Annals of Mathematics Studies
Number 183

Mumford-Tate Groups and Domains

Their Geometry and Arithmetic

Mark Green, Phillip Griffiths, and
Matt Kerr

PRINCETON UNIVERSITY PRESS
PRINCETON AND OXFORD
2012

Published by Princeton University Press, 41 William Street,
Princeton, New Jersey 08540

In the United Kingdom: Princeton University Press, 6 Oxford Street,
Woodstock, Oxfordshire OX20 1TW

press.princeton.edu

Library of Congress Cataloging-in-Publication Data

Green, M. (Mark)
Mumford-Tate groups and domains : their geometry and arithmetic / Mark Green, Phillip Griffiths, Matt Kerr.
p. cm. – (Annals of mathematics studies ; no. 183)
Includes bibliographical references and index.
ISBN 978-0-691-15424-4 (hardcover) – ISBN 978-0-691-15425-1 (pbk.)
1. Mumford-Tate groups. 2. Hodge theory. 3. Geometry, Algebraic. I. Griffiths, Phillip, 1938- II. Kerr, Matthew D., 1975- III. Title.
QA564.G634 2012
516.3'5–dc23

2011037621

British Library Cataloging-in-Publication Data is available

This book has been composed in LaTeX.

The publisher would like to acknowledge the authors of this volume for providing the camera-ready copy from which this book was printed.

Printed on acid-free paper ∞

Printed in the United States of America

10 9 8 7 6 5 4 3 2 1

Contents

Mumford-Tate Groups and Domains

Introduction

Mumford-Tate groups are the fundamental symmetry groups in Hodge theory. They were introduced in the papers [M1] and [M2] by Mumford. As stated there the purpose was to interpret and extend results of Shimura and Kuga ([Sh1], [Sh2], and [Ku]). Since then they have played an important role in Hodge theory, both in the formal development of the subject and in the use of Hodge theory to address algebro-geometric questions, especially those that are arithmetically motivated. The informative sets of notes by Moonen [Mo1] and [Mo2] and the recent treatment in [PS] are two general accounts of the subject.

We think it is probably fair to say that much, if not most, of the use of Mumford-Tate groups has been in the study of abelian varieties or, what is essentially the same, polarized Hodge structures of level one[1] and those constructed from this case. The papers [De1], [De2], and [De3] formulated the definitions and basic properties of Mumford-Tate groups in what is now the standard way, a formulation that provides a setting in which Mumford-Tate groups were particularly suited for the study of Shimura varieties, which play a central role in arithmetic geometry. Noteworthy is the use of Mumford-Tate groups and Shimura varieties in Deligne's proof [DMOS] that Hodge classes are absolute in the case of abelian varieties, and their role in formulating conjectures concerning motivic Galois groups (cf. [Se]). See [Mi2] for a useful and comprehensive account and [R] for a recent treatment of Shimura varieties, and [Ke] for a Hodge-theoretic approach.

As will be explained, the perspective in this monograph is in several ways complementary to that in the literature. Before discussing these, we begin by noting that Chapter I is an introductory one in which we give the basic definitions and properties of Mumford-Tate groups in both the case of Hodge structures and of mixed Hodge structures. Section II.A is also introductory where we review the definitions of period domains and their compact duals as well as the canonical exterior differential system on them.

[1]Level one means a Tate twist of a polarized Hodge structure of *effective weight one*. In general for $n \geqq 1$, a Hodge structure has *effective weight* n if the non-zero Hodge (p, q) components with $p + q = n$ have $p \geqq 0$, $q \geqq 0$. When no confusion seems likely, we shall omit the term effective.

As will be shown, Mumford-Tate groups M will be reductive algebraic groups over $\mathbb{Q}$ such that the derived or adjoint subgroup of the associated real Lie group $M_{\mathbb{R}}$ contains a compact maximal torus. In order to keep the statements of the results as simple as possible, we will emphasize the case when $M_{\mathbb{R}}$ itself is semi-simple. The extension to the reductive case will be usually left to the reader.

Before turning to a discussion of the remaining contents in this monograph, we first note that throughout we shall use the notation V for a $\mathbb{Q}$-vector space and $Q : V \otimes V \to \mathbb{Q}$ for a non-degenerate form satisfying $Q(u, v) = (-1)^n Q(v, u)$ where n is the weight of the Hodge structure under consideration. In many cases there will be given a lattice $V_{\mathbb{Z}}$ with $V = V_{\mathbb{Z}} \otimes \mathbb{Q}$.

One way in which our treatment is complementary is that we have used throughout the interpretation of Mumford-Tate groups in the setting of period domains D and their compact duals $\check{D}$. The latter are rational, homogeneous varieties defined over $\mathbb{Q}$. Variations of Hodge structure are integral manifolds of a canonical exterior differential system (EDS), defined on all of $\check{D}$, and also on quotients of D by discrete subgroups. This leads to a natural extension of the definition of the Mumford-Tate group M_φ of a Hodge structure $\varphi \in D$ to the Mumford-Tate group $M_{F^\bullet}$ associated the Hodge filtration given by a point $F^\bullet \in \check{D}$, and to the Mumford-Tate group $M_{(F^\bullet, E)}$ of an integral element $E \subset T_{F^\bullet}\check{D}$ of the EDS. Both of these extensions will be seen to have important geometric and arithmetic implications.

A second complementary perspective involves the emphasis throughout on *Mumford-Tate domains* D_{M_φ} (cf. Section (II.B)), defined as the orbit of the point $\varphi \in D$ by the group $M_\varphi(\mathbb{R})$ of real points of the Mumford-Tate group M_φ.[2] One subtlety, discussed in Section IV.G, is that the Mumford-Tate domain depends on its particular representation as a homogeneous complex manifold. The same underlying complex manifold may appear in multiple, and quite different, ways as a Mumford-Tate domain.

For later reference we note that Mumford-Tate domains will have compact duals, which are rational, homogeneous varieties that as homogeneous varieties are defined over a number field.

We shall denote by $M_\varphi(\mathbb{R})^0$ the identity component of $M_\varphi(\mathbb{R})$ in the classical topology and by $D^0_{M_\varphi}$ the component of D_{M_φ} through φ. To a point $\varphi \in D$, i.e., a polarized Hodge structure V_φ on V, is associated the algebra of *Hodge tensors* $\mathrm{Hg}^{\bullet,\bullet}_\varphi \subset T^{\bullet,\bullet} := \bigoplus_{k,l} V^{\otimes k} \otimes \check{V}^{\otimes l}$.

[2]Later in this monograph when discussing the geometry of Mumford-Tate domains, when confusion seems unlikely, we shall not distinguish between the $\mathbb{R}$-algebraic group $M_{\mathbb{R}}$ and its real points $M(\mathbb{R})$ and the corresponding real Lie group $M_{\mathbb{R}}$.

For reasons discussed below, it is our opinion that the classical Noether-Lefschetz loci (cf. Section (II.C)), defined traditionally by the condition on $\varphi \in D$ that a vector $\zeta \in V$ be a Hodge class, should be replaced by the *Noether-Lefschetz locus* NL_φ associated to $\varphi \in D$, where by definition

$$\mathrm{NL}_\varphi = \left\{ \psi \in D : \mathrm{Hg}_\varphi^{\bullet,\bullet} \subseteq \mathrm{Hg}_\psi^{\bullet,\bullet} \right\}.[3]$$

We will then prove the

(II.C.1) THEOREM: *The component $D^0_{M_\varphi}$ of the Mumford-Tate domain D_{M_φ} is the component of NL_φ through $\varphi \in D$.*

An application of this result is the estimate given in theorem (III.C.5) for the codimension of the Noether-Lefschetz locus, in the extended form suggested above, in the parameter space of a variation of Hodge structure. This estimate seems to be unlike anything appearing classically; it illustrates both the role of Mumford-Tate groups and, especially, the integrability condition in the EDS in "dimension counts."

For a simple first illustration of this, since $\check{D}$ is a projective variety defined over $\mathbb{Q}$ we may speak of a $\mathbb{Q}$*-generic point* $F^\bullet \in \check{D}$, meaning that the $\mathbb{Q}$-Zariski closure of $F^\bullet$ is $\check{D}$. In the literature there are various criteria, some of them involving *genericity* of one kind or another, that imply that M_φ is equal to the $\mathbb{Q}$-algebraic group $G = \mathrm{Aut}(V, Q)$. We show that, except when the weight $n = 2p$ is even and the only non-zero Hodge number is $h^{p,p}$, if $F^\bullet$ is a $\mathbb{Q}$-generic point of $\check{D}$, then the Mumford-Tate group $M_{F^\bullet}$ is equal to G. A converse will also be discussed. These issues will also be addressed in a more general context in Section VI.A (cf. (VI.A.5)).

A remark on terminology: For Hodge structures of weight one, what we are here calling Mumford-Tate domains have been introduced in [M2] and used in [De2], [De3]. For reasons to be explained in Section II.B, we shall define *Shimura domains* to be the special case of Mumford-Tate domains where M_φ can be described as the group fixing a set of Hodge tensors in degrees one and two.[4] There are then strict inclusions of sets

$$\begin{pmatrix} \text{period} \\ \text{domains} \end{pmatrix} \subset \begin{pmatrix} \text{Shimura} \\ \text{domains} \end{pmatrix} \subset \begin{pmatrix} \text{Mumford-Tate} \\ \text{domains} \end{pmatrix}.$$

[3]In the classical weight $n = 1$ this point of view is taken in the original papers [M1] and [M2] on the subject, as well as in [De2].

[4]The degree of $t \in V^{\otimes^k} \otimes \check{V}^{\otimes^l}$ is $k + l$.

We remark that a Shimura domain and a Mumford-Tate domain may be considered as period domains with additional structures. When that additional data is trivial we have the traditional notion of a period domain.[5]

Another result relating Mumford-Tate groups and period domains, the structure theorem stated below, largely follows from results in the literature (cf. [Schm1], [A1]) and the use of Mumford-Tate domains. To state it, we consider a global variation of Hodge structure (cf. Section (III.A))

$$\Phi : S \to \Gamma\backslash D$$

where S is smooth and quasi-projective. We assume that the $\mathbb{Q}$-vector space V has an integral structure $V_{\mathbb{Z}}$ and for $G_{\mathbb{Z}} = G \cap \mathrm{Aut}(V_{\mathbb{Z}})$ we denote by $\Gamma \subset G_{\mathbb{Z}}$ the monodromy group. As explained below, we consider Φ up to finite data, which in effect means that we consider Φ up to isogeny, meaning that we can replace S by a finite covering and take the induced variation of Hodge structure. We also denote by M_{Φ} the Mumford-Tate group associated to the variation of Hodge structure. It is also a reductive $\mathbb{Q}$-algebraic group, and we denote by

$$M_{\Phi} = M_1 \times \cdots \times M_{\ell} \times A$$

the almost product decomposition of M_{Φ} into its $\mathbb{Q}$-simple factors M_i and abelian part A. We also denote by $D_i \subset D$ the $M_i(\mathbb{R})$-orbit of a lift to D of the image $\Phi(\eta)$ of a very general point $\eta \in S$. Thus D_i is a Mumford-Tate domain for M_i. Then we have the

(III.A.1) THEOREM: (i) *The D_i are homogeneous complex submanifolds of D.* (ii) *Up to finite data, the monodromy group splits as an almost direct product $\Gamma = \Gamma_1 \times \cdots \times \Gamma_k$, $k \leqq l$, where for $1 \leqq i \leqq k$ the $\mathbb{Q}$-Zariski closure $\overline{\Gamma_i^{\mathbb{Q}}} = M_i$.* (iii) *Up to finite data, the global variation of Hodge structure is given by*

$$\Phi : S \to \Gamma_1 \backslash D_1 \times \cdots \times \Gamma_k \backslash D_k \times D'$$

where $D' = D_{k+1} \times \cdots \times D_{\ell}$ is the part where the monodromy is trivial.

A consequence of the proof will be that

> *The tensor invariants of Γ coincide with those of the arithmetic group $M_{1,\mathbb{Z}} \times \cdots \times M_{k,\mathbb{Z}}$ where $M_{i,\mathbb{Z}} = M_i \cap G_{\mathbb{Z}}$.*

---

[5]The motivation for the terminology is that for us this case needs to be distinguished from the general case, when the algebra of Hodge tensors is not generated in degrees one and two. In weight one it is the case originally introduced by Shimura in the 1960's. The somewhat subtle distinctions in terminology will be explained when we discuss what is meant in this work by the "classical and non-classical" cases.

It is known (cf. [De-M1], [De5]) that Γ need not be an arithmetic group,[6] i.e., a group commensurable with $M_{1,\mathbb{Z}} \times \cdots \times M_{k,\mathbb{Z}}$. However, from the point of view of its tensorial invariants it is indistinguishable from that group.

Because of this result and the arithmetic discussion in Chapters V–VIII it is our feeling that *Mumford-Tate domains are natural objects for the study of global variations of Hodge structure.* In particular, the Cattani-Kaplan-Schmid study of limiting mixed Hodge structures in several variables [CKS] and the recent Kato-Usui construction [KU] of extensions, or partial compactifications, of the moduli space of equivalence classes of polarized Hodge structures might be carried out in the context of Mumford-Tate domains. A previously noted subtlety here is that as a complex homogeneous manifold, the same complex manifold D may have several representations $D = G_{\mathbb{R},i}/H_i$ as a homogeneous space. It is reasonable that the extension of the Kato-Usui theory will depend on the particular $\mathbb{Q}$-algebraic group G_i. For this reason, as well as for material needed later in this monograph, in Section I.C we give a brief introduction to the Mumford-Tate groups associated to mixed Hodge structures.

Classically there is considerable literature (cf. [Mo1] and [Mo2]) on the question: What are the possible Mumford-Tate groups of polarized Hodge structures whose corresponding period domain is Hermitian symmetric?[7] In those works the question, "What are the possible Mumford-Tate groups?" is also posed.

In Chapter IV for general polarized Hodge structures we discuss and provide some answers to the questions:

(i) *Which semi-simple $\mathbb{Q}$-algebraic groups M can be Mumford-Tate groups of polarized Hodge structures?*[8]

and, more importantly,

(ii) *What can one say about the different realizations of M as a Mumford-Tate group?*

(iii) *What is the relationship among the corresponding Mumford-Tate domains?*

To address these questions, we use a third aspect in which this study differs from previous ones in that we invert the first question. For this we use the notion

[6]So far as we know, all non-arithmetic monodromy groups are subgroups of $\mathrm{SU}(n, 1)$. A question of which we are not aware if there is an answer to is whether Γ is arithmetic in case the real Lie group associated to $\overline{\Gamma^{\mathbb{Q}}}$ has no simple factors of real rank one. In this regard we note the paper [Kl4]

[7]In this regard we call attention to the papers [Z1] and [Z2] where this question is addressed.

[8]A variant of this question over $\mathbb{R}$ is treated in [Simp1] — see footnote 12.

of a *Hodge representation* (M, ρ, φ), which is given by a reductive $\mathbb{Q}$-algebraic group M, a representation

$$\rho : M \to \mathrm{Aut}(V, Q),$$

and a circle

$$\varphi : \mathbb{U}(\mathbb{R}) \to M(\mathbb{R}),$$

where $\mathbb{U}(\mathbb{R})$ is a maximal compact subgroup of the real algebraic group $\mathbb{S} =: \mathrm{Res}_{\mathbb{C}/\mathbb{R}}\,\mathbb{G}_m$,[9] such that $(V, Q, \rho \circ \varphi)$ gives a polarized Hodge structure.[10] For any Hodge representation, the circle $\varphi(\mathbb{U}(\mathbb{R}))$ lies in a maximal *compact* torus $T \subset M(\mathbb{R})$ whose compact centralizer H_φ is the subgroup of $M(\mathbb{R})$ preserving the polarized Hodge structure $(V, Q, \rho \circ \varphi)$.

We shall say that a representation $\rho : M \to \mathrm{Aut}(V)$ *leads to a Hodge representation* if there is a Q and φ such that $(V, Q, \rho \circ \varphi)$ is a Hodge representation. We define a *Hodge group* to be a reductive $\mathbb{Q}$-algebraic group M that has a Hodge representation. In Chapter IV our primary interest will be in the case where M is semi-simple. The other extreme case when M is an algebraic torus will be discussed in Chapter V.

Given a Hodge representation (M, ρ, φ) when M is semi-simple, we observe that there is an associated Hodge representation $(M_a, \mathrm{Ad}, \varphi)$ where the polarized Hodge structure on $(\mathfrak{m}, B, \mathrm{Ad}\,\varphi)$ is induced from the inclusion $\mathfrak{m} \subset \mathrm{End}_Q(V)$, noting that the Cartan-Killing form B is induced by Q. Here, M_a is the adjoint group, which is a finite quotient of M by its center. For the conjugate $\varphi_m = m^{-1}\varphi m$ by a generic $m \in M_a(\mathbb{R})$, it is shown that M_a is the Mumford-Tate group of $(\mathfrak{m}, B, \mathrm{Ad}\,\varphi_m)$. Thus, at least up to finite coverings,[11] the issue is to use the standard theory of roots and weights to give criteria to have a Hodge representation, and then to apply these criteria in examples.

Because the real points $M(\mathbb{R})$ map to

$$M(\mathbb{R}) \xrightarrow{\rho} \mathrm{Aut}(V_{\mathbb{R}}, Q),$$

[9] See Section I.A for a discussion of the Deligne torus.

[10] Without essential loss of generality, we assume as part of the definition that the induced representation $\rho_* : \mathfrak{m} \to \mathrm{End}(V, Q)$ is injective.

[11] The issue of finite coverings and the related faithfulness of Hodge representations is interesting in its own right; it will be analyzed in the text. The point is that irreducible Hodge representations of M will be parameterized by pairs (λ, φ) where λ is the highest weight of an irreducible summand of the $\mathfrak{m}_{\mathbb{C}}$-module $V_{\mathbb{C}}$. Given λ there are conditions on the lattice of groups with Lie algebra $\mathfrak{m}$ that λ be the highest weight of a representation of a particular M in the lattice. To be a Hodge representation will impose conditions on the pair (λ, φ).

the issue arises early on of what the *real form* $M(\mathbb{R})$ can be.[12] The first result is that:

(1) If (M, ρ, φ) is a Hodge representation, then as noted above $M(\mathbb{R})$ contains a compact real maximal torus T with $\dim T = \operatorname{rank}_{\mathbb{C}} M_{\mathbb{C}}$. We will abbreviate this by saying that $M(\mathbb{R})$ contains a compact maximal torus. Furthermore,

(2) If M is semi-simple, then the condition (1) is sufficient; indeed, the adjoint representation of M leads to a Hodge representation.

The natural starting point for an analysis of Hodge representations is the well-developed theory of complex representations of complex semi-simple Lie algebras. To pass from the theory of irreducible complex representations of a complex semi-simple Lie algebra to the theory of irreducible real representations of real semi-simple Lie groups,[13] there are three elements that come in:

(i) The theory of real forms of complex simple Lie algebras. These were classified by Cartan, and there is now the beautiful tool of Vogan diagrams with an excellent exposition in [K].

(ii) By Schur's lemma, irreducible real representations $V_{\mathbb{R}}$ of a real Lie algebra $\mathfrak{m}_{\mathbb{R}}$ break into three cases depending on whether

[12]In [Simp1] a variant of this question is studied and solved. Namely Simpson defines the notion of a *Hodge type*, which by a result in his paper is given by the pair consisting of a real, semi-simple Lie group $M_{\mathbb{R}}$ together with a circle S^1 in $M_{\mathbb{R}}$ whose centralizer $Z_{M_{\mathbb{R}}}(S^1) = H$ is a compact subgroup of $M_{\mathbb{R}}$. For a Hodge group (M, φ) as defined in this monograph, the associated real Lie group $M(\mathbb{R})$ and image of $\varphi : S^1 \to M(\mathbb{R})$ give a Hodge type. He shows that for a Hodge group the adjoint representation gives a real polarized Hodge structure. As discussed in Section IV.F, a Hodge type gives a homogeneous complex manifold and these are exactly those that are discussed in [GS].

Simpson also proves an existence theorem ([Simp1], Theorem 3 and Corollary 46), which roughly stated implies that if X is a compact Kähler manifold and $\rho : \pi_1(X) \to M_{\mathbb{R}}$ is a homomorphism that is $\mathbb{R}$-Zariski dense in $M_{\mathbb{R}}$, then $M_{\mathbb{R}}$ is the real Mumford-Tate group of a complex variation of Hodge structure and therefore $M_{\mathbb{R}}$ is of Hodge type. This is a wonderful existence result, saying informally that certain Hodge-theoretic data is "motivic over $\mathbb{R}$." We note also the interesting papers [Kl2] and [Kl3].

[13]If M is a $\mathbb{Q}$-simple algebraic group, then the group $M(\mathbb{R})$ will be a semi-simple — but not necessarily simple — real Lie group. An example of this is $\operatorname{Res}_{k/\mathbb{Q}} \mathrm{SL}_2(k)$ where $k = \mathbb{Q}(\sqrt{-d})$ with d a positive rational number. Then $\mathrm{SL}_2(k)(\mathbb{R}) \cong \mathrm{SL}_2(\mathbb{R}) \times \mathrm{SL}_2(\mathbb{R})$. For ease of exposition, we shall assume that $M(\mathbb{R})$ is also simple. The analysis can, in a straightforward fashion, be extended to the general case.

Of course we ultimately need to consider representations of M that are defined over $\mathbb{Q}$. This will be discussed in Section IV.A.

$$\mathrm{End}_{\mathfrak{m}_{\mathbb{R}}}(V_{\mathbb{R}}) = \begin{cases} \mathbb{R} & \text{(Real case)} \\ \mathbb{C} & \text{(Complex case)} \\ \mathbb{H} & \text{(Quaternionic case).} \end{cases}$$

Which case we end up in is determined by the weight λ associated to the representation and by the real form, as encoded in its Vogan diagram.[14] The possible invariant bilinear forms Q on V depend on which case we are in. In the real case, Q is unique up to a scalar and its parity — symmetric or alternating — is determined, but in the complex and quaternionic cases, invariant Q's of both parities exist. Perhaps the most delicate point in our analysis in Chapter IV is (IV.E.4), a theorem in pure representation theory, which we need in order to deal with the parity of Q in the real case; this result allows us to distinguish between real and quaternionic representations.

(iii) Given a real form $\mathfrak{m}_{\mathbb{R}}$, there is a one-to-one correspondence

$$\begin{array}{c}\text{Connected real Lie} \\ \text{groups having Lie} \\ \text{algebras } \mathfrak{m}_{\mathbb{R}}\end{array} \iff \begin{array}{c}\text{Subgroups } M_{P'}, \\ P \supseteq P' \supseteq R.\end{array}$$

Here P and R are respectively the weight and root lattice. In the case of interest to us when $\mathfrak{t} \subseteq \mathfrak{m}_{\mathbb{R}}$, the associated maximal torus in $M_{P'}(\mathbb{R})$ is

$$T \cong \mathfrak{t}/\Lambda \quad \text{where} \quad \Lambda \cong \mathrm{Hom}(P', \mathbb{Z}).$$

The representation of $\mathfrak{m}_{\mathbb{C}}$ having highest weight λ lives on $M_{P'}(\mathbb{R})$ if, and only if, $\lambda \in P'$. Note that the center

$$Z(M_{P'}(\mathbb{R})) \cong P'/R, \qquad \pi_1(M_{P'}(\mathbb{R})) \cong P/P'.$$

The disparity between P and R requires some analysis when $\lambda \in P$ but $\lambda \notin R$.

Because Mumford-Tate groups are $\mathbb{Q}$-algebraic groups, the issue of describing in terms of dominant weights the irreducible representations over $\mathbb{Q}$ of the simple $\mathbb{Q}$-algebraic groups whose maximal torus is anisotropic arises. Here again there is a highly developed theory. For those simple $\mathbb{Q}$-algebraic groups whose maximal torus is anisotropic the theory simplifies significantly, and those aspects needed for this work are summarized in part III of Section IV.A. The upshot is that for the purposes of Mumford-Tate groups as discussed in this monograph, one may focus the detailed root-weight analysis on the real case.

The outline of the steps to be followed in our analysis of Hodge representations is given in (IV.A.3). To state the result, we need to introduce some notation

[14]The weight λ associated to the representation will be explained (footnote 15).

that will be explained in the text. Given a real, simple Lie group $M(\mathbb{R})$ with Lie algebra $\mathfrak{m}_{\mathbb{R}}$ having Cartan decomposition

$$\mathfrak{m}_{\mathbb{R}} = \mathfrak{k} \oplus \mathfrak{p},$$

where $\mathfrak{k}$ is the Lie algebra of a maximal compact subgroup $K \subset M(\mathbb{R})$ containing a compact maximal torus T, there is the root lattice $R \subset i\check{\mathfrak{t}}$ and weight lattice $P \subset i\check{\mathfrak{t}}$ with $R \subset P$. We then define a map $\psi : R \to \mathbb{Z}/2\mathbb{Z}$ by

$$\psi(\alpha) = \begin{cases} 0 & \text{if } \alpha \text{ is a compact root} \\ 1 & \text{if } \alpha \text{ is a non-compact root.} \end{cases}$$

Since the Cartan involution is a well-defined Lie algebra homomorphism, it follows that ψ is a homomorphism. We next define a homomorphism

$$\Psi : R \to \mathbb{Z}/4\mathbb{Z}$$

by $\Psi =$ "2ψ" — i.e., $\Psi(\alpha) = 0$ for compact roots and $\Psi(\alpha) = 2$ for non-compact roots. The reason for working both "mod 2" and "mod 4" will be explained in the remark.

Given a choice of positive, simple roots there is defined a Weyl chamber $\mathbf{C}$ and weights $\lambda \in P \cap \mathbf{C}$. To each $\lambda \in \mathbf{C}$ there is associated an irreducible $\mathfrak{m}_{\mathbb{C}}$-module W^{λ}, and one may define the irreducible $\mathfrak{m}_{\mathbb{R}}$-module $V_{\mathbb{R}}$ associated to λ.[15] This $\mathfrak{m}_{\mathbb{R}}$-module will be seen to have an invariant form Q. As noted, in some cases there is more than one invariant form Q and we must choose it based on the result. One of our main results is then:

(IV.E.2) THEOREM: *Assume that M is a simple $\mathbb{Q}$-algebraic group that contains an anisotropic maximal torus. Assume that we have an irreducible representation*

$$\rho : M \to \operatorname{Aut}(V)$$

defined over $\mathbb{Q}$. We let δ be the minimal positive integer such that $\delta\lambda \in R$. Then ρ leads to a Hodge representation if, and only if, there exists an integer m such that

$$\Psi(\delta\lambda) \equiv \delta m \pmod 4.$$

Implicit in this result is that the invariant form Q may be chosen to be defined over $\mathbb{Q}$. This and the related results and computation of examples are expressed in terms of congruences mod 2 and mod 4. The reason for the "mod

[15]If $\operatorname{Res}_{\mathbb{C}/\mathbb{R}} W^{\lambda}$ denotes the irreducible $\mathfrak{m}_{\mathbb{R}}$-module obtained by restriction of scalars, then $\operatorname{End}_{\mathfrak{m}_{\mathbb{R}}}(\operatorname{Res}_{\mathbb{C}/\mathbb{R}} W^{\lambda})$ is a division algebra and the definition of $V_{\mathbb{R}}$ depends on whether this algebra is $\mathbb{R}$, $\mathbb{C}$ or $\mathbb{H}$.

2" is the sign in $Q(u,v) = \pm Q(v,u)$, and the reason for the "mod 4" is that the 2nd Hodge-Riemann bilinear relations

$$i^{p-q}Q(v,\overline{v}) > 0, \qquad v \in V^{p,q}$$

depend on $p - q$ mod 4.

As an illustration of the application of this analysis, we have the following (the notations are given in the appendix to Chapter IV): The only real forms of simple Lie algebras that give rise to Hodge representations of odd weight are

$\mathrm{su}(2p, 2q),\ p + q$ even, compact forms included
$\mathrm{su}(2k+1, 2l+1)$
$\mathrm{so}(4p+2, 2q+1),\ \mathrm{so}^*(4k)$
$\mathrm{sp}(2n, \mathbb{R})$ [16]
EV and EVII (real forms of E_7).

A complete list of the real forms having Hodge representations is given in the table after Corollary (IV.E.3).

At the end of Section IV.E, in the subject titled **Reprise**, we have summarized the analysis of which pairs (M, λ) give faithful Hodge representations and which pairs give odd weight Hodge representations. It is interesting that there are $\mathbb{Q}$-forms of some of the real, simple Lie groups that admit Hodge representations but do not have faithful Hodge representations.

In Section IV.B we turn to the adjoint representation. Here the Cartan-Killing form B gives an invariant form Q, and as a special case of theorem (IV.E.2) the criteria (IV.B.3) to have a Hodge representation may be easily and explicitly formulated in terms of the compact and non-compact roots relative to a Cartan decomposition of the Lie algebra. A number of illustrations of this process are given. The short and direct proof of this result is also given.

An interesting question of Serre (see 8.8 in [Se]) is whether G_2 is a motivic Galois group. This has recently been settled by Dettweiler and Reiter [DR] using a slight modification of the definition that replaces motives by motives for motivated cycles.[17] One has also the related (or equivalent, assuming the Hodge conjecture) question of whether G_2 is the Mumford-Tate group of a motivic Hodge structure. This too follows from [DR], as will be explained in a forthcoming work of the third author with G. Pearlstein [KP2].

[16] Our convention for the symplectic groups is that $2n$ is the number of variables.

[17] See in addition the interesting works [GS], [HNY], [Yun], and [Ka].

In this section we also give another one of the main results in this work, which provides a converse to the observation above about the adjoint representation:

THEOREM: *Given a representation* $\rho : M \to \mathrm{Aut}(V, Q)$ *defined over* $\mathbb{Q}$ *and a circle* $\varphi : \mathbb{U}(\mathbb{R}) \to M(\mathbb{R})$, $(V, \pm Q, \rho \circ \varphi)$ *gives a polarized Hodge structure if, and only if,* $(\mathfrak{m}, B, \mathrm{Ad}\,\varphi)$ *gives a polarized Hodge structure.*

Given a representation $\rho : M \to \mathrm{Aut}(V, Q)$, we identify $\rho_*(\mathfrak{m}) \subset \mathrm{End}_Q(V)$ with $\mathfrak{m}$. Then, as noted above, if $\rho \circ \varphi$ gives a polarized Hodge structure, there is induced on $(\mathfrak{m}, B) \subset \mathrm{End}_Q(V)$ a polarized sub-Hodge structure. The interesting step is to show that conversely a polarized Hodge structure $(\mathfrak{m}, B, \mathrm{Ad}\,\varphi)$ induces one for $(V, \pm Q, \rho \circ \varphi)$. This requires aspects of the structure theory of semi-simple Lie algebras. The proof makes use of the explicit criteria (IV.B.3) that $\mathrm{Ad}\,\varphi$ for a co-character $\varphi : \mathbb{U}(\mathbb{R}) \to T \subset M(\mathbb{R})$ give a polarized Hodge structure on $(\mathfrak{m}, B)$.

From our analysis we have the following conclusion: *Mumford-Tate groups are exactly the* $\mathbb{Q}$*-algebraic groups* M *whose associated real Lie groups* $M(\mathbb{R})$ *have discrete series representation in* $L^2(M(\mathbb{R}))$ (cf. [HC1], [HC2], [Schm2], [Schm3]). The discrete series, and the limits of discrete series, are of arithmetic interest as the infinite components of cuspidal automorphic representations in $L^2(M(\mathbb{Q})\backslash M(\mathbb{A})$. This potential connection between arithmetic issues of current interest and Hodge theory seems to us unlikely to be accidental.

In Section IV.F we establish another fundamental result:

(IV.F.1) THEOREM: (i) *The subgroup* $H_\varphi \subset M(\mathbb{R})$ *that stabilizes the polarized Hodge structure associated to a Hodge representation* (M, ρ, φ) *is compact and is equal to the subgroup that stabilizes the polarized Hodge structure associated to the polarized Hodge structure* $(\mathfrak{m}, \mathrm{Ad}, \varphi)$.

(ii) *Under the resulting identification of the two Mumford-Tate domains with the homogeneous complex manifold* $M(\mathbb{R})/H_\varphi$*, the infinitesimal period relations coincide.*

This suggests introducing the concept of a *Hodge domain* $D_{\mathfrak{m},\varphi}$, which is a homogeneous complex manifold $M(\mathbb{R})_a/H_\varphi$ where H_φ is the compact centralizer of a circle $\varphi : \mathbb{U}(\mathbb{R}) \to M(\mathbb{R})_a$. Thus a Hodge domain is equivalent to the data (M, φ) where φ satisfies the conditions in (IV.B.3). We observe that a Hodge domain carries an invariant exterior differential system corresponding to the infinitesimal period relation associated to the polarized Hodge structure $(\mathfrak{m}, B, \mathrm{Ad} \circ \varphi)$.[18] *A given Hodge domain may appear, as a complex manifold,*

[18] In the literature there is a reference to Griffiths-Schmid domains, defined to be homogeneous complex manifolds of the form $M_\mathbb{R}/H$ where $M_\mathbb{R}$ is a real, generally non-compact

in many different ways as a Mumford-Tate domain.[19] Moreover, there will in general be many relations among various Hodge domains.[20] A particularly striking illustration is the realization of each of the two G_2-invariant exterior differential systems on a 5-manifold found by E. Cartan and the Lie-Klein correspondence between them (cf. [Ca] and [Br]). Other interesting low dimensional examples will also be analyzed in detail at the end of Section IV.F.

A fourth aspect of this work is our emphasis throughout on the properties of Mumford-Tate groups in what we call the *non-classical case*. Here we want to make an important and somewhat subtle point of terminology. By the *classical case* we shall mean the case where (i) the Hodge domain $D_{M_\varphi} = M(\mathbb{R})_a/K$ is Hermitian symmetric; thus, K is a maximal compact subgroup of the adjoint group $M(\mathbb{R})_a$ whose center contains a circle $S^1_0 = \{z \in \mathbb{C}^* : |z| = 1\}$; (ii) only $z, 1, z^{-1}$ appear as the characters of $\mathrm{Ad}\, S^1_0$ acting on $\mathfrak{m}_\mathbb{C}$, and (iii) $\mathrm{Ad}(i)$ is the Cartan involution.[21] These are the Hermitian symmetric domains that may be equivariantly embedded in Siegel's upper-half space. The inverse limit of quotients $\Gamma\backslash D_{M_\varphi}$ by arithmetic groups are essentially the complex points of components of Shimura varieties, about which there is a vast and rich theory (cf. [Mi2] and [R] for general references).

The non-classical case itself separates into two classes. The first is when D_{M_φ} is Hermitian symmetric,[22] but where condition (ii) is not satisfied. In this case the infinitesimal period relation (IPR) may or may not be trivial. When it is trivial, which we shall refer to as the *unconstrained case*, it is possible, but does not in general seem to be known, whether (speaking informally) D_{M_φ} is

---

Lie group and H is a compact subgroup that contains a compact maximal torus T; therefore, $\dim T = \operatorname{rank} \mathfrak{m}_\mathbb{C}$. Hodge domains, as discussed in this monograph, differ in three respects: First, $M_\mathbb{R}$ are the real points $M(\mathbb{R})$ of a $\mathbb{Q}$-algebraic group M. Secondly, the circle $\varphi : \mathbb{U}(\mathbb{R}) \to M(\mathbb{R})$ is included as part of the definition, the point being that several different φ's may give equivalent complex structures but inequivalent polarized Hodge structures on $(\mathfrak{m}, B)$. Third, the exterior differential system giving the infinitesimal period relation is also implied by the definition.

[19] Again we mention the subtlety that arises in the several representations of a fixed complex manifold D as *homogeneous* complex manifolds. For example, $D = \mathrm{U}(2,1)/T_\mathrm{U}$ is a Mumford-Tate domain for polarized Hodge structures of weight $n = 3$ and with Hodge numbers $h^{3,0} = 1$, $h^{2,1} = 2$. But $D = \mathrm{SU}(2,1)/T_\mathrm{SU}$ is not a Mumford-Tate domain for polarized Hodge structures of any odd weight. Here, T_U and T_SU are the respective maximal tori in $\mathrm{U}(2,1)$ and $\mathrm{SU}(2,1)$. This issue is discussed and illustrated in Section IV.G.

[20] These relations may or may not be complex analytic. Both cases are of interest, the non-holomorphic ones from the perspective of representation theory.

[21] There is a subtlety here in that the adjoint action of the circle $S^1_0 = Z(K)$ is the "square-root" of a character that gives a weight two polarized Hodge structure on $(\mathfrak{m}, B)$. This is explained in the remark at the end of Section IV.F, where the φ in writing $D_{M_\varphi} = M(\mathbb{R})/K$ will be identified.

[22] We will see that all Hermitian symmetric domains are Hodge domains.

"motivic" in the sense that it parametrizes Hodge structures that arise algebro-geometrically.[23] In any case, the quotient $\Gamma\backslash D_{M_\varphi}$ by an arithmetic group is a quasi-projective, complex algebraic variety. The story of its field of definition seems to be worked out in the most detail in the classical case when the Mumford-Tate domain, factored by an arithmetic group, represents the solution to a moduli problem.

There are also interesting cases where D_{M_φ} is Hermitian symmetric but the IPR is non-trivial. In this case it is automatically integrable; D_{M_φ} is therefore foliated and variations of Hodge structure lie in the leaves of the foliation. Thus we have the situation where on the one hand D_{M_φ} cannot be motivic, while on the other hand for an arithmetic subgroup $\Gamma \subset M$, $\Gamma\backslash D_{M_\varphi}$ is a quasi-projective algebraic variety.

The second possibility in the non-classical case is when the Hodge domain D_{M_φ} is not Hermitian symmetric, which implies that the IPR is non-trivial with again the resulting implication that D_{M_φ} cannot be motivic in the above sense. In general, when the IPR is non-trivial from the viewpoint of algebraic geometry, Hodge theory is a *relative* subject.[24] In writing this monograph we have consistently sought to emphasize and illustrate what is different in the non-classical case.

For clarity, we summarize this discussion: The *classical case* is when D_{M_φ} may be equivariantly embedded in the Siegel-upper-half space; equivalently, D_{M_φ} parametrizes abelian varieties whose algebra of Hodge tensors contains a given algebra.[25] The *non-classical cases* are the remaining D_{M_φ}'s.

A fifth way in which our treatment is complementary to much of the literature is the discussion of the arithmetic aspects of Mumford-Tate domains and Noether-Lefschetz loci in the non-classical case. Before summarizing some of what is in the more arithmetically oriented Chapters V–VIII, we wish to make a few general observations. For these we assume given a global variation of Hodge structure

$$\Phi : S(\mathbb{C}) \to \Gamma\backslash D$$

where $S(\mathbb{C})$ are the complex points of a smooth, quasi-projective variety S defined over a field k, D is a period domain, and Γ is the monodromy group. If

[23] In this regard we call attention to the very interesting recent paper [FL], where it is shown that in the unconstrained case D_{M_φ} is a Mumford-Tate domain parametrizing a variation of Hodge structure of Calabi-Yau type.

[24] By this we mean that the basic algebro-geometric objects are variations of Hodge structure $\Phi : S \to \Gamma\backslash D_{M_\varphi}$, rather than the quotients $\Gamma\backslash D_{M_\varphi}$ by themselves.

[25] This is the case when the arithmetic aspects, especially Galois representations, of automorphic representation, are most highly developed. The principal reason is that in this case algebraic geometry, specifically the action of Galois groups on l-adic cohomology, may be employed.

one assumes a positive answer to the well-known question, "Is the *spread* of a global variation of Hodge structure again a variation of Hodge structure?", then we can assume that k is a number field. Associated to a point $p \in S(\mathbb{C})$ there are then two fields:

(i) the field of definition $k(p)$ of $p \in S(\mathbb{C})$;

(ii) the field of definition of the Plücker coordinate of any lift of $\Phi(p)$ to $\check{D}$.

Under special circumstances there are three additional fields associated to p.

(iii) the field $\mathrm{End}(\Phi(p))$ of endomorphisms associated to any lift of $\Phi(p)$ to a point of D;

(iv) the field of definition of the period point $\Phi(p)$ associated to a point $p \in S(\mathbb{C})$;

(v) the field of definition of the variety X_p.

In (iii), if we assume that the polarized Hodge structure $\Phi(p)$ is simple, then $\mathrm{End}(\Phi(p))$ is a division algebra, and when it is commutative we obtain a field. In (iv), we assume that $\Gamma \backslash D$ is a quasi-projective variety defined over a number field. In (v) we assume that there is a family of smooth projective varieties $\mathcal{X} \to S$ defined over k, and X_p is the fibre over $p \in S(\mathbb{C})$, whose period map is Φ.

There is a vast and rich literature about the relationships among these fields in the classical case where D is a bounded symmetric domain equivariantly embedded in Siegel's upper half space. This is the theory of Shimura varieties of Hodge type (cf. [Mi1] and [R]). In the non-classical case very little seems to be known, and one of our objectives is to begin to clarify what some interesting issues and questions are, especially as they may relate to Mumford-Tate groups.[26]

If we have a global variation of Hodge structure whose monodromy group Γ is an arithmetic group, then there are arithmetic aspects associated to particular irreducible representations of $G(\mathbb{R})$ in $L^2(\Gamma\backslash G(\mathbb{R}))$, as well as their adelic extensions. Here we have little to say other than to recall, as noted, that in the non-classical case the discrete series representations of $G(\mathbb{R})$ correspond to *automorphic cohomology* on $\Gamma\backslash D$, rather than to automorphic forms as in the

[26] We may think of (ii) and (iii) as reflecting arithmetic properties "upstairs"; i.e., on D or $\check{D}$, and (i), (iv), (v) as reflecting arithmetic properties "downstairs." They are related via the transcendental period mapping Φ, and the upstairs arithmetic properties of $\Phi(p)$ when p is defined over a number field is a very rich subject. Aspects of this will be discussed in Section VI.D.

classical case (cf. [GS] and, especially, [WW]).[27] So far as we know, there is not yet even speculation about a connection between the arithmetic aspects of $L^2(\Gamma\backslash G(\mathbb{R}))$ and the various arithmetic aspects, as mentioned previously, of a global variation of Hodge structure.[28]

In the largely expository Chapter V we will discuss Hodge structures with a high degree of symmetry; specifically, Hodge structures with complex multiplication or *CM Hodge structures.* For any Hodge structure $(V, \widetilde{\varphi})$, defined by $\widetilde{\varphi} : \mathbb{S}(\mathbb{R}) \to \mathrm{Aut}(V_{\mathbb{R}})$ and not necessarily pure, the endomorphism algebra $\mathrm{End}(V, \widetilde{\varphi})$ reflects its internal symmetries. It is reasonable to expect, and is indeed the case as will be discussed in the text, that there is a relationship between $\mathrm{End}(V, \widetilde{\varphi})$ and the Mumford-Tate group of $(V, \widetilde{\varphi})$. For CM-Hodge structure $(V, \widetilde{\varphi})$ with Mumford-Tate group $M_{\widetilde{\varphi}}$, there is an equivalence between

- $V_{\widetilde{\varphi}}$ is a CM-Hodge structure;
- $M_{\widetilde{\varphi}}(\mathbb{R})$ is contained in the isotropy group $H_{\widetilde{\varphi}}$;
- $M_{\widetilde{\varphi}}$ is an algebraic torus; and
- $M_{\widetilde{\varphi}}(\mathbb{Q})$ is contained in the endomorphism algebra $E_{\widetilde{\varphi}} = \mathrm{End}(V_{\widetilde{\varphi}})$.

In case $V_{\widetilde{\varphi}}$ is a simple Hodge structure, $\mathrm{End}(V, \widetilde{\varphi})$ is a totally imaginary number field of degree equal to $\dim V$. There is a converse result in (V.3).

We broaden the notion of CM type by defining an *n-orientation of a totally imaginary number field* (OIF(n)) and construct a precise correspondence between these and certain important kinds of CM Hodge structures. In the classical case of weight $n = 1$, we recover the abelian variety associated to a CM type.

Next, we generalize the notion of the *Kubota rank* and *reflex field* associated to a CM Hodge structure $V_{\widetilde{\varphi}}$ to the OIF(n) setting. This may then be used to compute the dimension, rational points, and Lie algebra of the Mumford-Tate group of $V_{\widetilde{\varphi}}$. When the Kubota rank is maximal, the CM Hodge structure is non-degenerate. In the classical case the corresponding CM abelian variety A is non-degenerate, and by a result of Hazama and Murty [Mo1] all powers of A satisfy the Hodge conjecture.

Chapter VI is devoted to the arithmetic study of Mumford-Tate domains and Noether-Lefschetz loci, both in D and in $\check{D}$. We use the notation $\mathcal{Z} \subset \check{D}$ for the

[27] This issue is discussed further near the end of this introduction.

[28] Here we note the paper [GS] and that there are some promising special results [C1], [C2], [C3]. The authors would like to thank Wushi Goldring for bringing this work to our attention.

set of flags $F^\bullet$ that fail to give Hodge structures; i.e., for which at least one of the maps

$$F^p \oplus \overline{F}^{n-p+1} \to V_{\mathbb{C}}$$

fails to be an isomorphism. Then points in $\check{D}\backslash\mathcal{Z}$ give Hodge structures that may not be polarized. More precisely, $\check{D} \setminus \mathcal{Z}$ is a union of open $G(\mathbb{R})$-orbits, each connected component of which corresponds to *Hodge structures having possibly indefinite polarizations* in the sense that the Hermitian forms in the second Hodge representation bilinear relations are non-singular but may not be positive or negative definite.[29]

For $\varphi \in D$ a polarized Hodge structure with Mumford-Tate group $M_\varphi := M$, we are interested in the loci

$$\mathrm{NL}_M = \left\{ \begin{array}{c} Q\text{-polarized Hodge structures with} \\ \text{Mumford-Tate group contained in } M \end{array} \right\}$$

$$\check{\mathrm{NL}}_M = \left\{ \begin{array}{c} \text{flags } F^\bullet \in \check{D} \text{ with Mumford-} \\ \text{Tate group contained in } M \end{array} \right\}.$$

We show that no component of $\check{\mathrm{NL}}_M$ is contained in $\mathcal{Z}$, and the components of $\check{\mathrm{NL}}_M$ are indexed in terms of "*Mumford-Tate Hodge orientations*," extending to non-abelian Mumford-Tate groups the CM types/orientations discussed in Chapter V. Using this we give a computationally effective procedure to determine the components in terms of Lie algebra representations and Weyl groups.

A first consequence of this is that $M(\mathbb{C})$ acts transitively on each component of the locus $\check{\mathrm{NL}}_M$. Another result, which contrasts the classical and non-classical period domains, is that in the classical case NL_M is a single $M(\mathbb{R})$-orbit in D, whereas in the non-classical case this is definitely *not* true; there is more than one Q-polarized Hodge type.

Next, we observe that the components of NL-loci are all defined over $\overline{\mathbb{Q}}$, the reason being that each component contains CM-Hodge structures, which as points in $\check{D}$ are defined over $\overline{\mathbb{Q}}$, and each component is an orbit in $\check{D}$ of $M(\mathbb{C})$ and M is a $\mathbb{Q}$-algebraic subgroup of G. Thus the absolute Galois group $\mathrm{Gal}(\overline{\mathbb{Q}}/\mathbb{Q})$ (or equivalently $\mathrm{Gal}(\mathbb{C}/\mathbb{Q})$) acts on $\check{\mathrm{NL}}_M$ permuting the components. We then show that the normalizers of M in G are the groups stabilizing the NL-loci. In case M is nondegenerate and $\check{\mathrm{NL}}_M$ contains a simple nondegenerate CM Hodge structure, we show that:

[29]The orbit structure of the action of $M(\mathbb{R})$ on $\check{D}$ is extensively discussed in [FHW]; cf. the remarks at the end of this introduction.

- the action of the normalizer $N_G(M,\mathbb{C})$ on the components of $\breve{\mathrm{NL}}_M$ turns out to give a "continuous envelope" for the very discontinuous action of $\mathrm{Gal}(\mathbb{C}/\mathbb{Q})$; and
- the endomorphism algebra of a generic $\varphi \in \breve{\mathrm{NL}}_M$ is a field, and this field contains the field of definition of each component.

In Chapter VII we develop an algorithm for determining all Mumford-Tate subdomains of a given period domain. This result is then applied to the classification of all CM Hodge structures of rank 4 and when the weight $n = 1$ and $n = 3$, to an analysis of their Hodge tensors and endomorphism algebras, and the number of components of the Noether-Lefschetz loci. The result is that one has a complex but very rich arithmetic story; e.g., the results for the $n = 3$ case are summarized in the list given in Theorem (VII.F.1). We note in particular the intricate structure of the components of the Noether-Lefschetz loci in D and in its compact dual $\check{D}$, and the two interesting cases where the Hodge tensors are generated in degrees 2 and 4, and not just in degree 2 by $\mathrm{End}(V_\varphi)$. One application (cf. VII.H, type (i)) is that a particular class of period maps appearing in mirror symmetry never has image in a proper subdomain of D.

As an observation, it is generally understood that in Hodge theory or algebraic geometry — and especially at their interface — "special points" are of particular interest. These are points with symmetries that have an internal structure that relates arithmetic and geometry. We believe that the tables in Chapter VII give a good picture of the intricacies of the Hodge-theoretic special points. Of course, these relate directly to algebraic geometry when the weight $n = 1$, but when the weight $n = 3$ the story is in its earliest stages.[30]

In Chapter VIII we discuss some arithmetic aspects of the situation

$$\Phi : S(\mathbb{C}) \to \Gamma\backslash D$$

where S parametrizes a family $\mathcal{X} \to S$ of smooth, projective varieties defined over a number field k. We recall the notion of absolute Hodge classes (AH) and strongly absolute classes (SAH). There are well-known conjectures (here H denotes Hodge classes):

(i) H $\Rightarrow$ AH,

(ii) H $\Rightarrow$ SAH.

[30] The case considered here is when the Hodge structure is of "mirror-quintic-type." On the algebro-geometric side there is a vast literature, much of it motivated by the connection in physics.

We denote by NL a Noether-Lefschetz locus in $\Gamma\backslash D$, and show that (cf. [Vo])

(ii) *implies that* $\Phi^{-1}(\mathrm{NL})$ *is defined over a number field.*

Recalling that $\mathrm{NL} \subset \check{D}$ is defined over $\mathbb{Q}$, these observations may be thought of as relating "arithmetic upstairs" with "arithmetic downstairs" via the transcendental period map Φ. The particular case when the Noether-Lefschetz locus consists of isolated points was alluded to in the discussion of CM Hodge structures.

A related observation [A2] is that one may formulate a variant of the "Grothendieck conjecture" in the setting of period maps and period domains. Informally stated, Grothendieck's conjecture is that for a smooth projective variety X defined over a number field, all of the $\mathbb{Q}$-relations on the period matrix obtained by comparing bases for the algebraic de Rham cohomology group $\mathbb{H}^n_{\mathrm{Zar}}(\Omega^\bullet_{X(k)/k})$ and the Betti cohomology group $H^n_B(X(\mathbb{C}), \mathbb{Q})$ are generated by algebraic cycles on self-products of X. Again, informally stated the period domain analog of this is (cf. (VIII.A.8))

Let $p \in S(k)$ *and let* $\varphi \in D$ *be any point lying over* $\Phi(p)$. *Then* φ *is a very general point in the variety* $\check{D}_{M_\varphi}$.

Here, very general means that φ is a point of maximal transcendence degree in $\check{D}_{M_\varphi}$, which is a subvariety of $\check{D}$ defined over a number field. The relation between this and the above formulation in terms of period matrices arises when $\mathcal{X} \to S$ is a moduli space with the property that the fields of definition of X_p and of p are both number fields exactly when one of them is.

A final conjecture (VIII.B.1) may be informally stated as follows:

Let $\mathcal{E} \subset D$ *be a set of CM points and assume that the image* $\rho(\mathcal{E})$ *of* $\mathcal{E}$ *in* $\Gamma\backslash D$ *lies in* $\Phi(S(\mathbb{C}))$. *Then the* $\mathbb{C}$*-Zariski closure of* $\rho(\mathcal{E})$ *in* $\Phi(S(\mathbb{C}))$ *is a union of unconstrained Hermitian symmetric Mumford-Tate domains.*

A geometric consequence of this would be that if $S(\mathbb{C})$ contains a Zariski dense set of CM Hodge structures, then the corresponding Mumford-Tate domain in the structure theorem is *Hermitian symmetric* whose IPR is trivial.

Before turning to some concluding remarks in this introduction, we observe that Mumford-Tate groups may be said to lie at the confluence of several subjects:

(A) **Geometry**; specifically, algebraic geometry and variation of Hodge structure as realized by mappings to Mumford-Tate domains;

(B) **Representation theory**; specifically, the introduction and analysis of Hodge representations, leading among other things to the observation that Hodge groups turn out to be exactly the reductive $\mathbb{Q}$-algebraic groups whose associated semi-simple real Lie groups have discrete series representations; and

(C) **Arithmetic**; specifically, the rich honeycomb of arithmetically defined Noether-Lefschetz loci in a period domain.

It is certainly our sense that very interesting work should be possible at the interfaces among these areas.

We would like to mention several topics that are *not* discussed in this work and may be worth further study.

Extensions of moduli spaces of Γ-equivalence classes of polarized Hodge structures whose generic Mumford-Tate group is M.[31]

This means the following: Let M be a reductive $\mathbb{Q}$-algebraic group such that the pair (M, φ) gives a Hodge group together with a circle $\varphi : \mathbb{U}(\mathbb{R}) \to M(\mathbb{R})$, and let D_{M_φ} be the corresponding Mumford-Tate domain. Let $\Gamma \subset M$ be an arithmetic group and set

$$D_{M_\varphi}(\Gamma) = \Gamma \backslash D_{M_\varphi}.$$

We may think of $D_{M_\varphi}(\Gamma)$ as the *Γ-equivalence classes of polarized Hodge structures whose generic Mumford-Tate group is M*. A subtlety here is that D_{M_φ} may be realized in many different ways as a Mumford-Tate domain for polarized Hodge structures of different weights and different sets of Hodge numbers and, if $n > 1$, different Hodge orientations in the same period domain. What is common among all these realizations is that the infinitesimal period relations given by the Pfaffian system $I \subset T^* D_{M_\varphi}$ all coincide.

We let $S = (\Delta^*)^k \times \Delta^l$ be a punctured polycylinder and

$$\Phi : S \to D_{M_\varphi}(\Gamma)$$

a variation of Hodge structure whose monodromies are unipotent; i.e., if $\gamma_i \in \pi_1(S)$ is the circle around the origin in the ith factor of $(\Delta^*)^k$, then

$$\Phi_*(\gamma_i) = T_i \in \Gamma$$

is a unipotent element of M. Given any variation of Hodge structure as above, by a result of Borel the unipotency of monodromies may be achieved by passing

[31]This topic was already briefly alluded to above.

to a finite covering of S. By an *extension* we mean (i) a log-analytic variety $D^e_{M_\varphi}$ in which $D_{M_\varphi} \subset D^e_{M_\varphi}$ is a Zariski open set and to which the action of Γ extends to give a log-analytic quotient variety $D^e_{M_\varphi}(\Gamma) = \Gamma \backslash D^e_{M_\varphi}$; and (ii) setting $S^e = \Delta^k \times \Delta^l$, the above variation of Hodge structure extends to give

$$\begin{array}{ccc} S & \xrightarrow{\Phi} & D_{M_\varphi}(\Gamma) \\ \cap & & \cap \\ S^e & \xrightarrow{\Phi^e} & D^e_{M_\varphi}(\Gamma) \end{array}$$

where $\Phi^e : S^e \to D^e_{M_\varphi}(\Gamma)$ is a morphism of log-analytic varieties. The issue is:

Can the Kato-Usui theory [KU] be extended to the above situation?

The point is that since Hodge domains parametrize many different types of polarized Hodge structures, focusing on them may serve to isolate the essential algebraic group aspects of the Kato-Usui construction: the "input data" is (M, φ), and not any particular Hodge representation of M.[32]

Ordinary cohomology. This begins with the study of

$$H^*(D_{M_\varphi}(\Gamma), \mathbb{Q}).$$

There are a number of refinements to this.

(i) For a local system $\mathbb{V} \to D_{M_\varphi}(\Gamma)$ given by a Hodge representation of M, one may study $H^*(D_{M_\varphi}(\Gamma), \mathbb{V})$.

(ii) Tensoring with $\mathbb{C}$ and using

$$H^*(D_{M_\varphi}(\Gamma), \mathbb{V}_\mathbb{C}) \cong H^*_{DR}(D_{M_\varphi}(\Gamma), \mathbb{V}_\mathbb{C}),$$

one may study the *characteristic cohomology* (cf. [CGG])

$$H^*_I(D_{M_\varphi}(\Gamma), \mathbb{V}_\mathbb{C})$$

computed from the de Rham complex of C^∞ forms modulo the differential ideal generated by the C^∞ sections of $I \oplus \overline{I}$ where $\overline{I}$ is the complex conjugate sub-bundle to I in the complexification of the real tangent bundle. This is the cohomology that is relevant for variations of Hodge structure.

[32]This is carried out in a different context and for $\mathrm{SU}(2,1)$ in [C3]. Also, the example at the end of Section IV.A suggests a positive indication that the above question may be feasible.

(iii) We assume that Γ is *neat*, meaning that Γ acts freely on the Riemannian symmetric space X_M.[33] The fibering

$$D_M \to X_M$$

has rational, homogeneous projective varieties F as fibres, and therefore *additively*, but *not* multiplicatively, we have (for any coefficients, including a local system as above)

$$H^*(D_{M_\varphi}(\Gamma)) \cong H^*(\Gamma\backslash X_M) \otimes H^*(F).$$

Now two important points arise:

(a) Using that M is a $\mathbb{Q}$-algebraic group, for a suitable open compact subgroup $U \subset M(\mathbb{A})$ the *adelization*

$$M(\mathbb{Q})\backslash M(\mathbb{A})/U$$

of $\Gamma\backslash X_M$ may be defined and its cohomology is the subject of considerable arithmetic and representation-theoretic interest (see [Schw] for a recent survey). As a set, the adelization is the limit of $\Gamma\backslash D_M$'s or $\Gamma'\backslash X_M$'s over the congruence subgroups Γ' of Γ, or some variant of these constructions — e.g., taking

$$M(\mathbb{Q})\backslash M(\mathbb{A})/K_\varphi$$

where

$$K_\varphi = H_\varphi \times \prod K_p$$

is the restricted product where an element $\prod K_p$ is in $M(\mathbb{Z}_p)$ for almost all p.

(b) In the case discussed in [Schw], the cohomology that is of interest for the study of cuspidal automorphic representation is $H^*_{\text{cusp}}(M(\mathbb{Q})\backslash M(\mathbb{A})/K_\varphi, \mathbb{V})$. By virtue of the characterization of semi-simple Mumford-Tate groups as those semi-simple $\mathbb{Q}$-algebraic groups M such that $M(\mathbb{R})$ contains a compact maximal torus, which by Harish-Chandra [HC1], [HC2] is equivalent to $L^2(M(\mathbb{R}))$ containing non-trivial discrete series representations; it is essentially known that cuspidal cohomology may be defined in the context of Hodge domains and have a Lie algebra cohomological description of the sort

$$\bigoplus_{\pi} H^*(\mathfrak{m}_{\mathbb{R}}, H_\varphi; H_{\pi_\infty} \otimes V) \otimes H_{\pi_f}$$

(see [Schw, p. 258] for explanation of notations).

[33]This is always possible by passing to a subgroup of finite index in Γ. Here $X_M = M(\mathbb{R})/K$ where K is the maximal compact subgroup of $M(\mathbb{R})$ containing the isotropy group H_φ.

(iv) Combining (ii) and (iii), one may at least formally define the *cuspidal characteristic cohomology* as

$$\bigoplus_{\pi} H^*(\mathfrak{m}_{\mathbb{R}}, H_\varphi; \mathbf{w}, H_{\pi_\infty} \otimes V) \otimes H_{\pi_f}$$

where $\mathfrak{m}_{\mathbb{R}} = \mathfrak{k} \oplus \mathfrak{p}$ and $\mathbf{w} \subset \mathfrak{p}$ defines the infinitesimal period relation (see [CGG] for an explanation of the notation). Especially if it could be related to variations of Hodge structure defined over $\mathbb{Q}$ (or over a number field), this may be an object of interest to study.

Coherent cohomology. Here one begins with a holomorphic, homogeneous vector bundle $\mathcal{E} \to D_{M_\varphi}$ and considers the L^2-cohomology groups

$$H^*_{(2)}(D_{M_\varphi}, \mathcal{E})$$

and

$$H^*_{(2)}(\Gamma\backslash D_{M_\varphi}, \mathcal{E}).$$

Again it is exactly for Hodge domains that $H^*_{(2)}(D_{M_\varphi}, \mathcal{E})$ is well-understood, and varying φ and $\mathcal{E}$ realizes all the irreducible discrete series representations (see [Schm4] for an excellent overview). The groups $H^*_{(2)}(\Gamma\backslash D_{M_\varphi}, \mathcal{E})$ are less well-understood, although as shown in [WW] Poincaré series do give a nontrivial map

$$H^*_{(1)}(D_{M_\varphi}, \mathcal{E}) \to H^*_{(2)}(\Gamma\backslash D_{M_\varphi}, \mathcal{E}).$$

In the classical case when D_{M_φ} is Hermitian symmetric and the quotient $\Gamma\backslash D_{M_\varphi}$ is a component of the complex points of a Shimura variety, the adelization of $H^*(\Gamma\backslash D_{M_\varphi}, \mathcal{E})$ has been, and continues to be, the object of intense study (cf. [H], [Mi1], [Mi2], [BHR], and [Mor]). Once again there is a Lie algebra cohomological description of the cuspidal coherent cohomology ([H], [BHR]), and whose existence also is related to $M(\mathbb{R})$ having discrete series representations. This Lie algebra cohomological formulation makes sense for Hodge domains with H_φ replacing K.

In an interesting series of papers [C1], [C2], [C3] this automorphic cohomology is studied in detail for the Hodge domain

$$D_{M_\varphi} = \mathrm{SU}(2,1)/T$$

where the complex structure on D_{M_φ} is the one that does not fibre holomorphically or anti-holomorphically over an Hermitian symmetric space. There the automorphic cohomology in degrees one and two is studied and shown to correspond to automorphic representations whose archimedean component is a degenerate limit of discrete series, a phenomenon that cannot happen in the classical Shimura variety case.

Two further cases that may merit further detailed study are the period domains

$$\begin{cases} D_1 &= \mathrm{SO}(4,1,\mathbb{R})/\mathcal{U}(2) \\ D_2 &= \mathrm{Sp}(4,\mathbb{R})/\mathcal{U}(1)\times\mathcal{U}(1), \end{cases}$$

where in the second the complex structure is again the one that does not fibre holomorphically or anti-holomorphically over an Hermitian symmetric domain. In the first case there is no Hermitian symmetric domain with symmetry group $\mathrm{SO}(4,1)$, so the methods of [C1] and [C2] would not seem to apply. Each of these has algebro-geometric interest, the second because it arises in mirror symmetry.

Cycle spaces (cf. [FHW]). Cycle spaces are the following: In a Mumford-Tate domain $D_{M_\varphi} = M_\varphi(\mathbb{R})/H_\varphi$, the orbit $\mathcal{O}_\varphi := K\cdot\varphi$ of a point $\varphi\in D$ under the maximal compact subgroup $K\subset G(\mathbb{R})$ with $H_\varphi\subset K$ is a maximal compact, complex analytic subvariety of D_{M_φ}. The *cycle space*

$$\mathcal{U} = \left\{ g\cdot\mathcal{O}_\varphi : g\in M_\varphi(\mathbb{C}),\ g\cdot\mathcal{O}_\varphi\subset D_{M_\varphi}\right\}$$

is the set of translates of $\mathcal{O}_\varphi$ by those elements in the complex group that leaves the translate in D_{M_φ}.[34] The interest in cycle spaces began with the observation ([G1], [G2]) that $\dim\mathcal{O}_\varphi := d$ is the degree in which the cohomology $H^d(D_{M_\varphi},\mathcal{L})$ of suitable homogeneous line bundles was expected to occur, which in fact turned out to be the case [Schm5]. This suggested interpreting the somewhat mysterious group $H^d(D_{M_\varphi},\mathcal{L})$ as holomorphic sections of a vector bundle over $\mathcal{U}$ — a sort of Radon transform. For a discrete subgroup $\Gamma\subset M$ there is an "automorphic version" of this construction [WW], a variant of which has been used effectively in recent years (cf. [EGW], [Gi] and [C1], [C2], [C3]). The study of the cycle spaces themselves has turned out to be a rich subject with applications to representation theory; a recent comprehensive account is given in [FHW].

In concluding this introduction, we would like to observe that when we began this project we were of the view that Mumford-Tate groups and Mumford-Tate domains were primarily of interest because of their use in the period mappings arising from algebraic geometry. As our work has evolved, we have come to the point of view that the objects in the title of this monograph are arguably of equal interest in their own right, among other things for the rich arithmetic and representation-theoretic structure that they reveal.

[34]It may be shown that $\mathcal{U}$ is an open set in the Hilbert scheme associated to $\mathcal{O}_\varphi\subset\check{D}_{M_\varphi}$.

Finally, we would like to first thank the authors of [FH], [GW], and [K] — their books on Lie groups and representation theory were invaluable to this work. Secondly, we would like to thank Colleen Robles, J. M. Landsberg, and Paula Cohen for valuable suggestions and helpful edits. Lastly, we would like to express our enormous appreciation to Sarah Warren for her work in typing the manuscript through seemingly innumerable iterations and revisions.[35]

Notations (the terms to be explained in the text).

- Throughout, V will denote a rational vector space. For $k = \mathbb{R}$ or $\mathbb{C}$ we set $V_k = V \otimes_{\mathbb{Q}} k$.
- $\mathbb{S}$ and $\mathbb{U}$ will denote the *Deligne torus* and its maximal compact subgroup as defined in Section I.A.
- A *Hodge structure* on V will be denoted by $(V, \widetilde{\varphi})$, or frequently simply by $V_{\widetilde{\varphi}}$. A *polarized Hodge structure* will be denoted by (V, Q, φ) or (V_φ, Q).
- $E_{\widetilde{\varphi}}$ will denote the endomorphism algebra $\mathrm{End}(V, \widetilde{\varphi})$ of $(V, \widetilde{\varphi})$.
- *Polarized Hodge structures* on a $\mathbb{Q}$-vector space V with polarizing form Q will be denoted by φ. The space of all such, with given Hodge number $h^{p,q}$, is a *period domain* D. The period domain is acted on transitively by the real Lie group $G(\mathbb{R}) := \mathrm{Aut}(V_{\mathbb{R}}, Q)$, with isotropy group of $\varphi \in D$ denoted by H_φ, or just H if no confusion is possible.
- The *compact dual* $\check{D}$ is given by all filtrations $F^\bullet$ on $V_{\mathbb{C}}$ with $\dim F^p = f^p = \sum_{p' \geqq p} h^{p',q'}$ and satisfying the first Hodge-Riemann bilinear relations. It is a rational, homogeneous variety $G(\mathbb{C})/P$ defined over $\mathbb{Q}$ and with *Plücker embedding* $\check{D} \subset \mathbb{P}^N$. The *Plücker coordinate of* $F^\bullet \in \check{D}$ will be denoted by $[F^\bullet]$.[36]
- For $\varphi \in D$, $F^\bullet_\varphi$ will denote the corresponding point in $\check{D}$.
- The *Mumford-Tate group* of $\varphi \in D$ will be denoted by M_φ; that of $F^\bullet \in \check{D}$ will be denoted by $M_{F^\bullet}$. Sometimes, when no confusion is possible and to minimize notational clutter, we will set $M = M_\varphi$ or $M = M_{F^\bullet}$.

---

[35]The third author would also like to acknowledge partial support from EPSRC through First Grant EP/H021159/1.

[36]In a few circumstances we will omit the $^\bullet$ in $F^\bullet$, as including it would entail distracting notational clutter. Examples are $\mathrm{Hg}^{\bullet,\bullet}_{(F,W)}$, $\mathrm{Hg}^{\bullet,\bullet}_{\widehat{(F,W)}}$ and $\mathrm{Hg}^{\bullet,\bullet}_{(F,E)}$.

- *Integral elements* of the canonical exterior differential system on $\check{D}$ will be denoted by $E \subset T_{F^\bullet}\check{D}$. The corresponding Mumford-Tate group will be denoted by $M_{(F^\bullet,E)}$.

- Mumford-Tate domains will be denoted by D_{M_φ} or by just D_M if confusion is unlikely, for $\varphi \in D$, and by $\check{D}_{M_{F^\bullet}}$ or again $\check{D}_M$ for $F^\bullet \in \check{D}$.

- A *Hodge representation* is denoted
$$(M, \rho, \varphi)$$
where
$$\rho : M \to \mathrm{Aut}(V, Q)$$
is a representation and
$$\varphi : \mathbb{U}(\mathbb{R}) \to M(\mathbb{R})$$
is a circle such that $(V, Q, \rho \circ \varphi)$ gives a polarized Hodge structure.

- A *Hodge domain* is a homogeneous complex manifold $D_{\mathfrak{m}} = M_a(\mathbb{R})/H_\varphi$ where M is a reductive $\mathbb{Q}$-algebraic group and $\varphi : \mathbb{U}(\mathbb{R}) \to M(\mathbb{R})$ is a circle such that $(\mathfrak{m}, B, \mathrm{Ad} \circ \varphi)$ is a Hodge representation.

- $\mathcal{Z} \subset \check{D}$ will be the set of $F^\bullet = \{F^p\}$ such that some $F^p \otimes \overline{F}^{n-p+1} \to V_{\mathbb{C}}$ fails to be an isomorphism. The remaining points $F^\bullet \in \check{D}\backslash\mathcal{Z}$ will all give indefinitely polarized Hodge structures and in Chapters V–VII will frequently be denoted by φ where $\varphi : \mathbb{S}(\mathbb{U}) \to \mathrm{Aut}(V_{\mathbb{R}}, Q)$ defines the Hodge structure.

- We denote by G the $\mathbb{Q}$-algebraic group $\mathrm{Aut}(V, Q)$, and for any field $k \supseteq \mathbb{Q}$, $G(k)$ denotes the k-valued points of G. Similar notations will be used for other $\mathbb{Q}$-algebraic groups.

- K will denote a maximal compact subgroup of $G(\mathbb{R})$. As will be seen, a $\varphi \in D$ will determine a unique K with $H_\varphi \subset K$, sometimes denoted by K_φ.

- A Lie or algebraic group A is an *almost direct product* of subgroups A_i if the intersections $A_i \cap A_j$ are finite and the map $A_1 \times \cdots \times A_m \to A$ is finite and surjective. We denote by A^0 the identity components in either the Lie or algebraic group sense.

- We set $T^{k,l} = T^{k,l}(V) = (\otimes^k V) \otimes (\otimes^l \check{V})$ and $T^{\bullet,\bullet} = \bigoplus_{k,l} T^{k,l}$.

- For either $\varphi \in D$ or $F^\bullet \in \check{D}$, we denote by $\mathrm{Hg}_\varphi^{k,l}$ or $\mathrm{Hg}_{F^\bullet}^{k,l}$ the set of *Hodge classes* in $T^{k,l}$. $\mathrm{Hg}_\varphi^{\bullet,\bullet} = \bigoplus_{k,l} \mathrm{Hg}_\varphi^{k,l}$, and similarly $\mathrm{Hg}_{F^\bullet}^{\bullet,\bullet}$, will be the *algebra of Hodge tensors*. For M a group, we denote by $\mathrm{Hg}_M^{\bullet,\bullet}$ the algebra of M-Hodge tensors, as defined in Section IV.A.

- We denote by NL_φ, $\widecheck{\mathrm{NL}}_\varphi$, $\mathrm{NL}_{F^\bullet}$, NL_M, $\widecheck{\mathrm{NL}}_M$ the various *Noether-Lefschetz loci*, to be defined in Section II.C and Chapter VI.

- For a linear algebraic group $A \subset \mathrm{GL}(W)$, we denote by $A' \subset \mathrm{GL}(W)$ the subgroup defined by

$$A' = \left\{ \begin{array}{c} a \in \mathrm{GL}(W) : a \text{ fixes all } w \in T^{\bullet,\bullet}W \\ \text{that are fixed by } A \end{array} \right\}.$$

- For an algebraic torus T, we denote by $X^*(T)$, $X_*(T)$ the groups of characters, respectively co-characters of T.

- For $M \subset G$ a $\mathbb{Q}$-algebraic subgroup and $k = \mathbb{R}$ or $\mathbb{C}$, we denote by $N_G(M,k)$ the normalizer of $M(k)$ in $G(k)$.

- $W_M(T,\mathbb{R})$, $W_M(T,\mathbb{R})^0$, and $W_M(T,\mathbb{C})$ will denote various Weyl groups, defined in Section VI.D.

- $\mathrm{End}(V,\varphi)$ or $\mathrm{End}(V_\varphi)$ will denote the algebra of endomorphisms of a Hodge structure V_φ.

- CM-Hodge structure will be the standard abbreviation for a *complex multiplication Hodge structure*.

- Given a number field F, for k equal to $\mathbb{R}$ or $\mathbb{C}$, $S_F(k)$ will denote the set of embeddings of F in k.

- F^c will denote the Galois closure of a field F. For (F,Π) an oriented CM field associated to a polarized CM-Hodge structure (see Section V.A), F' will denote the generalized *reflex field* (which depends on Π), and $R(F,\Pi)$ will denote the generalized *Kubota rank*.

- The notations $\mathrm{OIF}(n)$, $\mathrm{subOIF}(n)$, $\mathrm{OCMF}(n)$, $\widetilde{\mathrm{OCMF}}(n)$, WCMHS, and SCMHS all represent concepts associated to CM-Hodge structure; see Section VI.B.

- $\mathrm{Corr}(A,\mathbb{Q})$ will denote the $\mathbb{Q}$-correspondences of an abelian variety A; see Section V.E.

- A *Q-quasi-unitary basis* for a polarized Hodge structure $V_{\mathbb{C}} = \underset{p+q=n}{\oplus} V^{p,q}$ is given by bases $\omega_{p,i}$ for $V^{p,q}$ satisfying

  (i) $\omega_{n-p,i} = \bar{\omega}_{p,i}$;

  (ii) $(\sqrt{-1})^{2p-n} Q(\omega_{p,i}, \bar{\omega}_{p,j}) = \delta^i_j$.

- Nilpotent orbits will be denoted by Φ^{nilp}.

- The notations and terminology from the theory of Lie groups and Lie algebras and their representations are collected in the appendix to Chapter IV, where most of the discussion involving Lie theory takes place.

- The canonical exterior systems generated by the *infinitesimal period relation* (IPR) will be denoted by $I \subset T\check{D}$.

- *Variations of Hodge structure* will be denoted by

$$\Phi : S \to \Gamma\backslash D.$$

- The *Noether-Lefschetz locus of a variation of Hodge structure* will be denoted by

$$\mathrm{NL}_{s_0}(S) \subset S.$$

Chapter I

Mumford-Tate Groups

I.A HODGE STRUCTURES

Let V be a $\mathbb{Q}$-vector space. For $k = \mathbb{R}$ or $\mathbb{C}$ we set $V_k = V \otimes k$ and $\mathrm{GL}(V)_k = \mathrm{GL}(V_k)$. There are three definitions of a *Hodge structure of weight* n, here given in historical order. In the first two definitions, we assume that n is positive and the p, q's in the definitions are non-negative. In the third definition, which will be the one used primarily in this monograph, n and p, q are arbitrary integers.

Definition (i): A *Hodge structure of weight* n is given by a *Hodge decomposition*

$$\begin{cases} V_{\mathbb{C}} = \underset{p+q=n}{\oplus} V^{p,q} \\ V^{q,p} = \bar{V}^{p,q}. \end{cases}$$

Definition (ii): A *Hodge structure of weight* n is given by a *Hodge filtration*

$$\begin{cases} F^n \subset F^{n-1} \subset \cdots \subset F^0 = V_{\mathbb{C}} \\ F^p \oplus \bar{F}^{n-p+1} \xrightarrow{\sim} V_{\mathbb{C}}. \end{cases}$$

These are equivalent by

$$F^p = \underset{p' \geqq p}{\oplus} V^{p',n-p'} \qquad ((\mathrm{i}) \implies (\mathrm{ii}))$$

$$V^{p,q} = F^p \cap \bar{F}^q \qquad ((\mathrm{ii}) \implies (\mathrm{i})).$$

For the third definition we shall use the *Deligne torus*

$$\mathbb{S} =: \mathrm{Res}_{\mathbb{C}/\mathbb{R}}\, \mathbb{G}_{m,\mathbb{C}}$$

where "Res" denotes the restriction of scalars à la Weil. A concrete, and in some ways more useful, description of $\mathbb{S}$ will be given below. It is an $\mathbb{R}$-algebraic group whose real points are

$$\mathbb{S}(\mathbb{R}) \cong \mathbb{C}^*.$$

Definition (iii)**:** A *Hodge structure of weight* n is given by a homomorphism of $\mathbb{R}$-algebraic groups

$$\widetilde{\varphi} : \mathbb{S}(\mathbb{R}) \to \mathrm{GL}(V)(\mathbb{R}) \tag{I.A.1}$$

such that for $r \in \mathbb{R}^* \subset \mathbb{S}(\mathbb{R})$

$$\widetilde{\varphi}(r) = r^n \,\mathrm{id}_V .$$

This is equivalent to definition (i) by

$$\begin{aligned} &\widetilde{\varphi}(z) = z^p \bar{z}^q \text{ on } V^{p,q} && ((\mathrm{i}) \implies (\mathrm{iii})) \\ &V^{p,q} = \{v \in V_{\mathbb{C}} : \widetilde{\varphi}(z)v = z^p \bar{z}^q v\} && ((\mathrm{iii}) \implies (\mathrm{i})). \end{aligned}$$

In other words, the action of $\widetilde{\varphi}(z)$ on $V_{\mathbb{C}}$ decomposes into eigenspaces $V^{p,q}$ as above, and this action is defined over $\mathbb{R}$ and is therefore given by a homomorphism (I.A.1). The *Weil operator* C is defined by

$$C = \widetilde{\varphi}(i)$$

and thus $C = i^{p-q}$ on $V^{p,q}$.

The alternate way one may formulate definition (iii) is the following: For $k = \mathbb{Q}$, $\mathbb{R}$ or $\mathbb{C}$ set

$$\mathbb{S}(k) = \left\{ \begin{pmatrix} a & -b \\ b & a \end{pmatrix} : \begin{matrix} a^2 + b^2 \neq 0 \\ a, b \in k \end{matrix} \right\}$$

$$\mathbb{U}(k) = \left\{ \begin{pmatrix} a & -b \\ b & a \end{pmatrix} : \begin{matrix} a^2 + b^2 = 1 \\ a, b \in k \end{matrix} \right\}.$$

Then

$$\mathbb{S}(\mathbb{R}) \cong \mathbb{C}^* \text{ via } \begin{pmatrix} a & -b \\ b & a \end{pmatrix} \to a + ib$$

$$\mathbb{S}(\mathbb{C}) \cong \mathbb{C}^* \times \mathbb{C}^* \text{ via } \begin{pmatrix} a & -b \\ b & a \end{pmatrix} \to (a + ib, a - ib)$$

and $\mathbb{U}(\mathbb{R}) \cong S^1$ is the maximal compact subgroup of $\mathbb{S}(\mathbb{R})$. We set

$$\varphi = \widetilde{\varphi}|_{\mathbb{U}(\mathbb{R})}.$$

Then (I.A.1) gives

$$\varphi : \mathbb{U}(\mathbb{R}) \to \mathrm{SL}(V)(\mathbb{R}) \tag{I.A.2}$$

where

$$\varphi(z)v = z^{p-q}v, \qquad v \in V^{p,q}.$$

This formulation, which is most useful when studying polarized Hodge structures, determines the Hodge decomposition but not the weight, for which one needs the scaling factor.[1]

The *Tate structure* $\mathbb{Q}(1)$ is defined by $V = 2\pi i\mathbb{Q} \subset \mathbb{C}$ and

$$\widetilde{\varphi}(z) = z^{-1}\bar{z}^{-1}.$$

Thus V is of pure Hodge type $(-1,-1)$. It is also sometimes referred to as the *Hodge-Tate structure.* We denote by $\mathbb{Q}(p)$ the pth tensor power of $\mathbb{Q}(1)$ and note that $\mathbb{Q}(-1) \cong \widetilde{\mathbb{Q}(1)}$.

In this paper we shall use both the (I.A.1) and (I.A.2) versions of definition (iii), and therefore shall denote a Hodge structure by $(V, \widetilde{\varphi})$ and (V, φ) or, when no confusion is likely, simply by $V_{\widetilde{\varphi}}$ and V_{φ}.

Remark: In the literature there are two sign conventions concerning the weight. Ours is this: For a complex torus $X = \mathbb{C}^g/\Lambda$ where $\mathbb{C}^g$ has coordinates $z^1, \ldots, z^g$ and $\Lambda \cong \mathbb{Z}^{2g}$ is a lattice, a differential form

$$\Psi = \sum_{|I|+|J|=n} dz^I \wedge d\bar{z}^J$$

defines a class in $H^n_{DR}(X) \cong H^n(X, \mathbb{C})$. Under a homothety $z^i \to \lambda z^i$, $\lambda \in \mathbb{R}^*$, we have

$$\Psi \to \lambda^n \Psi.$$

Thus, we will say that the Hodge structure on $H^n(X, \mathbb{Q})$ given by specifying that $H^n(X, \mathbb{C})^{(p,q)}$ is represented by Ψ as above where $|I| = p$, $|J| = q$ has weight n. In [Mo1] the weight, as defined there and following Deligne's convention [DMOS], is $-n$.

For later use it will be convenient to extend the action of φ to $\mathbb{S}(\mathbb{C}) \cong \mathbb{C}^* \times \mathbb{C}^*$ as given above where

$$\widetilde{\varphi}(z, w) = z^p w^q \text{ on } V^{p,q}.$$

This agrees with the above via $\mathbb{S}(\mathbb{R}) \subset \mathbb{S}(\mathbb{C})$ given by $z \to (z, \bar{z})$. Note that $\mathbb{U}(\mathbb{C}) \cong \mathbb{C}^*$, and the circle $U(\mathbb{R}) \cong S^1$ in particular maps into $\mathbb{S}(\mathbb{C})$ by $z \to$

[1]We have used the notation $\mathrm{GL}(V)(\mathbb{R})$ and $\mathrm{SL}(V)(\mathbb{R})$ for the $\mathbb{R}$-points of the $\mathbb{Q}$-algebraic groups $\mathrm{GL}(V)$ and $\mathrm{SL}(V)$. From now on we shall generally use the simpler notations $\mathrm{GL}(V_{\mathbb{R}})$ and $\mathrm{SL}(V_{\mathbb{R}})$.

(z, z^{-1}). The diagonal inclusion $\mathbb{G}_m \subset \mathbb{S}$ maps $\mathbb{C}^* \cong \mathbb{G}_m(\mathbb{C})$ (and $\mathbb{R}^* \cong \mathbb{G}_m(\mathbb{R})$) into $\mathbb{C}^* \times \mathbb{C}^*$ by $\alpha \mapsto (\alpha, \alpha)$.

Hodge structures of weight n are sometimes called *pure Hodge structures*, and the term Hodge structure then refers to a direct sum of pure Hodge structures.

Definition: A *Hodge structure* is given by a $\mathbb{Q}$-vector space and a representation

$$\widetilde{\varphi} : \mathbb{S}(\mathbb{R}) \to \mathrm{GL}(V_{\mathbb{R}})$$

with the following property: the restriction of $\widetilde{\varphi}$ to $\mathbb{G}_m$ should be a morphism of $\mathbb{Q}$-algebraic groups, yielding a homomorphism of rational points

$$\widetilde{\varphi} : \mathbb{Q}^* \to \mathrm{GL}(V).$$

Here, $G_m \cong \mathbb{Q}^*$ is the subgroup $\{\left(\begin{smallmatrix} a & 0 \\ 0 & a \end{smallmatrix}\right) : a \in \mathbb{Q}^*\}$ of $\mathbb{S}(\mathbb{Q})$. Over $\mathbb{Q}$ we have the weight space decomposition

$$V = \oplus V^{(n)}$$

where $\widetilde{\varphi}(r) = r^n$ on $V^{(n)}$, and over $\mathbb{C}$ we have

$$V_{\mathbb{C}} = \bigoplus_{p,q,n} V_{\mathbb{C}}^{(n)p,q}$$

where $\varphi(z) = z^p \bar{z}^q$, $p + q = n$, on $V_{\mathbb{C}}^{(n)p,q}$.

Hodge structures admit the standard operations of linear algebra, in particular $\otimes$ and Hom. We denote by $\check{V}$ the dual vector space of V and denote by $\check{V}_\varphi$ the Hodge structure of weight $-n$ induced on $\check{V}$ by a weight n Hodge structure V_φ.

A *sub-Hodge structure* is given by a $\mathbb{Q}$-vector subspace $U \subset V$ such that

$$\widetilde{\varphi}(\mathbb{S}(\mathbb{R})) : U(\mathbb{R}) \to U(\mathbb{R});$$

i.e., the action of $\widetilde{\varphi}(\mathbb{S}(\mathbb{R}))$ leaves $U(\mathbb{R})$ invariant. When one decomposes V into a direct sum of pure Hodge structures, there will be a corresponding induced decomposition of U as a direct sum of pure sub-Hodge structures.

Polarizations are most conveniently defined using definition (ii). Let

$$Q : V \otimes V \to \mathbb{Q}$$

be a non-degenerate form with

$$Q(u, v) = (-1)^n Q(v, u) \qquad u, v \in V.$$

We shall denote by $G = \mathrm{Aut}(V, Q)$ the $\mathbb{Q}$-algebraic group associated to (V, Q), and by $G(\mathbb{R}), G(\mathbb{C})$ the real and complex points of G.

Definition: A *polarized Hodge structure of weight* n is given by (V, Q, φ) where φ is as in (I.A.2) and where the *Hodge-Riemann bilinear relations*

$$\begin{cases} Q(F^p, F^{n-p+1}) = 0 \\ Q(u, C\overline{u}) > 0 \qquad\qquad 0 \neq u \in V_{\mathbb{C}} \end{cases}$$

are satisfied.

The bilinear relations are equivalent to

$$\begin{cases} Q(V^{p,q}, V^{p',q'}) = 0 \qquad p' \neq n-p \\ i^{p-q} Q(V^{p,q}, \overline{V}^{p,q}) > 0. \end{cases}$$

In this work we shall mainly but not exclusively be concerned with polarized Hodge structures, which is why we shall frequently work with φ instead of $\widetilde{\varphi}$.

When working with Hodge structures arising from geometry there will usually be given a lattice $H_{\mathbb{Z}} \subset H_{\mathbb{R}}$ such that $H = H_{\mathbb{Z}} \otimes \mathbb{Q}$.

I.B MUMFORD-TATE GROUPS

Mumford-Tate groups, sometimes abbreviated MT-groups, are the natural symmetry groups that encode information about the $\mathbb{Q}$-structure and the Hodge structure.

Definitions: (i) The *Mumford-Tate group* $M_{\widetilde{\varphi}}$ *associated to a Hodge structure* $(V, \widetilde{\varphi})$ is the $\mathbb{Q}$-algebraic closure of

$$\widetilde{\varphi} : \mathbb{S}(\mathbb{R}) \to \mathrm{GL}(V_{\mathbb{R}}).$$

(ii) The *Mumford-Tate group* M_{φ} *associated to a Hodge structure* (V, φ) *of weight* n is the $\mathbb{Q}$-algebraic closure of

$$\varphi : \mathbb{U}(\mathbb{R}) \to \mathrm{SL}(V_{\mathbb{R}}).$$

Thus

$$\begin{cases} M_{\widetilde{\varphi}}(\mathbb{R}) \supset \widetilde{\varphi}(\mathbb{S}(\mathbb{R})) \\ M_{\varphi}(\mathbb{R}) \supset \varphi(\mathbb{U}(\mathbb{R})), \end{cases}$$

and in each case the Mumford-Tate groups are the intersection of the $\mathbb{Q}$-algebraic subgroups of $\mathrm{GL}(V)$ whose real points contain $\widetilde{\varphi}(\mathbb{S}(\mathbb{R}))$, respectively $\varphi(\mathbb{U}(\mathbb{R}))$. In the literature, M_{φ} is usually called the *special Mumford-Tate group* or "Hodge group." Because of the centrality of M_{φ} in this monograph, and because the term "Hodge group" will be used in Chapter IV in another context, we will simply refer to both $M_{\tilde{\varphi}}$ and M_φ as Mumford-Tate groups and let the subscripts $\tilde{\varphi}$ and φ designate which one we are referring to.

We may define $M_{\tilde{\varphi}}$ and M_φ for a general Hodge structure $\tilde{\varphi} : \mathbb{S}(\mathbb{R}) \to \mathrm{GL}(V_{\mathbb{R}})$ (i.e., a $\mathbb{Q}$-direct sum of pure weight Hodge structures) in exactly the same way. As long as $(V, \tilde{\varphi})$ is not pure of weight zero, the $\mathbb{Q}$-closure definition implies that

$M_{\tilde{\varphi}}$ is the semi-direct product of its subgroups M_φ and $\mathbb{G}_{m,\mathbb{Q}}$.

If (V, Q, φ) is a weight n polarized Hodge structure, then φ preserves $Q \in \check{V} \otimes \check{V}$ and thus $M_\varphi \subset G = \mathrm{Aut}(V, Q)$.

We shall now formulate and prove the basic property for M_φ in the pure case; subsequently we shall do the same for $M_{\tilde{\varphi}}$. In each case the basic property is an answer to the question:

> *What are the defining equations for the $\mathbb{Q}$-algebraic group M_φ, respectively $M_{\tilde{\varphi}}$?*

For this we let

$$\begin{cases} T^{k,l} = V^{\otimes^k} \otimes \check{V}^{\otimes^l}, & k, l \geqq 0 \\ T^{\bullet,\bullet} = \underset{k,l \geqq l}{\oplus} T^{k,l} \end{cases}$$

be the *tensor spaces* and *tensor algebra* of V. Recall that for a pure Hodge structure of even weight $n = 2p$, the *Hodge classes* are defined by

$$\mathrm{Hg}(V_\varphi) = V \cap V^{p,p}\,.$$

Setting $\mathrm{Hg}_\varphi^{k,l} = \mathrm{Hg}(T_\varphi^{k,l})$, we then have

$$\begin{cases} \mathrm{Hg}_\varphi^{k,l} \subset T^{k,l} \\ \mathrm{Hg}_\varphi^{\bullet,\bullet} = \underset{k,l \geqq 0}{\oplus} H_\varphi^{k,l} \end{cases}$$

consisting of the *Hodge classes in $T^{k,l}$* and the *algebra of Hodge tensors* in $T^{\bullet,\bullet}$.

(I.B.1) BASIC PROPERTY (I): *M_φ is the subgroup of G fixing* $\mathrm{Hg}_\varphi^{\bullet,\bullet}$.

This result provides our answer to the previous question. The proof will be given in several steps.

Step one: *If $t \in \mathrm{Hg}_{\varphi}^{k,l}$, then M_{φ} fixes t.*

PROOF. Since t is rational, fixing it defines a $\mathbb{Q}$-algebraic subgroup $G(t)$ of $\mathrm{GL}(V)$. If t is of Hodge type (p,p), where $n(k-l)=2p$, then $\varphi(z)t = z^{p-p}t$. It follows that

$$\varphi(\mathbb{U}(\mathbb{R})) \subseteq G(t)(\mathbb{R}),$$

and by the minimality of M_{φ} we conclude that $M_{\varphi} \subset G(t)$. □

Step two: *If M_{φ} stabilizes the line $\mathbb{Q}t$ spanned by $t \in T^{k,l}$, then $t \in \mathrm{Hg}_{\varphi}^{k,l}$ is a Hodge tensor.*

PROOF. Since $\varphi(\mathbb{U}(\mathbb{C})) \subset M_{\varphi}(\mathbb{C})$, if M_{φ} stabilizes $\mathbb{Q}t$, then t will be an eigenvector for $\varphi(z)$ for a general $z \in \mathbb{U}(\mathbb{R})$. Hence t is of pure Hodge type, and since it is rational it must be a Hodge tensor. □

We then have

CHEVALLEY'S THEOREM: *Let M be a closed $\mathbb{Q}$-algebraic subgroup of* $\mathrm{GL}(V)$. *Then M is the stabilizer of a line L in a finite direct sum $\overset{m}{\underset{i=1}{\oplus}} T^{k_i,k_i}$.*

A proof of this result will be given below.

COMPLETION OF THE PROOF OF (I.B.1). Denoting by $\mathrm{Fix}(*)$ the $\mathbb{Q}$-algebraic subgroup of $\mathrm{GL}(V)$ defined by fixing pointwise a set $*$ of tensors in $T^{\bullet,\bullet}$, by step one we have

$$M_{\varphi} \subseteq \mathrm{Fix}(\mathrm{Hg}_{\varphi}^{\bullet,\bullet}).$$

For the reverse inclusion, by Chevalley's theorem there exists

$$\tau = (t_1, \ldots, t_m) \in \overset{m}{\underset{i=1}{\oplus}} T^{k_i,k_i}$$

such that $M_{\varphi} \subset \mathrm{GL}(V)$ is defined by stabilizing the line $\mathbb{Q}\tau$. In particular, M_{φ} stabilizes each line $\mathbb{Q}t_i \subset T^{k_i,k_i}$ and by step two each t_i is Hodge. Thus

$$\mathrm{Fix}(\mathrm{Hg}_{\varphi}^{\bullet,\bullet}) \subseteq \bigcap_i \mathrm{Fix}(t_i) \subseteq \mathrm{Stab}(\mathbb{Q}\tau) = M_{\varphi}. \qquad \square$$

PROOF OF CHEVALLEY'S THEOREM. From the open embedding $\mathrm{GL}(V) \hookrightarrow \mathrm{End}(V)$ of algebraic varieties comes the injection of coordinate rings

$$\mathbb{Q}[\mathrm{GL}(V)] \hookleftarrow \mathbb{Q}[\mathrm{End}(V)] \cong \underset{k\geq 0}{\oplus} \mathrm{Sym}^k(\mathrm{End}(V)^{\vee}) \subseteq \underset{k\geq 0}{\oplus} T^{k,k}.$$

The action of $\mathrm{GL}(V)$ on itself by conjugation extends to the adjoint action on $\mathrm{End}(V)$, and these induce compatible actions on the coordinate rings that, moreover, are compatible with the action of $\mathrm{GL}(V)$ on $\oplus T^{k,k}$. Write $S := \mathbb{Q}[\mathrm{End}(V)]$. If we choose a basis for V, we may think of S as polynomials $P(X^i_j)$ in the matrix entries of $X \in \mathrm{End}(V)$.

The stabilizer of the ideal

$$I(M) \subseteq \mathbb{Q}[\mathrm{GL}(V)]$$

of M, viewed as a subvariety of $\mathrm{GL}(V)$, is the largest algebraic subgroup contained in the Zariski closure of M. But since M is Zariski closed, and a subgroup, the stabilizer is just M. Note that S and

$$I := I(M) \cap S$$

inherit a nonnegative $\mathrm{GL}(V)$-invariant grading from the above injection of coordinate rings, and M is also the subgroup of $\mathrm{GL}(V)$ stabilizing I in S. This follows from the above because $\mathbb{Q}[\mathrm{GL}(V)] = \mathbb{Q}[\mathrm{End}(V)][\frac{1}{\det}]$ and det is nonvanishing on M.

Since M is an algebraic variety in $\mathrm{GL}(V)$, $I(M)$ is finitely generated; the same goes for I, the ideal of the Zariski closure of M in $\mathrm{End}(V)$. Let $\{P_i\} \subset I^{\leq k}$ be a generating set for I, and set $d := \dim(I^{\leq k})$. Since $I^{\leq k}$ generates I and the action is compatible with products, M is the stabilizer of $I^{\leq k}$ in $S^{\leq k}$, and this is, by linear algebra, the same as the stabilizer of

$$L := \wedge^d I^{\leq k} \subseteq \wedge^d S^{\leq k}$$

where $\wedge^d S^{\leq k} \subset \bigoplus_{\ell \geq 0} T^{\ell,\ell}$. □

Remark: The basic idea behind this argument is very simple:

(i) If M is a $\mathbb{Q}$-algebraic group acting on a finite dimensional $\mathbb{Q}$-vector space W, then stabilizing a subspace $\mathfrak{U} \subset W$ is the same as stabilizing the line $\Lambda^d \mathfrak{U} \subset \Lambda^d W$ where $\dim \mathfrak{U} = d$;

(ii) In the coordinate ring $\mathbb{Q}[X^i_j]$, M is exactly the stabilizer of the ideal that defines M;

(iii) By finite generation, (ii) will hold on a finite dimensional subspace of $\mathbb{Q}[X^i_j]$.

We now give the analog of (I.B.1) for the Mumford-Tate group $M_{\tilde{\varphi}}$. For this we denote by $\mathrm{Hg}^{\bullet,\bullet}_{\tilde{\varphi}}$ the algebra of Hodge tensors in $T^{\bullet,\bullet}$ when we write $V_{\tilde{\varphi}}$ as a $\mathbb{Q}$-direct sum of pure Hodge structures and expand everything out.

BASIC PROPERTY (II): *$M_{\tilde{\varphi}}$ is the subgroup of* $\mathrm{GL}(V)$ *fixing* $\mathrm{Gr}_0^W \mathrm{Hg}_{\tilde{\varphi}}^{\bullet,\bullet}$.

Here, the notation $\mathrm{Gr}_0^W \mathrm{Hg}_{\tilde{\varphi}}^{\bullet,\bullet}$ refers to the weight zero part $\bigoplus_i \mathrm{Hg}_{\tilde{\varphi}}^{k_i,k_i}$ of $\mathrm{Hg}_{\tilde{\varphi}}^{\bullet,\bullet}$.

PROOF. The argument in step one in (I.B.1) applies here to show that $M_{\tilde{\varphi}}$ fixes $\mathrm{Gr}_0^W \mathrm{Hg}_{\tilde{\varphi}}^{\bullet,\bullet}$; that is

$$M_{\tilde{\varphi}} \subseteq \mathrm{Fix}\, \mathrm{Gr}_0^W \mathrm{Hg}_{\tilde{\varphi}}^{\bullet,\bullet}.$$

To show the reverse inclusion, by the argument in step two we have

$$\mathrm{Fix}\, \mathrm{Gr}_0^W \mathrm{Hg}_{\tilde{\varphi}}^{\bullet,\bullet} \subseteq \bigcap_i \mathrm{Fix}(t_i) = M_{\tilde{\varphi}},$$

where the t_i are of weight zero since they belong to T^{k_i,k_i}. □

For later use we note the following.

(I.B.2) PROPOSITION: (i) *If $(V, \tilde{\varphi})$ is a Hodge structure, not necessarily of pure weight, and $W \subset V$ is a subspace invariant under $M_{\tilde{\varphi}}$, then W is a sub-Hodge structure.* (ii) *If (V, φ) is a Hodge structure of weight n and $W \subset V$ is a sub-space invariant under M_{φ}, the W is a sub-Hodge structure.*

PROOF OF (i). From $M_{\tilde{\varphi}}(W) \subseteq W$ we infer that

$$\tilde{\varphi}(\mathbb{S}(\mathbb{R})) : W_{\mathbb{R}} \to W_{\mathbb{R}}$$

and then $(W, \tilde{\varphi})$ is a sub-Hodge structure of $(V, \tilde{\varphi})$.

PROOF OF (ii). This is the same argument. □

Remark: In (ii), if $(V, \tilde{\varphi})$ is a Hodge structure not necessarily of pure weight and $W \subset V$ is subspace invariant under M_{φ}, it does not follow that W is a sub-Hodge structure. We may take $V = V_1 \oplus V_2$ and $\tilde{\varphi} = \varphi_1 \oplus \varphi_2$ where (V_i, φ_i) is a Hodge structure of weight $2p_i$ with $p_1 \neq p_2$, and for W the line

$$\mathbb{Q}(t_1 \oplus t_2)$$

where $t_i \in \mathrm{Hg}(V_i, \varphi_i)$. Since this line is not invariant under scaling, it is not a sub-Hodge structure.

To generalize (I.B.2) to subspaces of the $T^{k,l}$, we will use the following important property of MT-groups: Let

$$\rho : \mathrm{GL}(V) \to \mathrm{GL}(V_\rho)$$

be a representation of $\mathrm{GL}(V)$. Then a Hodge structure $(V, \tilde{\varphi})$ on V induces a Hodge structure $V_{\rho(\tilde{\varphi})} := (V_\rho, \rho \circ \tilde{\varphi})$ on V_ρ, and we have

(I.B.3) *$M_{\rho(\widetilde{\varphi})}$ is the image of $M_{\widetilde{\varphi}}$ under the natural map $\rho : \mathrm{GL}(V) \to \mathrm{GL}(V_\rho)$.*

PROOF OF (I.B.3). For an algebraic group X defined over $\mathbb{Q}$ and a subgroup $Y \subset X(\mathbb{C})$, we denote by

$$\overline{Y}^{\mathbb{Q}} \subset X$$

the intersection of all $\mathbb{Q}$-algebraic subgroups $Z \subset X$ which have $Y \subset Z(\mathbb{C})$.

Now let $f : X_1 \to X_2$ be a morphism of algebraic groups defined over $\mathbb{Q}$ and $Y \subset X_1(\mathbb{C})$. We observe that

$$f(\overline{Y}^{\mathbb{Q}}) = \overline{f(Y)}^{\mathbb{Q}}. \tag{I.B.4}$$

Applying this when

$$\begin{cases} X_1 = \mathrm{GL}(V) \text{ and } X_2 = \mathrm{GL}(V_\rho) \\ f = \rho \\ Y = \widetilde{\varphi}(\mathbb{S}(\mathbb{R})) \end{cases}$$

we find that (I.B.3) follows from the above where

- $f(Y) = (\rho \circ \widetilde{\varphi})\mathbb{S}(\mathbb{R})$
- $\overline{Y}^{\mathbb{Q}} = M_{\widetilde{\varphi}}$
- $f(\overline{Y}^{\mathbb{Q}}) = \rho(M_{\widetilde{\varphi}}) = M_{\rho(\widetilde{\varphi})}$,

the last step following from (I.B.4). □

Now let $W \subset T^{k,l}$ be a linear subspace. Then the generalization of (I.B.2) is that

(I.B.5) *W is a sub-Hodge structure if, and only if,*

$$M_{\widetilde{\varphi}}(W) \subseteq W.$$

PROOF OF (I.B.5). If W is a sub-Hodge structure, then $W_{\mathbb{C}}$ is a sum of eigenspaces for $\widetilde{\varphi}(\mathbb{S}(\mathbb{R}))$ and thus

$$\widetilde{\varphi}(\mathbb{S}(\mathbb{R}))W_{\mathbb{C}} \subseteq W_{\mathbb{C}}.$$

We then have

$$\begin{aligned} M_{\widetilde{\varphi}}(W) &= \overline{\widetilde{\varphi}(\mathbb{S}(\mathbb{R}))}^{\mathbb{Q}}(W) \\ &= \overline{\widetilde{\varphi}(\mathbb{S}(\mathbb{R}))W_{\mathbb{C}}}^{\mathbb{Q}} \subseteq \overline{W}_{\mathbb{C}}^{\mathbb{Q}} = W, \end{aligned}$$

where the first step is the definition of $M_{\tilde{\varphi}}$. Thus, if W is a sub-Hodge structure then $M_{\tilde{\varphi}}(W) \subseteq W$.

For the converse, if $M_{\tilde{\varphi}}(W) \subseteq W$, then $\tilde{\varphi}(\mathbb{S}(\mathbb{R}))W_{\mathbb{C}} \subseteq W_{\mathbb{C}}$ and so $W_{\mathbb{C}}$ is a direct sum of $\tilde{\varphi}(\mathbb{S}(\mathbb{R}))$-eigenspaces. Thus W is a sub-Hodge structure. □

The same property then automatically extends to W in finite direct sums of $T^{k,l}$'s and hence to arbitrary representations of M, which are subquotients of these.

Remark: Our terminology is taken from [Mo1], [Mo2]. An alternate approach to the definition of MT-groups is in terms of "fixing things" as in the basic property (I.B.1). This has the advantage (at least for M_{φ}) of better generalizing to the case of non-split mixed Hodge structures, discussed in Section I.C.

As a corollary, we have

(I.B.6) *If (V, Q, φ) is a polarized Hodge structure then $M_{\tilde{\varphi}}$ and M_{φ} are reductive.*

PROOF. It is enough to consider $W \subset T^{k,l}$ invariant under $M_{\tilde{\varphi}}$. Using (I.B.5) and Q, we have $T^{k,l} = W \oplus W^{\perp}$ as a direct sub of Hodge structures invariant under $M_{\tilde{\varphi}}$. Reductivity of M_{φ} then follows from that of $M_{\tilde{\varphi}}$ since it is a normal subgroup. □

Finally, we note the following:

(I.B.7) *Given a morphism $\sigma : V_{\tilde{\varphi}} \to V'_{\tilde{\varphi}'}$ of Hodge structures, denote by $\sigma(V)_{\sigma(\tilde{\varphi})}$ the image of σ. It is sub-Hodge structure of $V'_{\varphi'}$. Then we have surjective maps*

$$M_{\tilde{\varphi}} \twoheadrightarrow M_{\sigma(\tilde{\varphi})} \twoheadleftarrow M_{\varphi'}.$$

We note that σ does *not* induce a homomorphism of Mumford-Tate groups.

I.C MIXED HODGE STRUCTURES AND THEIR MUMFORD-TATE GROUPS

In this section we shall define and discuss the Mumford-Tate groups for mixed Hodge structures (MHS), including both the general and polarized cases. A central topic of general interest is the mixed Hodge structures that arise when a family of polarized Hodge structures degenerates. In this case, one obtains the polarized limiting mixed Hodge structure (LMHS, the polarization being understood). Moreover, the degeneration of a family of polarized Hodge structures is, in a precise sense, approximated by a corresponding nilpotent orbit Φ^{nilp}.

The definition of Mumford-Tate groups for variations of Hodge structure $\Phi : (\Delta^*)^r \to \mathcal{M}_\Gamma$ given in Chapter III is compatible with nilpotent orbits, in the sense that one has

(I.C.1) $$M_\Phi \supseteq M_{\Phi^{\text{nilp}}}$$

since global Hodge tensors are unchanged by the process of constructing Φ^{nilp}. As nilpotent orbits are the points of the boundary components in the Kato-Usui extensions, or partial compactifications, of period domains [KU], discrete-group quotients of Mumford-Tate domains should admit log-analytic completions in the same sense. Although we shall not pursue this line of thought in this monograph,[2] it might be kept in mind as motivation for some of what is discussed below.

Thus, we will consider MT groups $M_{(F,W)}$ of mixed Hodge structures; when these arise as limit mixed Hodge structures, an inclusion related to (I.C.1) holds. Namely, let $\{T_i^u\}$ be the unipotent parts in the Jordan decompositions of the monodromy operators $\{T_i\}$ associated to the variation of Hodge structure Φ (Section II.C), $\{N_i^u\}$ their logarithms, and $W_\bullet^{\text{lim}} := W_\bullet(N_1 + \cdots + N_r)$ the monodromy weight filtration. Then $M_{(F^\bullet_{\text{lim}}, W_\bullet^{\text{lim}})}$ is contained in the centralizer of the $\{T_i^u\}$ in M_Φ. In particular, as morphisms of mixed Hodge structure, the $\{N_i\}$ are Hodge tensors for the limiting mixed Hodge structure and hence must be fixed by the Mumford-Tate group.

We will prove a basic result (Theorem (I.C.6), generalizing (I.B.1)), which may simplify some proofs and aid in the computation of these groups. Unless the weight filtration splits over $\mathbb{Q}$, Mumford-Tate groups of mixed Hodge structures are not reductive; indeed, isolating the unipotent radical as done in Lemma (I.C.9) plays a key rôle in the proof.

Recall that a *mixed Hodge structure* is given by a $\mathbb{Q}$-vector space V endowed with an increasing filtration $W_\bullet V$ together with a decreasing filtration $F^\bullet V_\mathbb{C}$, which induces the structure of a weight i Hodge structure on each graded piece $\mathrm{Gr}_i^W V$. Associated to such a MHS is the *Deligne bigrading*

$$V^{p,q} := \{W_{p+q} \cap F^p\} \cap \left\{ W_{p+q} \cap \bar{F}^q + \sum_{i \geq 2} W_{p+q-i} \cap \bar{F}^{q-i+1} \right\}$$

of $V_\mathbb{C}$.

(I.C.2) PROPOSITION: (i) [De4] *$V^{\bullet,\bullet}$ is the unique bigrading of $V_\mathbb{C}$ satisfying*

- $V_\mathbb{C} = \underset{(p,q)}{\oplus} V^{p,q}$;

[2]It will be treated in a forthcoming work of M. Kerr and G. Pearlstein.

- $F^a V_{\mathbb{C}} = \underset{(p,q):p \geq a}{\oplus} V^{p,q}$;
- $W_n V_{\mathbb{C}} = \underset{(p,q):p+q \leq n}{\oplus} V^{p,q}$;
- $\overline{V}^{b,a} \equiv V^{a,b} \mod \underset{(p,q):p<a,q<b}{\oplus} V^{p,q}$.

(ii) *$V^{\bullet,\bullet}$ is compatible with morphisms of MHS and with the basic operators of tensor, dual, sub, and quotient.*

This yields a natural extension of (iii) in (I.A.1) to the mixed case: recalling the isomorphism $\mathbb{S}(\mathbb{C}) \cong \mathbb{C}^* \times \mathbb{C}^*$, define a morphism of $\mathbb{C}$-algebraic groups

$$\widetilde{\varphi} : \mathbb{S}(\mathbb{C}) \to \mathrm{GL}(V)_{\mathbb{C}} \tag{I.C.3}$$

by the rule

$$\widetilde{\varphi}(z, w)\big|_{V^{p,q}} = \text{multiplication by } z^p w^q.$$

One can also define $\varphi : \mathbb{U}(\mathbb{C}) \to \mathrm{GL}(V)_{\mathbb{C}}$ by composing $\widetilde{\varphi}$ with the inclusion $\mathbb{C}^* \cong \mathbb{U}(\mathbb{C}) \subset \mathbb{S}(\mathbb{C})$ given by $z \mapsto (z, z^{-1})$. For the proofs below, it is convenient to define an "extended mixed Hodge representation"

$$\widehat{\varphi} : \mathbb{S}(\mathbb{C}) \to \mathrm{GL}(V_{\mathbb{C}}) \times \mathbb{G}_{m,\mathbb{C}} \tag{I.C.4}$$

by

$$\widehat{\varphi}(z, w) := (\widetilde{\varphi}(z, w), zw).$$

This is really nothing but (I.C.3) with V replaced by $V \oplus \mathbb{Q}(-1)$.

As before, we have the tensor spaces $T^{k,l} := V^{\otimes k} \otimes \check{V}^{\otimes l}$, and $\mathrm{GL}(V) \times \mathbb{G}_m$ acts on $T^{k,l}(p) := T^{k,l} \otimes \mathbb{Q}(p)$ by $(A, \alpha)\tau \otimes \zeta := (A\tau) \otimes \alpha^{-p}\zeta$. They give rise to spaces of Hodge tensors by

$$\mathrm{Hg}^{k,l;p}_{(F,W)} := (T^{k,l})^{p,p} \cap T^{k,l}$$
$$\mathrm{Hg}^{k,l;p}_{\widehat{(F,W)}} := (T^{k,l}(p))^{0,0} \cap T^{k,l}(p).$$

(I.C.5) **Definition:**

(i) The *Mumford-Tate groups* $M_{(F,W)}$ (resp. $M_{\widetilde{(F,W)}}, M_{\widehat{(F,W)}}$) of the MHS on V are the largest subgroups of $\mathrm{SL}(V)$ (resp. $\mathrm{GL}(V)$, $\mathrm{GL}(V) \times \mathbb{G}_m$) pointwise fixing $\mathrm{Hg}^{k,l;p}_{(F,W)}$ (for all k, l, p) (resp. $\mathrm{Hg}^{k,l;0}_{(F,W)}$ (for all k, l), $\mathrm{Hg}^{k,l;p}_{\widehat{(F,W)}}$ (for all k, l, p)).

(ii) Denote by M_φ (resp. $M_{\widetilde{\varphi}}, M_{\widehat{\varphi}}$) the $\mathbb{Q}$-closures of the images of φ (resp. $\widetilde{\varphi}, \widehat{\varphi}$).

In Section II.A we will define the space of Hodge tensors for general flags $F^\bullet$ in the compact dual of the period domain. For those $F^\bullet$ that define pure Hodge structures, the definition will agree with the one just given.

The rest of this section is devoted to the proof of

(I.C.6) THEOREM: *We have* $M_{\widetilde{(F,W)}} = M_{\widetilde{\varphi}}$ *and* $M_{\widehat{(F,W)}} = M_{\widehat{\varphi}}$*, but in general*[3] $M_{(F,W)} \supsetneq M_{\varphi}$.

We begin by observing that by Proposition (I.C.2)(ii), if V_0 is a subquotient MHS of V, then the mixed Hodge representation for V_0, denoted by $\widehat{\varphi}_0$, factors through $\widehat{\varphi}$ in the sense of having the diagram

$$\begin{array}{ccc} \mathbb{S}(\mathbb{C}) & \xrightarrow{\widehat{\varphi}} & M_{\widehat{\varphi},\mathbb{C}} \\ & \searrow^{\widehat{\varphi}_0} & \downarrow \\ & & \mathrm{GL}(V_0)_{\mathbb{C}} \times \mathbb{G}_{m,\mathbb{C}} \,. \end{array}$$

Hence, arguing as in the proof of (I.B.3), we have a canonical surjective morphism

$$M_{\widehat{\varphi}} \twoheadrightarrow M_{\widehat{\varphi}_0}$$

defined over $\mathbb{Q}$. A similar argument works for the $\mathbb{Q}$-direct sum of subquotients

$$V^{\mathrm{split}} := \bigoplus_i \mathrm{Gr}_i^W V, \tag{I.C.7}$$

using the diagram

$$\begin{array}{ccc} \mathrm{GL}(V)_{\mathbb{C}} \times \mathbb{G}_{m,\mathbb{C}} & \xrightarrow[\cong]{\Xi} & \mathrm{GL}(V^{\mathrm{split}})_{\mathbb{C}} \times \mathbb{G}_{m,\mathbb{C}} \\ \cup & & \cup \\ \mathrm{Aut}(V, W_\bullet) \times \mathbb{G}_{m,\mathbb{C}} & \xrightarrow{\xi} & (\times_i \mathrm{GL}(\mathrm{Gr}_i^W V))_{\mathbb{C}} \times \mathbb{G}_{m,\mathbb{C}} \\ & \nwarrow^{\widehat{\varphi}} \quad \mathbb{S}(\mathbb{C}) \quad \nearrow_{\widehat{\varphi}_{\mathrm{split}}} & \end{array}$$

(I.C.8) **Remark:** Here Ξ, which is induced by the Deligne splitting, is only defined over $\mathbb{C}$; on the other hand, ξ — which is induced by taking the $W_\bullet$-graded pieces — and both vertical inclusions are defined over $\mathbb{Q}$. We note that $\mathrm{Ker}(\xi)$ is unipotent.

[3]The exception here is the case of $\mathbb{Q}$-split MHS, i.e., "general Hodge structures" as explained.

Taking $\mathbb{Q}$-closures, and again arguing as in (I.B.3), we have proved the

(I.C.9) LEMMA: *There is a canonical short-exact sequence of algebraic groups over* $\mathbb{Q}$

$$0 \to \mathfrak{U} \to M_{\widehat{\varphi}} \xrightarrow{\pi} M_{\widehat{\varphi}}^{\text{split}} \to 0$$

where $\mathfrak{U}$ *is unipotent and* $M_{\widehat{\varphi}}^{\text{split}}$ *is the* $\mathbb{Q}$*-closure of* $\widehat{\varphi}_{\text{split}}$.

Next let $M \subset \mathrm{GL}(V) \times \mathbb{G}_m$ be any closed $\mathbb{Q}$-algebraic subgroup.

(I.C.10) **Definition:** We let M' be the largest subgroup of $\mathrm{GL}(V) \times \mathbb{G}_m$ fixing all tensors $\tau \in T^{k,l}(p)$ (over all k, l, p) fixed by M.

Obviously $M \subset M'$ and M' is defined over $\mathbb{Q}$. Now recall the following from [DMOS], part (i) of which just restates Chevalley:

(I.C.11) PROPOSITION: (i) *M is the stabilizer of a $\mathbb{Q}$-line in some multitensor representation* $\bigoplus_i T^{k_i,l_i}(p_i)$.

(ii) *If M is reductive, or if every $\mathbb{Q}$-character of M extends to a $\mathbb{Q}$-character of* $\mathrm{GL}(V) \times \mathbb{G}_m$ *then* $M = M'$.

We are interested in the case $M = M_{\widehat{\varphi}}$, which is not reductive. Here is why we want to use (ii):

(I.C.12) LEMMA: $M'_{\widehat{\varphi}} = M_{\widehat{(F,W)}}$.

PROOF (SIMILAR TO [DMOS]). For $\tau \in T^{k,l}(p)$,

$$\tau \text{ is of type } (0,0) \quad \Leftrightarrow \quad \tau \text{ is fixed by } \widehat{\varphi}(\mathbb{S}(\mathbb{C}))$$

and we claim that this is

(I.C.13) $$\Leftrightarrow \quad \tau \text{ is fixed by } M_{\widehat{\varphi}}.$$

To see " $\Longrightarrow$ " in (I.C.13), note that $M_{\widehat{\varphi}}$ is the intersection of all $\mathbb{Q}$-subgroups $L \subseteq \mathrm{GL}(V) \times \mathbb{G}_m$ with $L_{\mathbb{C}} \supseteq \widehat{\varphi}(\mathbb{S}(\mathbb{C}))$. For any rational τ fixed by $\widehat{\varphi}(\mathbb{S}(\mathbb{C}))$, there is a $\mathbb{Q}$-equation equivalent to fixing it, which defines such an L.

Comparing Definitions (I.C.10) and (I.C.5)(i) the Lemma is now clear. $\square$

PROOF OF THEOREM (I.C.6). We will show that every $\mathbb{Q}$-character $\chi : M_{\widehat{\varphi}} \to \mathbb{G}_m$ extends to some character $\widehat{\chi} : \mathrm{GL}(V) \times \mathbb{G}_m \to \mathbb{G}_m$, also defined over $\mathbb{Q}$.

Recall that $\mathbb{S}(\mathbb{C}) \cong \mathbb{C}^* \times \mathbb{C}^*$, and consider the diagram

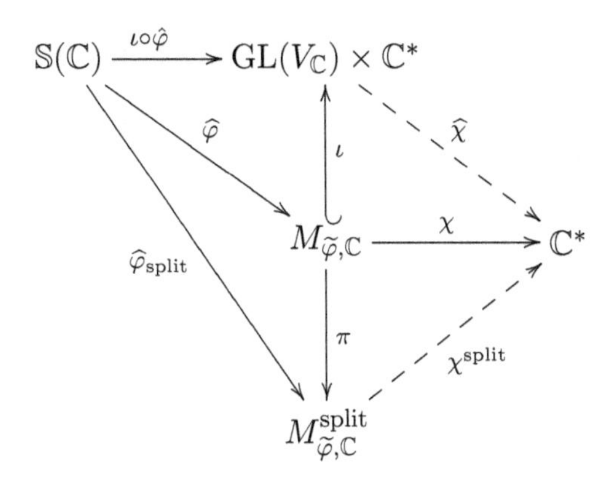

in which χ^{split} and $\widehat{\chi}$ are yet to be defined and where $\iota \circ \widehat{\varphi}$ had previously been denoted $\widehat{\varphi}$. We make the observations:

(i) Any homomorphism of algebraic groups $\mathbb{C}^* \times \mathbb{C}^* \to \mathbb{C}^*$, in particular $\chi \circ \widehat{\varphi}$, must be of the form $(z, w) \mapsto z^m w^n$ (for $m, n \in \mathbb{Z}$).

(ii) Since a unipotent group, such as $\ker(\pi)$, has no nontrivial characters, χ factors through a $\mathbb{Q}$-character

$$\chi^{\mathrm{split}} : M^{\mathrm{split}}_{\widehat{\varphi}} \to \mathbb{G}_m.$$

(iii) As V^{split} is a direct sum of pure Hodge structures, $\widehat{\varphi}_{\mathrm{split}}$ is defined over $\mathbb{R}$, hence so is $\chi^{\mathrm{split}} \circ \widehat{\varphi}_{\mathrm{split}} = \chi \circ \widehat{\varphi}$.

By (iii), $\chi \circ \widehat{\varphi}$ sends the real points $\mathbb{S}(\mathbb{R}) = \{(z, \overline{z}) \in \mathbb{C}^* \times \mathbb{C}^*\}$ to the real points $\mathbb{G}_m(\mathbb{R}) \cong \mathbb{R}^*$; therefore in (i) we must have $m = n$; i.e., $(\chi \circ \widehat{\varphi})(z, w) = (zw)^n$.

Now define the $\mathbb{Q}$-character $\widehat{\chi} : \mathrm{GL}(V) \times \mathbb{G}_m \to \mathbb{G}_m$ by

$$\widehat{\chi}(g, t) := t^n.$$

We have

$$\begin{aligned}(\widehat{\chi} \circ \iota \circ \widehat{\varphi})(z, w) &= \widehat{\chi}(\varphi(z, w), zw) \\ &= (zw)^n \\ &= (\chi \circ \widehat{\varphi})(z, w),\end{aligned}$$

i.e., $\widehat{\chi} \circ \iota = \chi$ on $\widehat{\varphi}(\mathbb{S}(\mathbb{C}))$. Since $\widehat{\chi} \circ \iota$ and χ are both defined over $\mathbb{Q}$, they must also agree on the $\mathbb{Q}$-closure $M_{\widehat{\varphi}}$ of $\widehat{\varphi}(\mathbb{S}(\mathbb{C}))$. Therefore χ extends and

Proposition (I.C.11)(ii) gives that $M_{\widehat{\varphi}} = M'_{\widehat{\varphi}}$, which in turn is equal to $M_{\widehat{(F,W)}}$ by Lemma (I.C.12).

To show that $M_{\widetilde{\varphi}} = M_{\widetilde{(F,W)}}$, we observe that the projection $\mathrm{GL}(V) \times \mathbb{G}_m \twoheadrightarrow \mathrm{GL}(V)$ sends $M_{\widehat{\varphi}} \twoheadrightarrow M_{\widetilde{\varphi}}$ and $M_{\widehat{(F,W)}} \twoheadrightarrow M_{\widetilde{(F,W)}}$.

Finally, for a counterexample to $M_{\varphi} = M_{(F,W)}$, let V be a non-split Hodge-Tate MHS, so that $M_{(F,W)}$ is a nontrivial unipotent subgroup of $\mathrm{GL}(V)$. But since $V_{\mathbb{C}} = \bigoplus_{p} V^{p,p}$, φ is trivial and so $M_{\varphi} = \{1\}$. □

(I.C.14) **Remark:** One easily shows that the map $M_{\widehat{\varphi}} \twoheadrightarrow M_{\widetilde{\varphi}}$ is an isogeny if V is graded-polarizable and an isomorphism if V is a limiting mixed Hodge structure, or more generally, some $T^{k,l}$ of V has a $\mathbb{Q}(1)$ as a subquotient. In the literature, $M_{\widehat{\varphi}}$ is referred to as the *extended MT group*.

Chapter II

Period Domains and Mumford-Tate Domains

II.A PERIOD DOMAINS AND THEIR COMPACT DUALS

The principal general reference is [C-MS-P]. We refer also to [CGG] for a survey and some more recent developments. We assume given (V, Q) as in Section I.A, together with a set of *Hodge numbers* $h^{p,n-p} = h^{n-p,p}$ with $\sum_{p=0}^{n} h^{p,n-p} = \dim V$. We also set $f^p = \sum_{p' \geqq p} h^{p',n-p'}$.

Definition: The *period domain* D associated to the above data is the set of polarized Hodge structures (V, Q, φ) with $\dim V^{p,q} = h^{p,q}$.

The natural symmetry group acting on D is the group $G(\mathbb{R})$ of real points of the $\mathbb{Q}$-algebraic group $G = \operatorname{Aut}(V, Q)$. Elementary linear algebra shows that $G(\mathbb{R})$ operates transitively on D, so that upon choice of a reference Hodge structure $\varphi \in D$ we have an identification

$$D = G(\mathbb{R})/H_{\varphi},$$

where the *isotropy group* H_{φ} is the subgroup of $G(\mathbb{R})$ that preserves the Hodge decomposition $V_{\mathbb{C}} = \underset{p+q=n}{\oplus} V^{p,q}$ corresponding to φ. It is easy to see that H_{φ} is compact and is the centralizer in $G(\mathbb{R})$ of the circle $\varphi(\mathbb{U}(\mathbb{R}))$. When confusion seems unlikely we shall simply write H for H_{φ}.

Note: To be precise, we should note that H_{φ} is an algebraic group defined over $\mathbb{R}$ and write $D = G(\mathbb{R})/H_{\varphi}(\mathbb{R})$. However, we shall use the traditional notation $D = G(\mathbb{R})/H_{\varphi}$ for period domains. We do remark that the group $H_{\varphi}(\mathbb{C})$ of $\mathbb{C}$-valued points of H_{φ} will be used in Chapter VI.

Group theoretic description of D. If we fix a reference point $\varphi_0 \in D$, then we have a set-theoretic identification

$$\text{(II.A.1)} \qquad \left\{ \begin{array}{c} \text{real representations} \\ g\varphi_0 g^{-1} = \varphi_g : \mathbb{U}(\mathbb{R}) \to G(\mathbb{R}) \\ \text{conjugate to } \varphi_0 \end{array} \right\} \leftrightarrow D.$$

It is given by the map

$$g\varphi_0 g^{-1} \to gH_{\varphi_0} \in G(\mathbb{R})/H_{\varphi_0}.$$

If $\varphi_g = \varphi_{\tilde{g}}$, then we have

$$(g^{-1}\tilde{g})\varphi_0(g^{-1}\tilde{g})^{-1} = \varphi_0,$$

which using the fact that $H_{\varphi_0} = Z(\varphi_0)$ is the centralizer of φ_0 gives

$$\tilde{g} = gh, \qquad h \in H_{\varphi_0}.$$

Thus, the above mapping is well defined. A similar argument shows that it is injective, and it is obviously surjective.

In the situation in which we have a lattice $V_{\mathbb{Z}}$ we set $G_{\mathbb{Z}} = \mathrm{Aut}(V_{\mathbb{Z}}, Q)$ and recall that an *arithmetic group* is a subgroup $\Gamma \subset G$ that is commensurable with $G_{\mathbb{Z}}$; i.e., coincides with $G_{\mathbb{Z}}$ up to a finite group.

The Hodge structure φ induces a Hodge structure of weight zero on $\mathrm{Hom}(V, V)$ where

$$\mathrm{Hom}(V_{\mathbb{C}}, V_{\mathbb{C}})^{-i,i} = \left\{X : V^{p,q} \to V^{p-i,q+i}\right\}.$$

This Hodge structure is polarized, and the Lie algebra $\mathfrak{g}$ of G is a polarized sub-Hodge structure. We denote by

$$\mathrm{Ad}\,\varphi : \mathbb{U}(\mathbb{R}) \to \mathrm{GL}(\mathfrak{g}_{\mathbb{R}})$$

the induced Hodge structure. Writing

$$\mathfrak{g}_{\mathbb{C}} = \oplus \mathfrak{g}^{-i,i}$$

we observe that

$$\left[\mathfrak{g}^{-i,i}, \mathfrak{g}^{-j,j}\right] \subseteq \mathfrak{g}^{-(i+j),i+j}. \tag{II.A.2}$$

The image of the map $\mathrm{Ad}\,\varphi$ leaves $\mathfrak{h}_{\mathbb{R}}$ invariant and thus induces a complex structure on the real tangent space

$$\mathfrak{g}_{\mathbb{R}}/\mathfrak{h}_{\mathbb{R}} \cong T_{\mathbb{R},\varphi}D.$$

Passing to the complexifications we have

$$\mathfrak{h}_{\mathbb{C}} = \mathfrak{g}^{0,0}$$

and using the above identification

$$\begin{cases} T_{\varphi}D^{(1,0)} &= \underset{i>0}{\oplus}\, \mathfrak{g}^{-i,i} \\ T_{\varphi}D^{(0,1)} &= \underset{i<0}{\oplus}\, \mathfrak{g}^{-i,i} = \overline{T_{\varphi}D^{(1,0)}}. \end{cases}$$

From (II.A.2) it follows that the resulting $G(\mathbb{R})$-invariant almost-complex structure on D given in each tangent space by $T_\varphi D^{(1,0)}$ is integrable. Therefore, D is a homogeneous complex manifold; we shall shortly give another, more explicit, description of this complex structure.

An important observation is that *the Lie algebra $\mathfrak{m}_\varphi$ of the Mumford-Tate group M_φ is a sub-Hodge structure of* $\mathfrak{g}$. It follows that the *orbit*

$$D_{M_\varphi} =: M_\varphi(\mathbb{R}) \cdot \varphi \subset D$$

of the group of real points of M_φ is a homogeneous, complex submanifold of D.

Indeed, we have that

$$\mathfrak{m}_{\varphi,\mathbb{C}} = \mathfrak{m}_\varphi^- \oplus \mathfrak{m}_\varphi^0 \oplus \mathfrak{m}_\varphi^+$$

where $\mathfrak{m}_\varphi^\pm = \mathfrak{m}_{\varphi,\mathbb{C}} \cap \mathfrak{g}^\pm$ and $\mathfrak{m}_\varphi^0 = \mathfrak{m}_{\varphi,\mathbb{C}} \cap \mathfrak{g}^{0,0}$. Then $\mathfrak{m}_\varphi^-$ defines an integrable $M_\varphi(\mathbb{R})$-invariant, integrable almost-complex structure on the orbit D_{M_φ}, for the same reasons as just given for the case of $G(\mathbb{R})$.

This simple observation will be of considerable significance later. The orbit D_{M_φ} will be called a *Mumford-Tate domain* (formal definition to be given in Section II.B). As will be seen in Chapter III, it is these rather than the period domains that are the natural objects for the study of variations of Hodge structure (cf. Theorem III.A.1).

We note that, M_φ being reductive, its group of real points $M_\varphi(\mathbb{R})$ will split into an almost product of simple factors M_i and an abelian part A. More precisely, the real Lie algebra

$$\mathfrak{m}_{\varphi,\mathbb{R}} = \bigoplus_i \mathfrak{m}_i \oplus \mathfrak{A}$$

where the $\mathfrak{m}_i$ are simple and $\mathfrak{A}$ is abelian. This decomposition may be non-trivial, even when M_φ is $\mathbb{Q}$-simple, as happens for $\mathrm{SL}_2(\mathbb{Q}(\sqrt{-d}))$. Moreover, the factors M_i may be either compact or non-compact. In the classical case an example with some compact factors was given by Mumford [M2]. Since a bounded symmetric domain contains no compact, complex submanifolds, in the classical case the orbits of the compact factors turn out to be just points. In the non-Hermitian symmetric case, the period domain will always contain homogeneous, compact complex submanifolds and the compact factors may have positive dimensional orbits. As previously noted, the Mumford-Tate group may itself be compact with a positive dimensional orbit. Below we shall see that variations of Hodge structure will meet these compact orbits only in points.

We now turn to the compact duals of period domains.

Definition: The *compact dual* $\check{D}$ is the set of flags $F^\bullet = \{F^n \subset \cdots \subset F^0 = V_{\mathbb{C}}\}$ where $\dim F^p = f^p$ and which satisfy the first Hodge-Riemann bilinear relation

$$Q(F^p, F^{n-p+1}) = 0.$$

Again, elementary linear algebra shows that the complex Lie group $G(\mathbb{C})$ operates transitively on $\check{D}$. Upon choice of a reference point $F^\bullet \in \check{D}$ we have

$$\check{D} = G(\mathbb{C})/P(\mathbb{C})$$

where $P(\mathbb{C})$ are the $\mathbb{C}$-valued points of a parabolic algebraic group P defined over $\mathbb{Q}$. It follows that

$$\check{D} \subset \prod_p \mathrm{Grass}(f^p, V_{\mathbb{C}})$$

is a homogeneous, rational projective variety. In fact, $\check{D}$ is defined over $\mathbb{Q}$ and the $\mathbb{Q}$-algebraic group G operates transitively on the set $\check{D}(\mathbb{Q})$ of $\mathbb{Q}$-rational points of $\check{D}$. Concretely, each $\mathrm{Grass}(f^p, V_{\mathbb{C}})$ is embedded in projective space via the *Plücker coordinates* of a subspace of $V_{\mathbb{C}}$ and $\check{D}(\mathbb{Q})$ may be thought of as the set of flags whose Plücker coordinates are represented by homogeneous coordinate vectors having rational entries.

It is clear that D is an open set in $\check{D}$. If $\varphi \in D$ we shall denote by $F^\bullet_\varphi \in \check{D}$ the corresponding flag. Choosing $F^\bullet_\varphi$ as our reference point for $\check{D}$, the isotropy group in $G(\mathbb{C})$ of $F^\bullet_\varphi$ is a parabolic subgroup P and we have

$$\begin{array}{ccll} \check{D} & = & G(\mathbb{C})/P & \\ \cup & & & \\ D & = & G(\mathbb{R})/H, & \qquad H = G(\mathbb{R}) \cap P(\mathbb{C}) \end{array}$$

where we have omitted the subscript φ on H and P. For the complex tangent space $T_{F^\bullet_\varphi}\check{D}$ we have

$$T_{F^\bullet_\varphi}\check{D} \cong \mathfrak{g}_{\mathbb{C}}/\mathfrak{p}.$$

Using the above notations we have

$$\mathfrak{p} \cong \bigoplus_{i \leqq 0} \mathfrak{g}^{-i,i}$$

and, as above, the complex tangent space

$$T_{F^\bullet_\varphi}\check{D} \cong \bigoplus_{i>0} \mathfrak{g}^{-i,i}.$$

For $F^\bullet \in \check{D}$, not necessarily in D, we may define the Mumford-Tate group $M_{F^\bullet} \subset G$ as follows: First we note that if $n = 2p$ then we may define

$$\mathrm{Hg}^p_{F^\bullet} = F^p \cap V.$$

Then, we may proceed to define the algebra of *Hodge tensors* as before

$$\mathrm{Hg}^{\bullet,\bullet}_{F^\bullet} \subset T^{\bullet,\bullet},$$

where we recall our notational convention that $T^{\bullet,\bullet} = \bigoplus_{k,l} T^{k,l}$ where $T^{k,l} = (\otimes^k V) \otimes (\otimes^l \check{V})$ is a $\mathbb{Q}$-vector space.[1]

Definition: The *Mumford-Tate group $M_{F^\bullet}$ of a point $F^\bullet$ in the compact dual* is defined to be the subgroup of G that fixes all the Hodge tensors.

If $F^\bullet = F^\bullet_\varphi \in D$, then $\mathrm{Hg}^{\bullet,\bullet}_{F^\bullet_\varphi} = \mathrm{Hg}^{\bullet,\bullet}_\varphi$ as previously defined. By the basic property (I.B.1) we then have

$$M_{F^\bullet_\varphi} = M_\varphi.$$

This extension of the definition of Mumford-Tate groups will be important for the arithmetic considerations that follow.

Remark on terminology. We recall the earlier remark where we noted that in the literature $M_{F^\bullet_\varphi}$ is called a *special Mumford-Tate group*. Since in this paper these are the primary objects of interest, we will use the above terminology.

Arithmetic observation. The group $\mathrm{Aut}(\mathbb{C}/\mathbb{Q})$ acts on $\check{D}$. For example, thinking of $\check{D}$ as embedded in a product of projective spaces by the Plücker

[1]Explicitly, the filtration $\{0\} \subset F^n(V_\mathbb{C}) \subset \cdots \subset F^0(V_\mathbb{C}) = V_\mathbb{C}$ induces a filtration

$$\{0\} \subset F^0(\check{V}_\mathbb{C}) \subset F^{-1}(\check{V}_\mathbb{C}) \subset \cdots \subset F^{1-n}(\check{V}_\mathbb{C}) \subset F^{-n}(\check{V}_\mathbb{C}) = \check{V}_\mathbb{C},$$

of $\check{V}_\mathbb{C}$ by $F^{1-q}(\check{V}_\mathbb{C}) := \mathrm{Ann}\, F^q(V_\mathbb{C})$. These two filtrations induce a filtration

$$\{0\} \subset F^{kn}(T^{k,\ell}_\mathbb{C}) \subset \cdots \subset F^{-\ell n}(T^{k,\ell}_\mathbb{C}) = T^{k,\ell}_\mathbb{C}$$

of $T^{k,\ell}_\mathbb{C} = V_\mathbb{C}^{\otimes k} \otimes \check{V}_\mathbb{C}^{\otimes \ell}$ by

$$F^r(T^{k,\ell}_\mathbb{C}) := \bigoplus_{(\Sigma p_i)-(\Sigma q_j)\geq r} F^{p_1}(V_\mathbb{C}) \otimes \cdots \otimes F^{p_k}(V_\mathbb{C}) \otimes F^{-q_1}(\check{V}_\mathbb{C}) \otimes \cdots \otimes F^{-q_r}(\check{V}_\mathbb{C}).$$

If $n(k-\ell) = 2q$, then define

$$\mathrm{Hg}^{k,\ell}_{F^\bullet} := T^{k,\ell} \cap F^q(T^{k,\ell}_\mathbb{C}).$$

Otherwise, if $n(k-\ell)$ is odd, define $\mathrm{Hg}^{k,\ell}_{F^\bullet} = \{0\}$.

coordinates of the F^p, then $\sigma \in \mathrm{Aut}(\mathbb{C}/\mathbb{Q})$ operates on these Plücker coordinates. For $F^\bullet \in \check{D}$, we denote by $F^\bullet_\sigma$ the image of $F^\bullet$ under σ. The action of $\mathrm{Aut}(\mathbb{C}/\mathbb{Q})$ is highly discontinuous, and does not preserve $D \subset \check{D}$. However we observe that

$$\begin{cases} \mathrm{Hg}^{\bullet,\bullet}_{F^\bullet_\sigma} = \mathrm{Hg}^{\bullet,\bullet}_{F^\bullet_\sigma} \\ M_{F^\bullet_\sigma} = M_{F^\bullet}, \end{cases}$$

with the consequence that Galois conjugation *will* preserve the "Mumford-Tate Noether-Lefschetz loci" defined in Section II.C, permuting their irreducible components.

We shall revisit the action of $\mathrm{Aut}(\mathbb{C}/\mathbb{Q})$ in Section VI.D.

Generic Mumford-Tate groups. As in Section I.B, for an algebraic variety X defined over $\mathbb{Q}$ and an algebraic subvariety $Y \subset X(\mathbb{C})$, we say that Y *is* $\mathbb{Q}$-*generic* if the $\mathbb{Q}$-Zariski closure of Y is equal to X; we write

$$\overline{Y}^{\mathbb{Q}} = X.$$

(II.A.3) THEOREM: *Except in the case that the weight $n = 2m$ is even and $F^m = V_{\mathbb{C}}$, if $F^\bullet \in \check{D}$ is a $\mathbb{Q}$-generic point, then the Mumford-Tate group $M_{F^\bullet} = G$.*

Remark: This is a special case of more general results that will be given in Section VI.A, and which will provide a context for it. There are a number of similar results in the literature where the term "generic" has various meanings (cf. [Mo1], [Mo2], and [PS]). For a polarized Hodge structure arising from geometry, "generic" here means the following: *If the Plücker coordinates of the period matrices of the Hodge filtration satisfy no polynomial relations over $\mathbb{Q}$, then the Mumford-Tate group is G.*

PROOF. We denote by $\mathrm{Hg}^{\bullet,\bullet} \subset T^{\bullet,\bullet} = \underset{k,l}{\oplus}(V^{\otimes k} \otimes \check{V}^{\otimes L})$ the tensors that are Hodge tensors for *all* $F^\bullet \in \check{D}$. Then

$$\text{(II.A.4)} \qquad F^\bullet \text{ is } \mathbb{Q}\text{-generic} \Rightarrow \mathrm{Hg}^{\bullet,\bullet}_{F^\bullet} = \mathrm{Hg}^{\bullet,\bullet}.$$

Indeed, the condition for $\zeta \in T^{\bullet,\bullet}$ to be a Hodge tensor for a point $F^\bullet \in \check{D}$ is an algebraic relation defined over $\mathbb{Q}$, and so $F^\bullet$ being $\mathbb{Q}$-generic implies that the condition $\zeta \in \mathrm{Hg}^{\bullet,\bullet}_{F^\bullet}$ implies no $\mathbb{Q}$-relations on the Plücker coordinates of $F^\bullet$ beyond those satisfied for all points in $\check{D}$.

We now define

$$M_{\check{D}} = \left\{ \begin{array}{c} \mathbb{Q}\text{-algebraic subgroup of } G \\ \text{that } \textit{pointwise} \text{ fixes } \mathrm{Hg}^{\bullet,\bullet} \end{array} \right\}.$$

From (II.A.4), for $F^\bullet$ a $\mathbb{Q}$-generic point of $\check{D}$ we have

$$M_{F^\bullet} = M_{\check{D}}.$$

Thus, we have to show: Except in the case noted in the theorem, we have

$$M_{\check{D}} = G. \tag{II.A.5}$$

For this, we first note that the subspace $\mathrm{Hg}^{\bullet,\bullet} \subset T^{\bullet,\bullet}$ is preserved by G. We then have the general

OBSERVATION: *If $B \subset A \subset \mathrm{GL}(W)$ are linear groups and $U \subset W$ is a linear subspace such that $A(U) \subseteq U$ and B the subgroup of A that fixes U pointwise, then B is a normal subgroup of A.*

It follows that $M_{\check{D}}$ is a normal subgroup of G. Since $G = \mathrm{Aut}(V, Q)$ is $\mathbb{Q}$-simple except in the case $\dim V = 2$ and Q is symmetric, it follows that excluding these cases we have either $M_{\check{D}} = G$ or $M_{\check{D}} = \{e\}$. Reflection shows that the latter case occurs only when $\mathrm{Hg}^{\bullet,\bullet} = T^{\bullet,\bullet}$, which is the exception noted in the theorem. □

(II.A.6) COROLLARY: *For a $\mathbb{Q}$-generic polarized Hodge structure $\varphi \in D$, the Mumford-Tate group $M_\varphi = G$.*

Mumford-Tate groups contained in the isotropy group. A natural question is

> *What are the conditions on $\varphi \in D$ implied by the assumption that $M_\varphi(\mathbb{R}) \subseteq H_\varphi$, where H_φ is the isotropy group of φ?*

In particular, the assumption implies that $M_\varphi(\mathbb{R})$ is compact. The converse is not true; see the example of $\mathrm{SU}(r+2)$ in Section IV.C. Anticipating a much fuller discussion in Chapter VI, we shall show that

(II.A.7) PROPOSITION: *Given a reductive $\mathbb{Q}$-algebraic subgroup $M \subset G$, the subvariety*

$$\{\varphi \in D : M(\mathbb{R}) \subset H_\varphi\}$$

is defined over $\mathbb{Q}$.

PROOF. We shall use the following elementary fact: If X and Y are algebraic varieties defined over $\mathbb{Q}$, and if $Z \subset X \times Y$ is a subvariety also defined over $\mathbb{Q}$, then the subvariety

$$\{y \in Y : X \times \{y\} \subset Z\}$$

of Y is defined over $\mathbb{Q}$. This is a result in elimination theory.

We apply this when $X = M, Y = \check{D}$ and

$$Z = \{(a, F^\bullet) \in M \times \check{D} : a(F^\bullet) = F^\bullet\}.$$

Then, when intersected with D, the subvariety of $\check{D}$ defined above is just $\{\varphi \in D : M \subset H_\varphi\}$. □

As mentioned in Chapter VI, (II.A.7) will be a special case of the result (VI.D.1) that *the orbits of Mumford-Tate groups are contained in subvarieties of $\check{D}$ that are defined over number fields.* If $M_\varphi \subseteq H_\varphi$ then the orbit of $M_\varphi(\mathbb{R})$ is just φ, which as a consequence of (II.A.7) must be a point in $\check{D}$ that is defined over a number field. This also follows from of a result of Borcea [Bo] that φ is a complex multiplication Hodge structure and M_φ is an algebraic torus (see (V.4)).

Mumford-Tate groups of integral elements. In Hodge theory, what we shall refer to as the *classical cases* are those when the

$$n = 1 \text{ or } n = 2 \text{ and } h^{2,0} = 1, \tag{II.A.8}$$

in which cases the period domain D, which is Hermitian symmetric, may be equivariantly embedded in the Siegel generalized upper half space. Except in the cases when D is Hermitian symmetric but not classical as above, which exist but seem to not be much studied — cf. the discussion following the proof of Theorem (IV.E.4) — D is far from having the complex analytic properties of a bounded symmetric domain [GS].[2] The usual function-theoretic properties of a bounded symmetric domain D and algebro-geometric properties of quotients $\Gamma\backslash D$, where Γ is an arithmetic subgroup of G, are then replaced by the property that variations of Hodge structure are integral manifolds of a canonical exterior differential system on the period domain D.[3] This exterior differential system is the restriction to D of a $G(\mathbb{C})$-invariant exterior differential system on $\check{D}$ generated by sections of a sub-bundle $I \subset T^*\check{D}$. Setting $I^\perp = W \subset T\check{D}$, with the usual identification

$$T_{F^p}\,\mathrm{Grass}(f^p, V_\mathbb{C}) \cong \mathrm{Hom}(F^p, V_\mathbb{C}/F^p),$$

we may define I by the condition that the fibre

$$W_{F^\bullet} \subset \oplus\,\mathrm{Hom}(F^p, F^{p-1}/F^p).$$

[2] A result of Carlson-Toledo is then when $n = 2$, $h^{2,0} \geqq 2$ and $h^{1,1} \geqq 3$, for $\Gamma \subset \mathrm{SO}(2h^{2,0}, h^{1,1})$ a co-compact discrete group acting freely on D the quotient $\Gamma\backslash D$ does not have the homotopy type of a compact Kähler manifold.

[3] Interestingly, there are cases where D is Hermitian but where the infinitesimal period relation (in the following definition) is non-trivial.

Symbolically, we may think of I as the *differential constraint*

$$dF^p \subseteqq F^{p-1},$$

which is equivalent to

$$Q(dF^p, F^{n-p+2}) = 0.$$

Definition: The sub-bundle $I \subset T^*\check{D}$ defined above will be called the *infinitesimal period relation* (IPR).

As previously noted, equivalent data to the Pfaffian system $I \subset T^*\check{D}$ is the dual $G(\mathbb{C})$-invariant distribution

$$I^{\perp} = W \subset T\check{D}.$$

At a point $\varphi \in D$ we have for the fibre of W

$$W_\varphi = \mathfrak{g}^{-1,1}. \tag{II.A.9}$$

Except in the cases when $\mathfrak{g}^{-i,i} = 0$ for $i \geqq 2$, this distribution is non-integrable.

The Pfaffian system I generates a differential ideal $\mathcal{I} \subset \Omega^{\bullet}_{\check{D}}$. If θ^α are 1-forms locally generating the vector bundle I, then $\mathcal{I}$ is generated over $\mathcal{O}_{\check{D}}$ by the θ^α and their exterior derivatives $d\theta^\alpha$.

Definition: An *integral element* for the differential ideal $\mathcal{I}$ is given by a linear subspace

$$E \subset T_{F^\bullet}\check{D}$$

such that for all $\psi \in \mathcal{I}$

$$\psi|_E = 0.$$

Any tangent space to an integral manifold is an integral element. Integral elements may be thought of as infinitesimal solutions to the exterior differential system. We refer to [CGG] for a recent discussion of the exterior differential system arising in Hodge theory.

We will now define the Mumford-Tate group $M_{(F^\bullet,E)}$ for an integral element E. The idea is that $M_{(F^\bullet,E)}$ will be the subgroup of G that fixes all the Hodge tensors that, to first order, remain Hodge tensors in the directions in E.

More formally, when $n(k-l) = 2m$ we set $T^{k,l} = V^{\otimes k} \otimes \check{V}^{\otimes l}$ and

$$\mathrm{Hg}^{k,l}_{F^\bullet} = F^m \cap T^{k,l}.$$

The action of $E \subset \bigoplus_p \mathrm{Hom}(F^p, F^{p-1}/F^p)$ induces a map

$$E \xrightarrow{\theta} \mathrm{Hom}\left(\mathrm{Hg}^{k,l}_{F^\bullet}, F^{m-1}T^{k,l}_{\mathbb{C}}/F^m T^{k,l}_{\mathbb{C}}\right)$$

and we set

$$\mathrm{Hg}^{k,l}_{(F^\bullet,E)} = \left\{ \zeta \in \mathrm{Hg}^{k,l}_{F^\bullet} : \theta(e)\cdot\zeta = 0 \text{ for all } e \in E \right\}$$

and

$$\mathrm{Hg}^{\bullet,\bullet}_{(F^\bullet,E)} = \underset{k,l}{\oplus} \mathrm{Hg}^{k,l}_{(F^\bullet,E)}.$$

Definition: The *Mumford-Tate group $M_{(F^\bullet,E)}$ of the integral element E* is defined to be the $\mathbb{Q}$-algebraic subgroup of G that fixes $\mathrm{Hg}^{\bullet,\bullet}_{(F^\bullet,E)}$.

It is known (cf. [CGG] and the references cited there) that when $F^\bullet = F^\bullet_\varphi \in D$ every integral element $E \subset T_\varphi D$ is tangent to local variations of Hodge structure

$$S \subset D.$$

We will also define the Mumford-Tate group M_Φ of a variation of Hodge structure $S \subset D$ to be the Mumford-Tate at a generic point. Then it will follow that:

$$M_{\varphi,E} = M_\Phi \textit{in case } \varphi \textit{ is a generic point of } S.$$

Here, "generic" means outside of a countable union of proper analytic subvarieties of S.

Finally, for later reference we note that in case $\varphi \in D$ so that we have the canonical identifications

$$\begin{array}{ccc} T_\varphi D & = & \underset{i>0}{\oplus} \mathfrak{g}^{-i,i} \\ \cup & & \cup \\ W_\varphi & = & \mathfrak{g}^{-1,1}, \end{array}$$

then it follows from (II.A.2) that

(II.A.10) *The subspace $E \subset \mathfrak{g}^{-1,1}$ is an integral element if, and only if, it is abelian Lie sub-algebra.*

Period domains were introduced as classifying spaces for polarized Hodge structures with given Hodge numbers. Among the interesting maps to them are those that arise from algebraic geometry. Locally these give integral manifolds of the exterior differential system generated by I. An obvious issue is whether period domains are "too big" — i.e., do they contain irrelevant information from an algebro-geometric perspective? One (to us) satisfactory response to that question, one that exhibits the fundamental role played by Mumford-Tate groups, will be given by Theorem (III.A.1).

A final general remark on integral elements: Given $F^\bullet \in \check{D}$ there is a graded vector space

$$\mathbb{M}_{F^\bullet} = \bigoplus_p (F^p/F^{p+1}).$$

It has a special structure because of the dualities

$$(\widetilde{F^p/F^{p+1}}) \cong F^{n-p}/F^{n-p+1}$$

arising from Q. For an integral element E there is a natural map

$$E \to \mathrm{Hom}_Q(\mathbb{M}_{F^\bullet}, \mathbb{M}_{F^\bullet}).$$

The condition that E satisfies the integrability conditions $d\theta|_E = 0$ for $\theta \in I$ then has the consequence:

(II.A.11) $\mathbb{M}_{F^\bullet}$ *is a graded* $\mathrm{Sym}^\bullet E$ *module.*

From examples it is known that this algebraic structure can be very rich and can contain significant geometric information. We will comment further on this.

II.B MUMFORD-TATE DOMAINS AND THEIR COMPACT DUALS

Let (V, Q, φ) be a polarized Hodge structure with Mumford-Tate group M and with associated period domain D.[4]

Definition: A *Mumford-Tate domain* $D_{M_\varphi} \subset D$ is given by the $M(\mathbb{R})$-orbit of $\varphi \in D$.

The basic properties of Mumford-Tate domains will be established in several places, especially in Section IV.F. Many of these will be based on the fact that the Lie algebra $\mathfrak{m} \subset \mathfrak{g}$ inherits a polarized sub-Hodge structure from that on $\mathfrak{g}$. From this, it follows that D_{M_φ} is a homogeneous complex manifold $M(\mathbb{R})/M(\mathbb{R}) \cap H_\varphi$ where $M(\mathbb{R}) \cap H_\varphi$ is the compact isotropy group. We note that if

$$M(\mathbb{R}) \sim M_S(\mathbb{R}) \times A(\mathbb{R})$$

is the almost product decomposition of $M(\mathbb{R})$ into its semi-simple and abelian parts, then $A(\mathbb{R})$ will be a compact torus contained in H_φ so that, setting $H_M = M_S(\mathbb{R}) \cap H_\varphi$ as a complex manifold we have

$$D_M = M_S(\mathbb{R})/H_M.$$

[4]To avoid notational clutter, when there is no danger of confusion we shall sometimes denote the Mumford-Tate group simply by M, rather than by M_φ.

However, as will be discussed and illustrated in Section IV.G, it is the representation of D_M as a *homogeneous* complex manifold parametrizing polarized Hodge structures whose generic point has M as Mumford-Tate group that is important. The *same* complex manifold may arise in *several ways* as a Mumford-Tate domain. We also observe that even though M is connected as a $\mathbb{Q}$-algebraic group, $M_S(\mathbb{R})$ may not be connected as a real Lie group and D_{M_φ} may have several components (cf. II.C and Chapter VI).

As noted, Mumford-Tate domains are the natural parameter spaces for all polarized Hodge structures on (V, Q) whose Mumford-Tate groups are a subgroup of M.

For the remainder of this section, we assume that M is semi-simple[5] and shall denote D_{M_φ} by D_M. As is the case for period domains, a Mumford-Tate domain D_M has a compact dual $\check{D}_M$ in which

$$D_M \subset \check{D}_M$$

is an open orbit of $M(\mathbb{R})$. In fact,

$$\check{D}_M = M(\mathbb{C})/P_M$$

where $P_M \subset M(\mathbb{C})$ is a parabolic subgroup with

$$P_M \cap M(\mathbb{R}) = H_M.$$

If $\varphi \in D_M$ is the reference point such that $M = M_\varphi$, then

$$\text{(II.B.1)} \qquad \begin{cases} \mathfrak{m}_{\mathbb{C}} &= \bigoplus\limits_k \mathfrak{m}^{-k,k} \\ \mathfrak{h}_{M,\mathbb{C}} &= \mathfrak{m}^{0,0} \end{cases}$$

and the Lie algebra of P_M is given by

$$\mathfrak{p}_M = \bigoplus_{k \leqq 0} \mathfrak{m}^{-k,k}.$$

Remark: The approach taken in this monograph is from the perspective of Hodge theory; the basic objects are polarized Hodge structures and their Mumford-Tate groups and domains, their algebra of endomorphisms, etc. An alternate, more Lie theoretic approach is to start with the rational, projective variety $\check{D}_M = M(\mathbb{C})/P_M$ and then consider the open orbits of the real form $M(\mathbb{R})$ of $M(\mathbb{C})$.[6] One of the main results (Theorem III.C.1) is that

[5] This discussion can be modified to apply to the case when M is reductive. An illustration of this is given in Section IV.G. We shall restrict here to the semi-simple case for expositional simplicity.

[6] This is the approach taken, e.g., in [GS], [W1], [W2], [WW], and [FHW].

> *Mumford-Tate domains are exactly the open $M(\mathbb{R})$-orbits in $\check{D}_M$ whose isotropy group is compact.*

In the Lie theoretic approach, the Hodge structure on $\mathfrak{m}_{\mathbb{C}}$ is replaced by the root space decomposition of $\mathfrak{m}_{\mathbb{C}}$ under the action of a Cartan sub-algebra. Anticipating the discussion in Chapter IV, and using the notations from the appendix to that chapter, here we will relate the two approaches.

Implicit in this discussion of the Hodge structure is the result given in Proposition (IV.A.2) that for a Mumford-Tate domain $D_M = M(\mathbb{R})/H$,

H contains a compact maximal torus T.

In fact, if we have

$$\varphi : \mathbb{U}(\mathbb{R}) \to M(\mathbb{R}), \tag{II.B.2}$$

then

$$H = Z_M\big(\varphi(\mathbb{U}(\mathbb{R}))\big) \tag{II.B.3}$$

is the centralizer in $M(\mathbb{R})$ of the circle $\varphi(\mathbb{U}(\mathbb{R}))$. As Cartan sub-algebra of $\mathfrak{m}_{\mathbb{C}}$ we take the complexification $\mathfrak{t}_{\mathbb{C}}$ of the Lie algebra $\mathfrak{t}$ of T. Under the adjoint action of $\mathfrak{t}_{\mathbb{C}}$ on $\mathfrak{m}_{\mathbb{C}}$ we have the root space decomposition

$$\mathfrak{m}_{\mathbb{C}} = \mathfrak{t}_{\mathbb{C}} \oplus \left(\underset{\alpha \in \mathfrak{r}}{\oplus} \mathfrak{m}_{\alpha} \right). \tag{II.B.4}$$

The roots $\alpha \in \mathfrak{r}$ are purely imaginary linear functions on $\mathfrak{t}$. They occur in conjugate pairs — thus

$$-\mathfrak{r} = \mathfrak{r},$$

and the root spaces $\mathfrak{m}_{\alpha}$ are 1-dimensional.

The torus

$$T = \mathfrak{t}/\Lambda$$

where Λ is a lattice in $\mathfrak{t}$, and the circle (II.B.2) is given by

$$\mathfrak{l}_{\varphi} \in \Lambda;$$

i.e., $\varphi(\mathbb{U}(\mathbb{R}))$ is the image of the line $\mathbb{R}\mathfrak{l}_{\varphi}$ under the exponential map $\mathfrak{t} \to T$.

The relation between the Hodge decomposition (II.B.1) and root space decomposition (II.B.4) is that on the one hand $\mathfrak{m}^{-k,k}$ is a sum of root spaces. On the other hand, the circle (II.B.2) acts on $\mathfrak{m}^{-k,k}$ by $\exp(-2ki)$. This then gives the basic relation

(II.B.5)
$$\mathfrak{m}^{-k,k} = \bigoplus_{\left\{\substack{\alpha\in\mathfrak{r} \\ \langle\alpha,\mathfrak{l}_\varphi\rangle=-2ki}\right.} \mathfrak{m}_\alpha .$$

If we set
$$\mathfrak{l}_\varphi^\perp = \{\alpha \in \mathfrak{r} : \langle \alpha, \mathfrak{l}_\varphi \rangle = 0\},$$
then

(II.B.6)
$$\mathfrak{h}_{M,\mathbb{C}} = \bigoplus_{\alpha\in\mathfrak{l}_\varphi^\perp} \mathfrak{m}_\alpha \oplus \mathfrak{t}_\mathbb{C}.$$

From (II.B.5) and (II.B.6) we infer the

(II.B.7) PROPOSITION: (i) *The complex tangent space*
$$T_\varphi D_M = \bigoplus_{\left\{\substack{\alpha\in\mathfrak{h}_\varphi^\perp \\ i\langle\alpha,\mathfrak{l}_\varphi\rangle>0}\right.} \mathfrak{m}_\alpha .$$

(ii) *The Lie algebra of* P_M *is given by*
$$\mathfrak{p}_M = \bigoplus_{i\langle\alpha,\mathfrak{l}_\varphi\rangle\leqq 0} \mathfrak{m}_\alpha .$$

For the remainder of this section, we will invert this discussion and let

- $M_\mathbb{R}$ be a real, semi-simple Lie group containing a compact maximal torus T,[7] and
- $H \subset M_\mathbb{R}$ be a compact, connected subgroup that is the centralizer of a circle in T given by $\mathfrak{l}_\varphi \in \Lambda$.

We will discuss the homogeneous complex structures on $M(\mathbb{R})/H$ where $H \subset M(\mathbb{R})$ is a compact subgroup containing T.

Case of $M(\mathbb{R})/T$. Identifying the tangent space at the identity coset with $\mathfrak{m}_\mathbb{R}/\mathfrak{t}$, an almost complex structure is given by
$$\begin{cases} J \in \mathrm{Hom}_T(\mathfrak{m}_\mathbb{R}/\mathfrak{t}, \mathfrak{m}_\mathbb{R}/\mathfrak{t}) \\ J^2 = -I. \end{cases}$$

[7] For this discussion we do not require that $M_\mathbb{R}$ be the real points of a $\mathbb{Q}$-algebraic group.

As irreducible representations of T

$$(\mathfrak{m}_{\mathbb{R}}/\mathfrak{t}) \otimes \mathbb{C} = \bigoplus_{\alpha \in \mathfrak{r}} \mathfrak{m}_\alpha$$

where $\mathfrak{m}_\alpha = \mathbb{C} e_\alpha$. Since $J(\mathfrak{m}_\alpha) = \mathfrak{m}_\alpha$ we have

$$J(e_\alpha) = \pm i e_\alpha.$$

Because $\overline{e}_\alpha = \pm e_{-\alpha}$ and J is real

$$J(e_{-\alpha}) = \pm i e_{-\alpha}.$$

We set

$$T^{1,0} = \text{span}\{e_\alpha : J(e_\alpha) = -ie_\alpha\}^8$$

so that

$$T^{0,1} = \overline{T^{1,0}} = \text{span}\{e_\alpha : J(e_\alpha) = ie_\alpha\}.$$

The integrability condition

$$\left[T^{1,0}, T^{1,0}\right] \subset T^{1,0}$$

implies that the subset of $\mathfrak{r}$ consisting of the roots α such that $J(e_\alpha) = -ie_\alpha$ is a semi-group, and this then is a set of positive roots $\mathfrak{r}^+_{\mathbf{C}}$ corresponding to a Weyl chamber $\mathbf{C}$. Thus

$$e_\alpha \in T^{1,0} \Longleftrightarrow \langle \alpha, \nu \rangle \geqq 0 \text{ for all } \nu \in \mathbf{C}.$$

Case of $M(\mathbb{R})/H$. Then

$$\begin{cases} J \in \text{Hom}_H(\mathfrak{m}_{\mathbb{R}}/\mathfrak{h}, \mathfrak{m}_{\mathbb{R}}/\mathfrak{h}) \\ J^2 = -I. \end{cases}$$

Let

$$\begin{aligned} \mathfrak{r}_{\mathfrak{h}} &= \text{span}_{\mathbb{R}}\{\alpha : e_\alpha \in \mathfrak{h}_{\mathbb{C}}\} \\ \mathfrak{r}_{\mathfrak{h}^\perp} &= \{\nu \in \mathfrak{r}_{\mathbf{C}} \otimes \mathbb{R} : \langle \nu, \mathfrak{h}_\alpha \rangle = 0 \text{ for all } e_\alpha \in \mathfrak{h}_{\mathbb{C}}\}.^9 \end{aligned}$$

Then

$$(\mathfrak{r}_{\mathfrak{h}})^\perp = \mathfrak{r}_{\mathfrak{h}^\perp}.$$

[8]In $\mathbb{C} = \mathbb{R}^2$ with coordinates (x, y) and the usual complex structure, $\partial_z = \frac{1}{\sqrt{2}}(\partial_x - i\partial_y)$ and $J(\partial_x) = \partial_y$, $J(\partial_y) = -\partial_x$ give $J(\partial_z) = -i\partial_z$.

[9]Remark that $\mathfrak{h}_\alpha \in \mathfrak{t}$ is the co-root associated to the root α.

The irreducible complex representations of H appearing in $(\mathfrak{m}_{\mathbb{R}}/\mathfrak{h}) \otimes \mathbb{C}$ have weights contained in linear spaces parallel to $\mathfrak{r}_{\mathfrak{h}}$. Since J is $\operatorname{Ad} H$-invariant we have

$$\begin{cases} J(e_\alpha) = \pm i e_\alpha \quad \text{for all } e_\alpha \notin \mathfrak{h}_{\mathbb{C}} \\ J(e_\alpha) = i e_\alpha \Leftrightarrow J(e_\beta) = i e_\beta \quad \text{when } \beta - \alpha \in \mathfrak{r}_{\mathfrak{h}} \\ J(e_\alpha) = i e_\alpha \Leftrightarrow J(e_{-\alpha}) = -i e_\alpha \quad \text{for all } e_\alpha \notin \mathfrak{h}_{\mathbb{C}}. \end{cases}$$

For the orthogonal projection

$$\mathfrak{r} \xrightarrow{\pi_{\mathfrak{h}^\perp}} \mathfrak{r}_{\mathfrak{h}}^\perp$$

integrability implies that

$$\pi_{\mathfrak{h}^\perp} \{\alpha \in \mathfrak{r} : e_\alpha \notin \mathfrak{h}_{\mathbb{C}} \text{ and } J(e_\alpha) = -i e_\alpha\}$$

is closed under addition, and thus this is a set of positive elements $\pi^+_{\mathfrak{h}^\perp}$ for the lattice $\pi_{\mathfrak{h}^\perp}(\mathfrak{r})$. We can now complete this to a set of positive roots $\mathfrak{r}^+$ for which the corresponding Weyl chamber $\mathbf{C}$ has the property

$$\mathbf{C} \cap \mathfrak{r}_{\mathfrak{h}^\perp} \neq (0).$$

Conversely, two Weyl chambers $\mathbf{C}, \mathbf{C}'$ determine the same set of positive elements in $\pi_{\mathfrak{h}^\perp}(\mathfrak{r})$ if, and only if,

$$\mathbf{C} \cap \mathfrak{r}_{\mathfrak{h}^\perp} = \mathbf{C}' \cap \mathfrak{r}_{\mathfrak{h}^\perp}.$$

This gives the

(II.B.8) **Conclusion:** *The homogeneous complex structures on $M_{\mathbb{R}}/H$ are in one-to-one correspondence with the set*

$$\left\{ \begin{array}{c} \text{Weyl chambers } \mathbf{C} \\ \text{such that } \mathbf{C} \cap \mathfrak{r}_{\mathfrak{h}^\perp} \neq (0) \end{array} \right\} \Big/ \sim_{\mathfrak{h}}$$

where

$$\mathbf{C} \sim_{\mathfrak{h}} \mathbf{C}' \Longleftrightarrow \mathbf{C} \cap \mathfrak{r}_{\mathfrak{h}^\perp} = \mathbf{C}' \cap \mathfrak{r}_{\mathfrak{h}^\perp}.$$

Finally, for later use we consider the situation of a pair H_1, H_2 of compact, connected subgroups of $M_{\mathbb{R}}$ with $T \subseteq H_1 \cap H_2$ and where we have invariant complex structures on $M_{\mathbb{R}}/H_1$ and $M_{\mathbb{R}}/H_2$ given by $(\mathfrak{h}_1^\perp)^+$ and $(\mathfrak{h}_2^\perp)^+$. We consider the diagram

(II.B.9)
$$\begin{array}{ccc} & M_{\mathbb{R}}/T & \\ \swarrow & & \searrow \\ M_{\mathbb{R}}/H_1 & & M_{\mathbb{R}}/H_2. \end{array}$$

(II.B.10) PROPOSITION: *There is a choice of invariant complex structures on* $M_{\mathbb{R}}/T$ *such that the maps in* (II.B.9) *are holomorphic if, and only if, there is a choice of positive roots* $\mathfrak{r}^+$ *for* $\mathfrak{m}_{\mathbb{C}}$ *such that* $\pi^+_{\mathfrak{h}^\perp} \subset \mathfrak{r}^+$ *for* $i = 1, 2$.

(II.B.11) COROLLARY: *If* H_1 *and* H_2 *are centralizers of circles* $\varphi_i : \mathbb{U}(\mathbb{R}) \to M_{\mathbb{R}}$ *given by* $\mathfrak{l}_{\varphi_i} \in \Lambda$, *and if there is a root* α *such that*

$$\begin{cases} i\,\langle \alpha, \mathfrak{l}_{\varphi_1} \rangle > 0 \\ i\,\langle \alpha, \mathfrak{l}_{\varphi_2} \rangle < 0, \end{cases}$$

then there is no complex structure on $M_{\mathbb{R}}/T$ *such that the mappings in* (II.B.9) *are both holomorphic.*[10]

II.C NOETHER-LEFSCHETZ LOCI IN PERIOD DOMAINS

Traditionally, Noether-Lefschetz (NL) loci have been described as follows:[11] Given a family $\{X_s\}_{s\in S}$ of smooth projective varieties with a smooth and quasi-projective parameter space S, a reference point $s_0 \in S$ and a Hodge class $\zeta \in \mathrm{Hg}^p(X_{s_0})$, the Noether-Lefschetz-locus is the subvariety $S_\zeta \subset S$ where ζ remains a Hodge class. The use of Mumford-Tate groups suggests an alternate, in some ways more satisfactory, approach.

Definitions: (i) Let $\varphi \in D$. Then the *Noether-Lefschetz locus* $\mathrm{NL}_\varphi \subset D$ is the set of $\psi \in D$ such that

$$\mathrm{Hg}^{\bullet,\bullet}_\varphi \subseteq \mathrm{Hg}^{\bullet,\bullet}_\psi.$$

(ii) Let $F^\bullet \in \check{D}$. Then we may give the same definition for the *Noether-Lefschetz locus* $\widecheck{\mathrm{NL}}_{F^\bullet} \subset \check{D}$.

Remarks: It is clear that $\widecheck{\mathrm{NL}}_{F^\bullet}$ is an algebraic subvariety, defined over $\mathbb{Q}$ but in general not irreducible, of $\check{D}$. We shall see in Chapter VI that the irreducible components of $\widecheck{\mathrm{NL}}_{F^\bullet}$ are defined over number fields. If $F^\bullet = F^\bullet_\varphi \in D$, then it is clear that

$$\mathrm{NL}_\varphi = \widecheck{\mathrm{NL}}_{F^\bullet} \cap D.$$

We shall also see that those components of $\mathrm{NL}_{F^\bullet}$ that intersect D are smooth.

[10] Recall that we have $\langle \alpha, \mathfrak{l}_\varphi \rangle = -2ki$ for an integer k. The root spaces that give the $(1,0)$ tangent space are those for which $k < 0$, which is the same sign as $i\,\langle \alpha, \mathfrak{l}_\varphi \rangle > 0$.

[11] This is an informal description; a precise definition will be given in Section III.A.

(II.C.1) THEOREM: *Let $\varphi \in D$. Then the $M_\varphi(\mathbb{R})^0$-orbit of φ is equal to the component of* NL_φ *passing through φ. We write this symbolically as*

$$D^0_{M_\varphi} = \mathrm{NL}^\circ_\varphi.$$

PROOF. We have noted that D_{M_φ} is a homogeneous complex submanifold of D. With the identification

$$T_\varphi D \cong \mathfrak{g}^- := \underset{i>0}{\oplus} \mathfrak{g}^{-i,i}$$ [12]

we have

$$T_\varphi D_{M_\varphi} = \mathfrak{m}^-_\varphi.$$

It is clear that set-theoretically

$$D_{M_\varphi} \subset \mathrm{NL}_\varphi,$$

and we have noted that NL_φ is a complex analytic subvariety of D. Since we are only considering the component of NL_φ passing through φ, to prove the theorem it is enough to show that the *Zariski* tangent spaces coincide; i.e.,

$$T_\varphi(\mathrm{NL}_\varphi) = \mathfrak{m}^-_\varphi.$$

LEMMA: *Define the subspace $\mathcal{A}_\mathbb{C} \subset \mathfrak{g}_\mathbb{C}$ by*

$$\mathcal{A}_\mathbb{C} = \left\{X \in \mathfrak{g}_\mathbb{C} : X(\zeta) = 0 \ \text{ for all } \ \zeta \in \mathrm{Hg}^{\bullet,\bullet}_\varphi\right\}.$$

Then $\mathcal{A}_\mathbb{C} = \mathfrak{m}_\mathbb{C}$.

PROOF. We observe that $\mathcal{A}_\mathbb{C}$ is a Lie sub-algebra of $\mathfrak{g}_\mathbb{C}$, and that it is defined over $\mathbb{Q}$ so that $\mathcal{A}_\mathbb{C} = \mathcal{A} \otimes \mathbb{C}$ where $\mathcal{A} \subset \mathfrak{g}$ is a Lie sub-algebra defined by the same conditions but where $X \in \mathfrak{g}$. It then follows from the basic property (I.B.1) that $\mathcal{A} = \mathfrak{m}_\varphi$. □

For $F^\bullet \in \check{D}$ the Zariski tangent space to $\mathrm{NL}_{F^\bullet}$ is computed as follows: If we think of

$$T_{F^\bullet}\check{D} \subset \underset{p}{\oplus} \mathrm{Hom}(F^p, V_\mathbb{C}/F^p)$$

and we consider Hodge classes

$$\zeta \in \mathrm{Hg}^{k,l}_{F^\bullet} = F^m \cap T^{k,l}$$

[12]A note on signs: For $i > 0$ elements in $\mathfrak{g}^{-i,i}$ shift the Hodge decomposition $V_\mathbb{C} = \oplus V^{p,q}_\varphi$ i places to the right and give the holomorphic tangent space $T_\varphi D$. This is the reason for the minus sign in $\mathfrak{g}^-$. It should not be confused with the direct sum of the negative root spaces in $\mathfrak{g}_\mathbb{C}$.

where $n(k-l)=2m$, then the Zariski tangent space

$$T_{F^\bullet}\mathrm{NL}_{F^\bullet}=\left\{\begin{array}{c}X\in T_{F^\bullet}\check{D}\subset\oplus\,\mathrm{Hom}(F^p,V_{\mathbb{C}}/F^p)\\ \text{such that } X(\zeta)=0 \text{ in } T^{k,l}_{\mathbb{C}}/F^mT^{k,l}_{\mathbb{C}}\\ \text{for all } \zeta \text{ as above.}\end{array}\right\}$$

In the case $F^\bullet=F^\bullet_\varphi\in D$ and $X\in\mathfrak{g}^-$ the condition in the brackets is the same as

$$X(\zeta)=0.$$

That is, and this is the key observation, with the identification $T_\varphi D\cong\mathfrak{g}^-$, since

$$\mathfrak{m}^{-i,i}:V^{p,q}_\varphi\to V^{p-i,q+i}_\varphi$$

the condition to be zero in the quotient $T^{k,l}_{\mathbb{C}}/F^mT^{k,l}_{\mathbb{C}}$ is the same as $X(\zeta)=0$ in $T^{k,l}_{\mathbb{C}}$. Thus by the lemma

$$T_{F^\bullet_\varphi}\mathrm{NL}_\varphi=\mathfrak{m}^-_\varphi$$

which completes the proof. □

(II.C.2) COROLLARY: *The fibre at φ of the normal bundle to NL_φ in D is given by $\mathfrak{g}^-/\mathfrak{m}^-_\varphi$. In particular*

$$\mathrm{codim}_D(\mathrm{NL}_\varphi)^0=\mathrm{codim}_{\mathfrak{g}^-}(\mathfrak{m}^-_\varphi).$$

Remark: If $n=2p$ and $\zeta\in\mathrm{Hg}^p_\varphi=V\cap F^p$, then the locus $D_\zeta\subset D$ where ζ remains a Hodge class is classically known to be smooth and of codimension $h^{p-1,p+1}+\cdots+h^{0,2p}$ (see [CGG]). If we now let ζ range over all Hodge tensors then it is not immediately clear how to compute the codimension of NL_φ in D. The only way we know how to do this is using Theorem (II.C.1).

It will be shown (Theorem (VI.A.3)) that

> *The subgroups of $G(\mathbb{R})$, respectively $G(\mathbb{C})$ leaving NL_φ, respectively $\mathrm{NL}_{F^\bullet}$ invariant are the normalizers respectively of M_φ and $M_{F^\bullet}$.*

Of particular arithmetic interest, to be discussed, are the finite group quotients of $N_G(M_\varphi)$ and $N_G(M_{F^\bullet})$ by their identity components, which act by permuting the components of the Noether-Lefschetz loci.

Remark: An extreme case is when the Noether-Lefschetz locus is a discrete set of points. This is equivalent to the identity component $M_\varphi(\mathbb{R})^0$ of the Mumford-Tate group being contained in the isotropy group H_φ. By a result of Borcea [Bo] and V.4, this is equivalent to M_φ being an algebraic torus. Among these are the polarized Hodge structures of CM type (Chapter V).

Discussion. The algebra $\mathrm{Hg}_\varphi^{\bullet,\bullet}$ of Hodge tensors has a number of properties.

(i) It decomposes under the irreducible representations $\{T_\rho\}$ of G on the tensor space. In fact, the projection $T^{k,l} \to T_\rho$ is itself a Hodge tensor for any $\varphi \in D$.

(ii) Using the polarization to identify V with $\check{V}$, we may consider

$$\mathrm{Hg}_\varphi^\bullet \subset T^\bullet = \bigoplus_k V^{\otimes k}.$$

We do not know whether or not it is finitely generated as an algebra. It is, however, *essentially finitely generated* in a sense that we now explain. For this we set

$$\mathrm{Hg}_\varphi^m = \mathrm{Hg}_\varphi^\bullet \cap \left(\bigoplus_{k \leqq m} V^{\otimes k} \right).$$

We then have

$$\subset \mathrm{Hg}_\varphi^m \subset \mathrm{Hg}_\varphi^{m+1} \subset \dots \tag{II.C.3}$$

and

$$\cup \mathrm{Hg}_\varphi^m = \mathrm{Hg}_\varphi^\bullet.$$

Definition: We define the Noether-Lefschetz locus $\mathrm{NL}_{\varphi,m} \subset D$ to be the subvariety of all $\psi \in D$ such that

$$\mathrm{Hg}_\varphi^m \subseteq \mathrm{Hg}_\psi^m.$$

Then

$$\supseteq \mathrm{NL}_{\varphi,m} \supseteq \mathrm{NL}_{\varphi,m+1} \supseteq$$

and

$$\bigcap_m \mathrm{NL}_{\varphi,m} = \mathrm{NL}_\varphi.$$

In the case the inclusion $\mathrm{NL}_{\varphi,m+1} \subset \mathrm{NL}_{\varphi,m}$ is strict, we may think of $\mathrm{NL}_{\varphi,m+1}$ as the subvariety — not just where new Hodge classes are added — *but where new Hodge classes are added that impose additional conditions on Noether-Lefschetz loci.*

Example: The first example is from Mumford [M2]. We take $n = 1$ so that D classifies weight one polarized Hodge structures. Then the polarization

$$Q \in \mathrm{Hg}(\wedge^2 V) \subset \mathrm{Hg}(V^{\otimes 2}).$$

For $\dim V = g = 1, 2, 3$ all of $\mathrm{Hg}_\varphi^\bullet$ is generated by Hg_φ^2 and the decomposition of $V^{\otimes k}$ into simple G-modules. However, when $g = 4$ Mumford gave an example of a φ where there is a class in Hg^4_φ not of this type and where [Mo1], [Mo2]

$$\mathrm{NL}_{\varphi,4} \underset{\neq}{\subset} \mathrm{NL}_{\varphi,2}.$$ [13]

Example: We refer to Proposition (IV.B.9) where it is shown that, for Mumford-Tate domains arising from the adjoint representation the polarizing form (in this case given by the Cartan-Killing form) and the Hodge tensor of degree three given by the Lie bracket, are sufficient to define the component through φ of the Noether-Lefschetz locus.

Since $\breve{\mathrm{NL}}_{F^\bullet_\varphi} \subset \check{D}$ is an algebraic subvariety, the intersection must be a finite one; i.e., there is an $m_0 = m_0(\varphi)$ such that $\mathrm{NL}_{\varphi,m} = N_{\varphi,m+1} = \cdots = \mathrm{NL}_\varphi$ for $m \geqq m_0$. If we define *effective generators* to be those that impose new, non-trivial Noether-Lefschetz conditions, which cut down either the dimension of components or decrease the number of components, then in this sense we may say that

$\mathrm{Hg}^\bullet_\varphi$ *is effectively finitely generated.*

Using the inclusions (II.C.3) we may define

$$M_{\varphi,m} = \left\{ \begin{array}{c} \text{subgroup of } G \text{ leaving} \\ \mathrm{Hg}^m_\varphi \text{ pointwise fixed} \end{array} \right\}.$$

Then with $\mathrm{NL}_{\varphi,m}$'s we have

$$\left\{ \begin{array}{ccccc} \cdots & \supseteq & M_{\varphi,m} & \supseteq & M_{\varphi,m+1} \quad \cdots \\ & \cap & M_{\varphi,m} & = & M_\varphi. \end{array} \right.$$

The proof of Theorem (II.C.1) carries over to show that for the orbit $\mathcal{O}(M_{\varphi,m}) := M_{\varphi,m}(\mathbb{R}) \cdot \varphi \subset D$ we have

(II.C.4) $\mathcal{O}(M_{\varphi,m})$ *is the component of* $\mathrm{NL}_{\varphi,m}$ *passing through* φ.

In Mumford's example we have

$$G = M_{\varphi,1} = M_{\varphi,2} = M_{\varphi,3} \underset{\neq}{\supset} M_{\varphi,4}.$$

To conclude this section, we recall that the orbit

$$D_{M_\varphi} =: M_\varphi(\mathbb{R}) \cdot \varphi$$

of φ is a *Mumford-Tate domain.*

[13] In Section VII.F we shall analyze this phenomenon, among others, in some interesting examples — see Theorem (VII.F.1) and the list following it.

Definitions: (i) We define the *height* $\mathrm{ht}(M_\varphi)$ by

$$\mathrm{ht}(M_\varphi) = \inf_{\mathfrak{g}} \left\{ \begin{array}{c} \max(\deg \zeta) \text{ where } \zeta \in \mathcal{H} \text{ and where} \\ \mathcal{H} \subset \mathrm{Hg}^\bullet \text{ is a set of Hodge tensors} \\ \text{that generate the equations of } \mathrm{NL}_\varphi \end{array} \right\}.$$

(ii) If $\mathrm{ht}(M_\varphi) \leq 2$ we shall call D_{M_φ} a *Shimura domain*, and say that M_φ and (V, φ) are *nondegenerate*.

Discussion. Since our Hodge structures are polarized, we have a Shimura domain exactly when the Hodge tensors that impose non-trivial Noether-Lefschetz conditions are generated by $\mathrm{End}(V, \varphi)$, the latter being the $\mathbb{Q}$-algebra of endomorphisms of the Hodge structure (V, φ). We also note that

> $\mathrm{ht}(M_\varphi) = 2$ *if, and only if,* $M_\varphi = Z_{\mathrm{GL}(V)}(\mathrm{End}(V, \varphi))$ *is the centralizer in* $\mathrm{GL}(V)$ *of* $\mathrm{End}(V, \varphi)$.

We have the strict inclusions

$$\begin{pmatrix} \text{period} \\ \text{domains} \end{pmatrix} \subset \begin{pmatrix} \text{Shimura} \\ \text{domains} \end{pmatrix} \subset \begin{pmatrix} \text{Mumford-Tate} \\ \text{domains} \end{pmatrix}.$$

As we shall see, Mumford-Tate domains are the most natural objects for variations of Hodge structure.

Example: For weight $n = 1$, in light of Mumford's example we have for the height of M_φ

rank V	maximum $\mathrm{ht}(M_\varphi)$
2, 4, 6	2
8	4

Chapter III

The Mumford-Tate Group of a Variation of Hodge Structure

Let $\Gamma \subset G$ be a discrete subgroup of $G(\mathbb{R})$. It acts properly discontinuously on D and the quotient

$$\mathcal{M}_\Gamma = \Gamma \backslash D$$

is a complex analytic variety. It is smooth outside of the images of the fixed points of Γ; locally these are quotients of a polycylinder in $\mathbb{C}^N$ by a finite group — i.e., they are *quotient singularities*. We shall refer to $\mathcal{M}_\Gamma$ as the *moduli space of* Γ*-equivalence classes of polarized Hodge structures.*

Definition: A *variation of Hodge structure* (VHS) is given by

$$\Phi : S \to \mathcal{M}_\Gamma \tag{III.1}$$

where S is a connected complex manifold and Φ is a locally liftable, holomorphic mapping that is an integral manifold of the canonical differential ideal $\mathcal{I}$.

To be locally liftable means that around each point of S there is a neighborhood $\mathcal{U}$ and a diagram

$$\begin{array}{ccc} & & D \\ & \overset{\varphi}{\nearrow} & \downarrow \\ \mathcal{U} & \xrightarrow{\Phi} & \mathcal{M}_\Gamma . \end{array}$$

The local lifting φ is assumed to be holomorphic and to satisfy

$$\varphi^*(\mathcal{I}) = 0.$$

These properties are invariant under the action of Γ and are thus well-defined for Φ.

Note: We are using the notation $\varphi : \mathcal{U} \to D$ and think of $\varphi(s) \in D$ for $s \in \mathcal{U}$. In other words, $\varphi : \mathcal{U} \to D$ denotes a family of Hodge structures in the sense of Section I.A, but here moving with a parameter s.

Upon choice of a base point $s_0 \in S$ and a point of D lying over $\Phi(s_0)$ we will have a diagram

$$\begin{array}{ccc} \widetilde{S} & \xrightarrow{\varphi} & D \\ \downarrow & & \downarrow \\ S & \xrightarrow{\Phi} & \mathcal{M}_\Gamma \end{array}$$

where $\widetilde{S}$ is the universal covering of S. At each point $\tilde{s} \in \widetilde{S}$ the space of Hodge tensors

$$\mathrm{Hg}^{\bullet,\bullet}_{\varphi(\tilde{s})} \subset T^{\bullet,\bullet}$$

is well defined. Elementary reasoning [Sch] shows that:

(III.2) *Outside of a countable union Z of proper analytic subvarieties of $\widetilde{S}$ the space $H^{\bullet,\bullet}_{\varphi(\tilde{s})}$ is a constant subspace $H^{\bullet,\bullet}_{\varphi}$ that is invariant under the action of Γ.*

A point $\eta \in S$ is said to be a *generic point for the variation of Hodge structure* if it lifts to a point $\tilde{\eta}$ in $\widetilde{S}\backslash Z$.

Definition: The *Mumford-Tate group M_Φ of the variation of Hodge structure* is the subgroup of G that fixes all the Hodge tensors $\mathrm{Hg}^{\bullet,\bullet}_{\varphi(\tilde{\eta})} \subset T^{\bullet,\bullet}$ where η is a generic point of S.

We will now phrase the above in another, more algebro-geometric way. We assume that there is a lattice $H_\mathbb{Z}$ and that Γ is a subgroup of $G_\mathbb{Z}$. Then once we have chosen the lift $\varphi(\tilde{\eta}) \in D$ of $\Phi(\eta) \in \mathcal{M}_\Gamma$, we may think of $\Phi(\eta)$ as a polarized Hodge structure on V. The space $\mathrm{Hg}^{\bullet,\bullet}_{\Phi(\eta)} \subset T^{\bullet,\bullet}$ is invariant under analytic continuation along paths in S, where we use the Gauss-Manin connection to identify the underlying vector spaces of the Hodge structures along the path. Since the polarizing form is preserved by monodromy and is positive or negative definite on $\mathrm{Hg}^{\bullet,\bullet}_{\Phi(\eta)}$, by our assumption that $\Gamma \subset G_\mathbb{Z}$ we may infer that monodromy acts by a finite group on this space. Passing to a finite cover, we may — and always will — assume that monodromy acts trivially. Setting $\mathrm{Hg}^{\bullet,\bullet}_{\Phi} =: \mathrm{Hg}^{\bullet,\bullet}_{\Phi(\eta)}$ and taking closed paths we then have that $\mathrm{Hg}^{\bullet,\bullet}_{\Phi}$ is invariant under the action of monodromy and gives a well-defined space of Hodge tensors at each point of S.

Associated to the variation of Hodge structure and lifting $\varphi(\tilde{\eta})$ is the *monodromy representation*

$$\rho : \pi_1(S) \to \Gamma,$$

and we may summarize the above by:

(III.3) *Associated to a variation of Hodge structure* (III.1) *is its Mumford-Tate group $M_\Phi \subset G$. The image $\rho(\pi_1(S))$ of the monodromy representation is a subgroup of M_Φ.*

In order to simplify notation, in what follows unless mentioned otherwise we will set

$$\Gamma = \rho(\pi_1(S))$$

and refer to it as the *monodromy group of the variation of Hodge structure.*

III.A THE STRUCTURE THEOREM FOR VARIATIONS OF HODGE STRUCTURES

In this section we will consider variations of Hodge structure (III.1) where S is a quasi-projective algebraic variety, which we shall refer to as *global variations of Hodge structure.* As discussed, we shall also consider *variations of Hodge structure up to finite data,* by which we mean those that allow S to be replaced by a finite covering $S' \to S$ and consider the variation of Hodge structure

$$\Phi' : S' \to \mathcal{M}_{\Gamma'}$$

induced from (III.1) and where Γ' is the associated monodromy group. Thus $\Gamma' \subset \Gamma$ is a subgroup of finite index.[1]

We denote by M_Φ the Mumford-Tate group of the variation of Hodge structure. By the general theory of algebraic groups over not necessarily algebraically closed fields of characteristic zero [Bor], for the Lie algebra we have a $\mathbb{Q}$-direct sum decomposition

$$\mathfrak{m}_\Phi = \mathfrak{m}_1 \oplus \cdots \oplus \mathfrak{m}_l \oplus \mathfrak{A}$$

where the $\mathfrak{m}_i$ are $\mathbb{Q}$-simple and $\mathfrak{A}$ is an abelian Lie algebra. Over $\mathbb{R}$ we then have

$$M_\Phi(\mathbb{R}) = M_1(\mathbb{R}) \times \cdots \times M_l(\mathbb{R}) \times A.$$

This decomposition is generally *not* the decomposition into $\mathbb{R}$-simple factors.

We denote by $\mathcal{O}(M_\Phi)$ the $M_\Phi(\mathbb{R})$-orbit of a generic point $\varphi(\tilde{\eta})$, and by $D_i \subset D$ the $M_i(\mathbb{R})$-orbit of $\varphi(\tilde{\eta})$.

(III.A.1) THEOREM: (i) *The D_i are complex submanifolds of D.* (ii) *Up to finite data, the monodromy group splits as an almost direct product*

$$\Gamma = \Gamma_1 \times \cdots \times \Gamma_k, \qquad k \leqq l$$

where $\overline{\Gamma}_i^{\mathbb{Q}} = M_i$.

[1] That is, *we shall consider variations of Hodge structure up to isogeny.*

Letting $D' = D_{k+1} \times \cdots \times D_l$ be the part where the monodromy is trivial, we have the

(III.A.2) COROLLARY: *Up to finite data, the global variation of Hodge structure is given by*

$$\Phi : S \to \Gamma_1 \backslash D_1 \times \cdots \times \Gamma_k \backslash D_k \times D'$$

where $\overline{\Gamma}_i^{\mathbb{Q}} = M_i$.

We may informally say that, *given the constraint imposed by the Mumford-Tate group of the variation of Hodge structure, the $\mathbb{Q}$-Zariski closure of the monodromy group is as large as it can be.*

PROOF. The argument proceeds in several steps.

Step one: *The decomposition $\mathfrak{m}_\Phi = \mathfrak{m}_1 \oplus \cdots \oplus \mathfrak{m}_l \oplus \mathfrak{A}$ is a decomposition into sub-Hodge structures.*

If we denote by

$$\psi : \mathbb{S}(\mathbb{R}) \to \mathrm{GL}(\mathfrak{m}_{\Phi,\mathbb{R}})$$

the Hodge structure on $\mathfrak{m}_{\Phi,\mathbb{R}}$ and by

$$\begin{cases} \psi_i : \mathbb{S}(\mathbb{R}) \to \mathrm{GL}(\mathfrak{m}_{i,\mathbb{R}}) \\ \psi_0 : \mathbb{S}(\mathbb{R}) \to \mathrm{GL}(\mathfrak{A}_{\mathbb{R}}), \end{cases}$$

the factors of ψ, we have to show that each $\mathfrak{m}_i$ and $\mathfrak{A}$ is closed under $\psi(\mathbb{S}(\mathbb{R}))$, which is (III.A.5), and that none of the ψ_i or ψ_0 is trivial. If one were trivial, then we would have a proper $\mathbb{Q}$-algebraic subgroup of M_Φ that contains the image $\psi(\mathbb{S}(\mathbb{R}))$, which is ruled out by the minimality of M_Φ.

Step two: *The D_i are complex submanifolds of D.*

We recall the earlier proof for $\mathcal{O}(M_\Phi)$; the same argument will apply to the M_i (the orbit of A is a point). The observations are that since $\mathfrak{m}_\Phi$ has a Hodge structure,

- the (real) tangent space to $\mathcal{O}(M_\Phi)$ at $\varphi(\tilde{\eta})$ is isomorphic to $\mathfrak{m}_{\Phi,\mathbb{R}}/\mathfrak{m}_{\Phi,\mathbb{R}} \cap \mathfrak{h}$;
- in the complexified tangent space, $\mathfrak{m}_\Phi^-$ gives an $M_\Phi(\mathbb{R})$-invariant almost-complex structure;
- this almost-complex structure is integrable.

Step three: *The image* $\varphi(\tilde{S})$ *lies in* $\mathcal{O}(M_\Phi)$.
For this it is sufficient to show that the integral element

$$E = \mathrm{Image}\{\widetilde{\varphi}_* : T_{\tilde{\eta}}(\tilde{S}) \to T_{\varphi(\tilde{\eta})}D\}$$

lies in $\mathfrak{m}_\Phi^-$. This follows from the discussion in Section II.A.
We now conclude the proof with the crucial

Step four (see [A1]): $\overline{\Gamma}^{\mathbb{Q}}$ *is a normal subgroup of the derived group* DM_Φ.
To establish this we review some standard results that are consequences of the following result of Schmid [Schm1]:[2]

> *Let* $v \in V_{\mathbb{C}}^{\Gamma}$ *be a vector fixed by the monodromy group. Then the Hodge* (p, q) *components* $v_s^{p,q} := \big(v(\Phi(s))\big)^{p,q}$ *are constant and fixed by* Γ.

A corollary of this is:

> *If* $v \in V_{\mathbb{C}}^{\Gamma}$ *is of pure Hodge type* (p, q) *at one point, then it is everywhere of type* (p, q).

An application of this corollary is the

RIGIDITY THEOREM: *Let*

$$\begin{cases} \Phi : S \to \Gamma \backslash D \\ \Phi' : S \to \Gamma' \backslash D' \end{cases}$$

be two variations of Hodge structure and $A \in \mathrm{Hom}(V, V')$ *with the properties:*

(i) *A intertwines the monodromy, in the sense that for* $\gamma \in \pi_1(S, s_0)$

$$A\rho(\gamma) = \rho'(\gamma)A$$

where ρ, ρ' *are the respective monodromy representations.*

(ii) *A is of Hodge type* $(0, 0)$ *at one point. Then A induces a morphism of the variations of Hodge structure, which we write as*

$$\Phi' = A\Phi.$$

A standard application of this is what is sometimes referred to as "the semi-simplicity of monodromy":

[2] A succinct proof, using the results in Section I.C, is given following this more leisurely argument.

Let $W \subset V$ be a Γ-invariant subspace that induces a sub-variation of Hodge structure. Then $Q_W := Q\big|_W$ is non-singular so that

$$V = W \oplus W^{\perp}$$

and $W^{\perp}$ is a sub-variation of Hodge structure, and the original variation of Hodge structure decomposes accordingly as a direct sum.

We note that if W consists of Hodge vectors, then Γ acts on W by a finite group. This is because of our assumption that $V = V_{\mathbb{Z}} \otimes \mathbb{Q}$ and that $\Gamma \subset \mathrm{GL}(V_{\mathbb{Z}})$. Then Q_W is a definite quadratic form and $\Gamma\big|_W$ is a subgroup of $\mathrm{Aut}(W_{\mathbb{Z}}, Q_W)$ where $W_{\mathbb{Z}} = W \cap V_{\mathbb{Z}}$. In particular, replacing V by $V^{\otimes^k} \otimes \check{V}^{\otimes^l}$, when

$$W = \mathrm{Hg}_{\Phi}^{k,l},$$

Γ acts by a finite group, and by passing to a subgroup of finite index we may assume that Γ fixes all the $\mathrm{Hg}_{\Phi}^{k,l}$ pointwise. This implies that $\Gamma \subset M_{\Phi}$ and hence

$$\overline{\Gamma}^{\mathbb{Q}} \subseteq M_{\Phi}.$$

To show that $\overline{\Gamma}^{\mathbb{Q}}$ is a normal subgroup of M_{Φ}, let Γ_ρ be the image of Γ under any tensor representation V_ρ. The theorem of the fixed part implies that $\ker(\overline{\Gamma}_{\rho}^{\mathbb{Q}} - I)$ is a subvariation of Hodge structure of V_ρ.[3] Next, by (II.B.5), M_{Φ} stabilizes this sub-variation. Therefore, given $g \in M_{\Phi}(\mathbb{R})$ and $\gamma \in \overline{\Gamma}^{\mathbb{Q}}$, $g\gamma g^{-1}$ fixes $\ker(\overline{\Gamma}_{\rho}^{\mathbb{Q}} - I)$ *pointwise*. Thus, by (I.C.10) and using the notations there, $g\gamma g^{-1} \in \Gamma'$. Since $\overline{\Gamma}_{\rho}^{\mathbb{Q}}$ is semi-simple and therefore reductive, by (I.C.11) we have $\overline{\Gamma}^{\mathbb{Q}} = \Gamma'$ and $g\gamma g^{-1} \in \overline{\Gamma}^{\mathbb{Q}}$. From this we may conclude that $\overline{\Gamma}^{\mathbb{Q}}$ is a normal subgroup of M_{Φ}.

To show that $\overline{\Gamma}^{\mathbb{Q}}$ is a normal subgroup of the derived group DM_{Φ}, we remark that the abelian part A of M_{Φ} is in the isotropy group of $\widetilde{\Phi}(\tilde{s}_0)$, and hence $A(\mathbb{R})$ is compact. Thus $\Gamma \cap A$ is a finite group, and again passing to a subgroup of finite index if necessary Γ will intersect A only in the identity. □

We may summarize one consequence of the previous argument as follows:

(i) Setting $\Gamma(\mathbb{Q}) = P$, *the Mumford-Tate group decomposes over* $\mathbb{Q}$

$$M_{\Phi} = P \times R.$$

Each of P and R decomposes further into $\mathbb{Q}$-simple factors, and we denote by D_P and D_R the corresponding period domains;

[3] By Schmid's theorem the (p, q) components of $\ker(\overline{\Gamma}_{\rho}^{\mathbb{Q}} - I)$ are constant and Γ_ρ-invariant; hence $\ker(\overline{\Gamma}_{\rho}^{\mathbb{Q}} - I)$ is a sub-variation of Hodge structure.

(ii) *The variation of Hodge structure is given by*

$$\Phi : S \to \Gamma\backslash D_P \times D_R$$

and is constant in the D_R-factor; (in particular, a variation of Hodge structure with a finite monodromy group is isotrivial); and

(iii) *The monodromy group Γ and the $\mathbb{Q}$-algebraic group P have the same tensor invariants on the $T^{k,l}$.*

Observing that P and the arithmetic group $P_{\mathbb{Z}} = P \cap G_{\mathbb{Z}}$ have also the same tensor invariants we may conclude that

(III.A.3) *The monodromy group has the same tensor invariants as an arithmetic group.*

As noted in the introduction (footnote 6), Γ need not be an arithmetic group, i.e., Γ need not be of finite index in $P_{\mathbb{Z}}$. This says that insofar the invariants of its representations on the $\mathbb{Q}$-vector spaces $T^{k,l}$ are concerned, it is indistinguishable from $P_{\mathbb{Z}}$.

Because of the structure theorem (III.A.1) it is our sense that *Mumford-Tate domains are the natural universal objects for the study of variations of Hodge structure.*

First, they are obviously minimal — if one is going to have $\Gamma\backslash D$ for which variations of Hodge structure are given by maps

$$\Phi : S \to \Gamma\backslash D$$

then the smallest D's are Mumford-Tate domains.

Secondly, they have good formal properties that result from

$$M_{\rho(\varphi)} = \rho(M_\varphi)$$

for any representation $\rho : V \to V_\rho$. For example, if we do not use Mumford-Tate domains, then given a period domain $D(V)$ corresponding to (V, Q), there is an embedding

$$D(V) \subset D(V \otimes V)$$

given by

$$(V, \varphi) \to (V \otimes V, \varphi \otimes \varphi).$$

The image is the Mumford-Tate domain in $D(V \otimes V)$ given by the orbit of the image of G in $\mathrm{GL}(V \otimes V)$ under the diagonal embedding

$$A \to A \otimes A.$$

It is clear that the variation of Hodge structure arising as the tensor product of one should map to a quotient of $D(V)$ and not of $D(V \otimes V)$.

We conclude this section by discussing the implications of the almost product decomposition

$$M_\varphi(\mathbb{R}) = M_1' \times \cdots \times M_l' \times M_{l+1}'$$

into its connected, $\mathbb{R}$-simple factors, M_{l+1}' being the abelian one. We denote by $D_{M_i'}$ the M_i'-orbit

$$M_i' \cdot \varphi \subset D$$

of φ.

(III.A.4) THEOREM: (i) *The* $D_{M_i'}$ *are* **complex** *submanifolds of* D. (ii) *In case* M_i' *is compact, the image of any local variation of Hodge structure* $\Phi : S \to D$ *meets* $D_{M_i'}$ *in points.*

PROOF. We begin by formalizing an observation made several times, including in step two in the proof of Theorem (III.A.1).

(III.A.5) LEMMA: *Let* k *be a field with* $\mathbb{Q} \subseteq k \subseteq \mathbb{R}$ *and* A *a* k*-algebraic subgroup of* G_k *that is normalized by* $M_{\varphi,k}$. *Then the Lie algebra* $\mathfrak{A}$ *of* A *is a* k*-sub Hodge structure of* $\mathfrak{g}_k$ *with respect to* $\operatorname{Ad}\varphi$.

PROOF. $\operatorname{Ad}\varphi$ factors through $\mathfrak{m}_{\varphi,k}$, by the definition of M_φ. Since $M_{\varphi,k}$ normalizes A, $\operatorname{Ad}\varphi(\mathfrak{A}) \subseteq \mathfrak{A}$. □

Using the lemma, (i) in the theorem follows by the argument just below (II.A.2).

We now turn to (ii). Recall that upon choice of a reference point $\varphi \in D$ we have

$$D = G(\mathbb{R})/H_\varphi$$

where H_φ is the compact isotropy group of φ. It is known (see below) that there is a *unique* maximal compact subgroup $K_\varphi \subset G(\mathbb{R})$ with $H_\varphi \subset K_\varphi$. Let M be a connected, compact $\mathbb{R}$-simple factor of $M_\varphi(\mathbb{R})$. Part (ii) in the theorem then will follow from the

(III.A.6) PROPOSITION: $M \subseteq K_\varphi$.

Indeed, by the lemma the orbit $\mathcal{O}(M)$ lies in a fibre of

$$G(\mathbb{R})/H_\varphi \to G(\mathbb{R})/K_\varphi, \tag{III.A.7}$$

and it is known [C-MS-P] the differential

$$\Phi_* : TS \to TD$$

has image in the orthogonal complement to the tangent spaces of the fibres of (III.A.7).

The proof of the proposition will proceed in three steps. For notational simplicity, we will omit the subscript φ on H_φ and K_φ.

Step one: For a simple Lie algebra $\mathfrak{A} \subset \mathfrak{g}_\mathbb{R}$, we denote by

$$B_\mathfrak{A}(x,y) = \operatorname{Tr} \operatorname{ad} x \operatorname{ad} y \qquad x, y \in \mathfrak{A}$$

the Cartan-Killing form. It is known that $B_\mathfrak{A}$ is $\operatorname{ad} \mathfrak{A}$ -invariant and non-degenerate, and that *any* $\operatorname{ad} \mathfrak{A}$-invariant symmetric form is a multiple of $B_\mathfrak{A}$.

Suppose now that we have a $\varphi \in D$ and that $\mathfrak{A}$ is an $\mathbb{R}$-sub-Hodge structure of the Hodge structure on $\mathfrak{g}(\mathbb{R})$ determined by φ. Then the polarizing form Q on V induces a non-degenerate, symmetric form $Q_\mathfrak{A}$ on the $\mathbb{R}$-sub-Hodge structure $\mathfrak{A} \subset \mathfrak{g}(\mathbb{R})$. It follows that

$$Q_\mathfrak{A} \sim B_\mathfrak{A},$$

where the symbol $\sim$ means that the two forms are proportional.

Step two: Using the Hodge structure on $\mathfrak{g}$ given by φ, we have as usual

$$\begin{aligned} \mathfrak{g}_\mathbb{C} &= \bigoplus_i \mathfrak{g}^{-i,i} \\ &= \left(\bigoplus_j \mathfrak{g}^{-2j,2j}\right) \oplus \left(\bigoplus_k \mathfrak{g}^{-2k-1,2k+1}\right) \\ &:= \mathfrak{g}_\mathbb{C}^{\text{even}} \oplus \mathfrak{g}_\mathbb{C}^{\text{odd}}. \end{aligned}$$

We think, for example, of $\mathfrak{g}_\mathbb{C}^{\text{even}} \subset \operatorname{Hom}(V_\mathbb{C}, V_\mathbb{C})$ as the elements that shift

$$V^{p,q} \to V^{p-2k,q+2k}$$

the $V^{p,q}$'s by an even number of steps. We note that

(i) the decomposition $\mathfrak{g}_\mathbb{C} = \mathfrak{g}_\mathbb{C}^{\text{even}} \oplus \mathfrak{g}_\mathbb{C}^{\text{odd}}$ is defined over $\mathbb{R}$;

(ii) the form $Q_{\mathfrak{g}(\mathbb{R})}$ is positive on $\mathfrak{g}_\mathbb{R}^{\text{even}}$ and negative on $\mathfrak{g}_\mathbb{R}^{\text{odd}}$; and

(iii) $\mathfrak{k} = \mathfrak{g}_\mathbb{R}^{\text{even}}$

where $K \subset G(\mathbb{R})$ is the *unique* maximal compact subgroup of $G(\mathbb{R})$ containing H.

Step three: To complete the proof, we use the assumption that M is simple and that $\mathfrak{m} \subset \mathfrak{g}(\mathbb{R})$ is an $\mathbb{R}$-sub-Hodge structure and step one to conclude that

$$B_{\mathfrak{m}} \sim Q_{\mathfrak{m}}.$$

Moreover, since $\mathfrak{m} \subset \mathfrak{g}(\mathbb{R})$ is a sub-Hodge structure, we have as in step two

$$\mathfrak{m} = \mathfrak{m}^{\text{even}} \oplus \mathfrak{m}^{\text{odd}}.$$

Now we use the assumption that M is compact to infer that $B_{\mathfrak{m}}$ has the same sign as $B_{\mathfrak{k}}$, and thus by (ii) in step two

$$\mathfrak{m} = \mathfrak{m}^{\text{even}}.$$

Then by (iii) in step (ii) we have that

$$\mathfrak{m} \subseteq \mathfrak{k},$$

which implies the proposition. □

Remark: We have earlier noted that in the classical case,[4] the orbit of any compact, real factor of the Mumford-Tate group is a point, the reason being simply that a bounded domain in $\mathbb{C}^N$ contains no positive dimensional complex analytic subvarieties. The result in (III.A.4) means that insofar as variations of Hodge structure are concerned the non-classical case behaves as in the classical one. This is consistent with the general principal that *Hodge theory in weight larger than one is a relative theory.*

Remark: There are cases where K is the real Lie group associated to a $\mathbb{Q}$-algebraic subgroup $K_{\mathbb{Q}}$ of G. Then $M \subseteq K_{\mathbb{Q}}$, and there are cases where equality holds. When D is a bounded symmetric domain these are just the CM points.

Considering the orbits D_{M_φ} in the case when D is a bounded, symmetric domain in some $\mathbb{C}^N$, we have noted that $D_{M_\varphi} \subset D$ is a homogeneous complex submanifold. This does not automatically imply the D_{M_φ} is Hermitian symmetric, as for $N \geqq 3$ there are homogeneous but non-Hermitian symmetric bounded domains in $\mathbb{C}^N$. However we have (cf. [De3])

(III.A.8) PROPOSITION: *In the classical case the orbit D_{M_φ} is a Hermitian symmetric domain.*

---

[4]Or, more generally, in the case where D is Hermitian symmetric and the IPR is trivial.

For the classification given by Satake [Sa] of embeddings of Hermitian symmetric domains in Siegel upper-half-space we observe the previously noted result:

(III.A.9) COROLLARY: *The real factors of a Mumford-Tate group in the weight* $n = 1$ *case are all classical groups* [Mo2].

PROOF OF THE PROPOSITION. The proposition follows from two observations. The first is that $\mathfrak{m}_\varphi \subset \mathfrak{g}$ is a sub-Hodge structure. Thus we have

$$\text{(III.A.10)} \qquad \operatorname{Ad}\varphi : \mathbb{U}(\mathbb{R}) \to \mathfrak{m}_{\varphi,\mathbb{R}} \subset \mathfrak{g}_\mathbb{R}$$

where $\mathbb{U}(\mathbb{R}) \cong \{z \subset \mathbb{C} : |z| = 1\}$ is a circle. Writing

$$\begin{array}{ccccccc} \mathfrak{m}_{\varphi,\mathbb{C}} & = & \mathfrak{m}_\varphi^{-1,1} & \oplus & \mathfrak{m}_\varphi^{0,0} & \oplus & \mathfrak{m}_\varphi^{1,-1} \\ \cap & & \cap & & \cap & & \cap \\ \mathfrak{g}_\mathbb{C} & = & \mathfrak{g}^{-1,1} & \oplus & \mathfrak{g}^{0,0} & \oplus & \mathfrak{g}^{1,-1} \end{array}$$

where in the Cartan decomposition

$$\mathfrak{g} = \mathfrak{k} \oplus \mathfrak{p}$$

we have an identification $T_\varphi D \cong \mathfrak{p}$ and the decomposition

$$\begin{array}{ccccc} \mathfrak{p}_\mathbb{C} & = & \mathfrak{p}^{(1,0)} & \oplus & \mathfrak{p}^{(0,1)} \\ & & \| & & \| \\ & & \mathfrak{g}^{-1,1} & & \mathfrak{g}^{1,-1} \end{array}$$

gives the complex structure.

Next, the symmetry $\sigma_\varphi : D \to D$ about $\varphi \in D$ has differential given by

$$\sigma_{\varphi,*} = \operatorname{Ad}\varphi(\sqrt{-1}).$$

From (III.A.10) we infer that σ_φ leaves $T_\varphi D_{M_\varphi}$ invariant, and it follows that $\sigma_\varphi : D_{M_\varphi} \to D_{M_\varphi}$ defines a symmetry on D_{M_φ}. ☐

We remark that Satake's list gives significantly more than (III.A.9) in that it severely restricts what the $\mathbb{R}$-simple factors of a Mumford-Tate group can be in the classical case.

In contrast to (III.A.9), in Section IV.C we will show that the exceptional simple group G_2 is the Mumford-Tate group of many polarized Hodge structures. For example there is one of weight four with Hodge numbers $h^{4,0} = 1$, $h^{3,1} = h^{2,2} = 4$.

III.B AN APPLICATION OF MUMFORD-TATE GROUPS

This concerns the following natural

QUESTION: *Given a Hodge structure V_φ, when are there Hodge structures V_{φ_i} such that*

(III.B.1) $$V_\varphi \subset \overset{i}{\otimes} V_{\varphi_i}$$

is a sub-Hodge structure?

To describe Schoen's necessary condition, we recall that the *level* $l(V_\varphi)$ of a Hodge structure of weight n is defined by

$$l(V_\varphi) = \max\left\{|p-q| : V_\varphi^{p,q} \neq 0\right\}.$$

If n is positive and if $l(V_\varphi) < n$, then V_φ is a Tate twist of a lower weight Hodge structure.

Next, we define

$$\begin{aligned} \mu(V_\varphi) &= \left(\tfrac{1}{2}\right) l(\mathfrak{m}_\varphi) \\ &= \max_{|i|}\left\{\mathfrak{m}_\varphi^{-i,i} \neq 0\right\} \end{aligned}$$

and

$$\tau(V_\varphi) = \max\left\{\mu\left(V'_\varphi\right) : V'_\varphi \subset V_\varphi \text{ is a sub-Hodge structure}\right\}.$$

The invariant $\mu(V_\varphi)$ measures the maximum shift $V^{p,q} \to V^{p+r,q-r}$ of an element of $\mathfrak{m}_\varphi \subset \operatorname{Hom}(V,V)$. The basic observation is that if we have (III.B.1), then

(III.B.2) $$\tau(\overset{i}{\otimes} V_{\varphi_i}) \leqq \max_i\left\{\tau\left(V_{\varphi_i}\right)\right\}.$$

To see this, we observe in general that for Hodge structures V_{φ_1} and V_{φ_2} we have

(i) $M_{\varphi_1+\varphi_2} \subset M_{\varphi_1} \times M_{\varphi_2}$ and

(ii) $M_{\varphi_1 \otimes \varphi_2} \subset M_{\varphi_1} \times M_{\varphi_2}$.

In (i) $M_{\varphi_1+\varphi_2}$ refers to the Mumford-Tate group of $V_{\varphi_1} \oplus V_{\varphi_2}$, and in (ii) $M_{\varphi_1\otimes\varphi_2}$ is the Mumford-Tate group of $V_{\varphi_1\otimes\varphi_2}$. In both cases, the projections

$$M_{\varphi_1} \times M_{\varphi_2} \to M_{\varphi_i} \qquad i = 1, 2$$

are surjective. But aside from the case of tensor constructions noted above, the inclusions (i), (ii) are in general not isomorphisms (cf. [Mo1] and [Mo2]).

It follows from (ii) that we have an inclusion

$$M_{\otimes\varphi_i} \subset \prod_i M_{\varphi_i}$$

that gives

$$\mathfrak{m}_{\otimes\varphi_i} \subset \bigoplus_i \mathfrak{m}_{\varphi_i}$$

leading to

(III.B.3) SCHOEN'S CRITERION: *If* (III.B.1) *holds, then*

$$\tau(V_\varphi) \leqq \max_i\{\tau(V_{\varphi_i})\}.$$

As a simple consequence, if φ is generic in D and $V_\varphi^{n,0} \neq 0$, then $\mathfrak{m}_\rho = \mathfrak{g}$ will have no sub-Hodge structures and (III.B.1) cannot be satisfied non-trivially — i.e., when the weights of the V_{φ_i} are strictly less than n.

Remark: We note that this discussion holds verbatim for general Hodge structures $V_{\tilde{\varphi}}$ using the Mumford-Tate groups $M_{\tilde{\varphi}}$. We have presented the discussion in the polarized case because the topic of this section is period domains.

Schoen's criterion was used to address the following geometric question:

(III.B.4) *When is a smooth n-dimensional projective variety X dominated by a product of lower dimensional, projective varieties X_i in the sense that there is a dominant rational map*

$$X_1 \times \cdots \times X_k \dashrightarrow X?$$

After resolving the singularities of the X_i and of the map, we may assume that the X_i are smooth and the map is a morphism. We then have an inclusion

$$H^n(X) \subset \bigoplus_{n_1+\cdots+n_k=n} \left(\overset{i}{\otimes} H^{n_i}(X_i)\right)$$

and we are in the situation (III.B.1) where, up to Tate twists, the weights of the V_{φ_i} are all less than the weight n of V_φ. Thus

(III.B.5) *For (V, φ) the Hodge structure on $H^n(X)$, if $\tau(V_\varphi) = n$ then the situation* (III.B.4) *does not occur.*

In general, for $H^n(X)$ the computation of the Mumford-Tate group and τ of it is not a simple matter. However, in the case X is a generic member of a family, Schoen devised a very nice *arithmetic* method of addressing this issue. Other methods appear in the literature [PS].

To briefly explain this, we shall denote by

$$E \subset \mathfrak{m}_{\varphi}^{-1,1}$$

the integral element given by the tangent space at X to the family. We want to find a method to show that

$$\mathfrak{m}_{\varphi}^{-n,n} \neq 0. \tag{III.B.6}$$

The simple idea of using the Lie algebra generated by E will not work because, as noted in (II.A.10), E is abelian.

Schoen's idea is to first note that $\mathrm{Aut}(\mathbb{C}/\mathbb{Q})$ acts on V; the orbit $\mathrm{Aut}(\mathbb{C}/\mathbb{Q})\cdot F_{\varphi}^{\bullet}$ in $\check{D}$ reflects the degree of *non-alignment* with the $\mathbb{Q}$-structure, or *transcendence*, of $F_{\varphi}^{\bullet} \in \check{D}$. Secondly, since $\mathfrak{m}_{\varphi} \subset \mathfrak{g}$ is defined over $\mathbb{Q}$ there is an induced action of $\mathrm{Aut}(\mathbb{C}/\mathbb{Q})$ on $\mathfrak{m}_{\varphi} \subset \mathrm{Hom}(V, V)$. To establish (III.B.6) it will thus suffice to find $\Psi \in E$ and $\sigma \in \mathrm{Aut}(\mathbb{C}/\mathbb{Q})$ such that

$$\left(\sigma \circ \Psi \circ \sigma^{-1}\right)^{-n,n} \neq 0.$$

Schoen does this by finding $\{X, E, \sigma\}$ with the properties

- $\dim X = 3$ and $h^{3,0}(X) = h^{2,1}(X) = 1$;
- $\sigma : H^{1,2}(X) \xrightarrow{\sim} H^{0,3}(X)$ and

  $\sigma^{-1} : H^{3,0}(X) \xrightarrow{\sim} H^{2,1}(X)$
- $\sigma \circ X \circ \sigma^{-1} : H^{3,0}(X) \xrightarrow{\sim} H^{0,3}(X)$,

which implies that $\tau(H^3(X)) = 3$. This leads him to the

Conclusion. *The generic mirror-quintic Calabi-Yau (CY) threefold does not satisfy* (III.B.4).

Of course, there are other circumstances where one may show (III.B.6); e.g., if

$$M_{\varphi} = G.$$

From [Sch] it follows that this is the case if X is generic in the family of hypersurfaces of degree $\geqq n + 2$ in $\mathbb{P}^{n+1}$.

We have included the discussion of Schoen's results because most of the geometric applications of Mumford-Tate groups seem to be in the classical case, or in situations derived from the classical case ([DMOS], [Mo1], [Mo2], [PS]). The above results are definitely outside those arising from the classical cases. Further applications of Mumford-Tate groups will be given in Chapter V.

III.C NOETHER-LEFSCHETZ LOCI AND VARIATIONS OF HODGE STRUCTURE

We consider a variation of Hodge structure

(III.C.1) $$\Phi : S \to \mathcal{M}_\Gamma.$$

Given a point $s_0 \in S$, we choose a lifting of $\Phi(s_0)$ to a point $\varphi(\widetilde{s_0}) \in D$ where $\widetilde{s_0} \in \widetilde{S}$ lies over s_0. Then we have a diagram

$$\begin{array}{ccccc} \tilde{s}_0 & \in & \widetilde{S} & \xrightarrow{\varphi} & D \\ \downarrow & & \downarrow & & \downarrow \\ s_0 & \in & S & \xrightarrow{\Phi} & \Gamma\backslash D \end{array}$$

where for $\tilde{s} \in \widetilde{S}$ lying over $s \in S$, $\varphi(\tilde{s}) \in D$ is a polarized Hodge structure lying over $\Phi(s) \in \Gamma\backslash D$. The *Noether-Lefschetz locus*

(III.C.2) $$\mathrm{NL}_{s_0}(S) \subset S$$

of the pair (Φ, s_0) consisting of the variation of Hodge structure given by Φ and the point $s_0 \in S$ is defined as follows:

(i) First, $\varphi^{-1}(\mathrm{NL}_{\tilde{s}_0})$ is a complex analytic subvariety of $\widetilde{S}$;

(ii) The projection of $\varphi^{-1}(\mathrm{NL}_{\tilde{s}_0})$ is independent of the lifting $\tilde{s}_0$ of s_0 and is a complex analytic subvariety of S.

It follows from the theorem of Cattani-Deligne-Kaplan [CDK] that

(III.C.3) *In the global case when S is a quasi-projective algebraic variety, $\mathrm{NL}_{s_0}(S)$ is an algebraic subvariety of S.*

Here, we are interested in the question

What is the codimension of $\mathrm{NL}_{s_0}(S)$ in S?

A first, obvious estimate is

(III.C.4) $$\operatorname{codim}_S \mathrm{NL}_{s_0}(S) \leqq \operatorname{codim}_D \mathrm{NL}_{\varphi(\widetilde{s_0})} = \dim\left(\mathfrak{g}^-/\mathfrak{m}^-\right).$$

However, this does not take into account that (III.C.1) is an integral manifold of the exterior differential system generated by $I \subset T^*D$, nor of the integrability conditions in that system. The same arguments as in Section V of [CGG], together with Theorem (II.C.1), lead to the following considerations: For $T^k = \otimes^k V$, assuming that the weight $kn = 2m$ is even we have for the integral element

$$E = \varphi_*(T_{\widetilde{s}_0}\widetilde{S}) \subset \mathfrak{g}^{-1,1}$$

the maps (cf. (II.A.11))

$$\mathrm{Sym}^{m-2} E \otimes (E \cap \mathfrak{m}^{-1,1}) : (T^k)^{2m,0} \to (T^k)^{m+1,m-1}.$$

Denote by σ_m the rank of this map.

(III.C.5) THEOREM: *Assume that* $\mathrm{NL}_{s_0}(S)$ *is reduced and irreducible. Without loss of generality assume also that* s_0 *is a smooth point of* $\mathrm{NL}_{s_0}(S)$*. Then we have*

$$\operatorname{codim}_S \mathrm{NL}_{s_0}(S) \leqq \dim(\mathfrak{g}^{-1,1}/\mathfrak{m}^{-1,1}) - \sum_m \sigma_m.$$

Remark: The first term on the RHS represents the part of the normal space to the Mumford-Tate domain in the period domain that satisfies the IPR, and the second term reflects the integrability conditions implied by this constraint. As noted in [CGG], in the geometric case and assuming the Hodge conjecture, in a family of algebraic varieties there are "many" more algebraic cycles than predicted by naïve dimension counts. It is also noted there that there are examples where the above estimate is an equality, so that no better general estimate is possible.

Discussion. It is easy to construct local integral manifolds $S \subset D$ of $\mathfrak{I}$ such that $\varphi \in S$ and the intersection

$$\mathrm{NL}_\varphi \cap S$$

is non-reduced. For instance, although we shall not give details here, one may use EDS methods to find examples where we

(i) construct a smooth integral curve $S_1 \subset \mathrm{NL}_\varphi$ of $\mathfrak{I}$;

(ii) "thicken" S_1 to a k_0-jet of integral surface $S_1^{(k_0)}$ lying in NL_φ;

(iii) construct an integral surface S containing $S_1^{(k_0)}$ and osculating to order exactly k_0 to NL_φ along S_1.

Examples arising from geometry — specifically from families of hypersurfaces in projective space — are discussed in [O1], [O2]. There, she considers a point $s_0 \in S$ and a class $\zeta \in V_{s_0}^{p,p} \subset V_c$; *$\zeta$ need not be rational.* The Noether-Lefschetz locus S_ζ is defined locally by the condition that the vector ζ remain in F_s^p. It is then shown that the non-reducedness of S_ζ is in a sense a generic phenomenon.

Following is a simple example illustrating how non-reducedness can arise algebro-geometrically.

Example: We consider smooth 4-folds of degree d

$$X \subset \mathbb{P}^5 = \mathbb{P}(\check{\mathfrak{U}})$$

where $\mathfrak{U} \cong \mathbb{C}^6$ with coordinates $z_1, \ldots, z_6$ and $P \in \operatorname{Sym}^d \mathfrak{U} =: \mathfrak{U}^d$ defines X. Denote by $J_P \subset \operatorname{Sym} \mathfrak{U} := \mathfrak{U}^\bullet$ the Jacobian ideal of P. Then it is well known that there is a natural identification

$$H^{2,2}(X)_{\text{prim}} = \mathfrak{U}^{3d-6}/J_P.$$

The tangent space T to the local moduli space of X is given by

$$T = \mathfrak{U}^d/J_P.$$

For $A \in \mathfrak{U}^{3d-6}$ and $B \in \mathfrak{U}^d$ we denote by $[A]$ and $[B]$ the corresponding classes when we quotient by J_P. Then

$$[B] \cdot [A] = 0 \;\text{ in }\; H^{1,3}(X) \Leftrightarrow AB = \sum C_i P_i \tag{III.C.6}$$

where $P_i = \partial_{z_i} P$. This is the condition that, to 1st order, the class $[A]$ remains of type $(2,2)$. We want to formulate the condition that it remain of type $(2,2)$ to 2nd order.

For this we let

$$\operatorname{div} C = \sum_i \partial_{z_i} C_i.$$

Choosing two directions B^1 and B^2, the condition that $[A]$ move to 2nd order — i.e., that $\nabla_{\partial_1 \partial_2}[A] = 0$ in $H^{1,3}(X)$ — is the following: $\partial_1[A] = \partial_2[A] = 0$ gives

$$AB^\alpha = \sum_i C_i^\alpha P_i \qquad \alpha = 1, 2$$

and

$$\nabla_{\partial_1 \partial_2}[A] = (\operatorname{div} C^1) B^2 + (\operatorname{div} C^2) B^1 \in \mathfrak{U}^{4d-6}/J_P.$$

There is no reason that this should vanish (see [O1], [O2] for the general formalism).

For a specific example, if

$$\begin{cases} A = z_1^{d-2}Q \\ B = z_1^2 R \end{cases}$$

where Q, R do not involve z_1, then

$$AB = \left(\frac{1}{d}\right)(z_1 QR)(dz_1^{d-1}).$$

Then

$$\operatorname{div} C = \left(\frac{1}{d}\right)(Q + z_1 \partial_{z_1}(QR) = \left(\frac{1}{d}\right) QR$$

and

$$\operatorname{div} CB = \left(\frac{1}{d}\right) z_1^2 QR^2$$

which need not be in J_P.

Chapter IV

Hodge Representations and Hodge Domains

Blanket assumption for this chapter: M will be a **semi-simple** *$\mathbb{Q}$-algebraic group.* In this case, the Lie algebra $\mathfrak{m}$ is a direct sum $\mathfrak{m} = \oplus \mathfrak{m}_i$ where the $\mathfrak{m}_i$ are simple Lie algebras over $\mathbb{Q}$. The group M is then an almost direct product of the $\mathbb{Q}$-simple algebraic groups M_i whose Lie algebra is $\mathfrak{m}_i$ [Bor].[1]

The basic questions underlying this section are

(i) *Which M can be Mumford-Tate groups of polarized Hodge structures?*

and, more fundamentally,

(ii) *What can one say about the different realizations of M as a Mumford-Tate group?*

(iii) *What can one say about the corresponding Mumford-Tate domains?*

In this chapter we shall use standard material from the structure theory of semi-simple Lie algebras and their representation theory. For reference, in the appendix to this chapter we have collected the facts and listed the notations that we shall use.

This chapter will be divided into three parts. The first, short part is introductory, giving basic definitions and one answer to the first question above.

The second part will use the representation theory of real, simple Lie groups to give the strategy for an answer over $\mathbb{R}$ to the second question. That is, Hodge representations as defined below are given by data V, ρ, Q, φ where V, ρ, Q are defined over $\mathbb{Q}$. In part two we will only require this data to be defined over $\mathbb{R}$. The strategy will be carried out in detail in Sections IV.B and IV.E. We shall also give a number of examples illustrating how the strategy is implemented in special cases.

[1] As noted previously, the most natural setting for Mumford-Tate groups is when M is reductive and where the center of the associated Lie group $M_{\mathbb{R}}$ is compact. Working in this generality in this chapter necessitates more complicated arguments and obscures the essential representation-theoretic ideas and calculations. For these reasons we decided to restrict to the semi-simple case.

In the third part to this chapter we will use the theory of representations of $\mathbb{Q}$-algebraic groups to give the method for giving an answer to the second question over $\mathbb{Q}$ once the answer over $\mathbb{R}$ has been given.

The third question will be addressed in Sections IV.F and IV.G.

IV.A PART I: HODGE REPRESENTATIONS

We recall our notation (V, Q) for a $\mathbb{Q}$-vector space V and non-degenerate bilinear form $Q : V \otimes V \to \mathbb{Q}$ with $Q(u, v) = (-1)^n Q(v, u)$ for $u, v \in V$.

Definition: A *Hodge representation* (M, ρ, φ) is given by a representation defined over $\mathbb{Q}$

$$\rho : M \to \mathrm{Aut}(V, Q) \tag{IV.A.1}$$

and a non-constant homomorphism of Lie groups

$$\varphi : \mathbb{U}(\mathbb{R}) \to M(\mathbb{R})$$

such that $(V, Q, \rho \circ \varphi)$ is a polarized Hodge structure of weight n.

Without essential loss of generality, we shall assume that the induced map

$$\rho_* : \mathfrak{m} \to \mathrm{End}_Q(V)$$

is injective. We note that the representation is then faithful on the finite quotient $M/\ker \rho$ of M.

It is understood that the Hodge structure is of pure weight n.

For notational simplicity, in this section we will denote $\mathbb{U}(\mathbb{R})$ by S^1 considered as the circle $z = e^{2\pi i \xi}$ in $\mathbb{C}$. We will also assume that the representation ρ is *irreducible*; any representation is a direct sum of such so it is sufficient to treat the irreducible case.

We will see (IV.A.9) that for a Hodge representation (M, ρ, φ), a generic conjugate $\mathrm{Ad}(\varphi_m)$, where $\varphi_m = m^{-1}\varphi m$ and $m \in M(\mathbb{R})$, of the adjoint representation

$$\mathrm{Ad} : M \to (\mathfrak{m}, B)$$

gives a *Mumford-Tate representation,* defined to be a Hodge representation whose image $\rho(M)$ is the Mumford-Tate group of $(V, Q, \rho \circ \varphi)$. Thus, up to isogeny to answer question (i) it is enough to determine which M have Hodge representations. We note that in any case, given a Hodge representation (IV.A.1), the image $\rho(M)$ contains the Mumford-Tate group of $(V, Q, \rho \circ \varphi)$.

We begin with the following

(IV.A.2) PROPOSITION: *If M has a Hodge representation* (IV.A.1), *then $M(\mathbb{R})$ contains a compact maximal torus T with $\varphi(S^1) \subset T$ with* $\dim(T) = \operatorname{rank}(M)$. *Moreover, the subgroup $H_\varphi \subset M(\mathbb{R})$, such that $\rho(H_\varphi)$ fixes the polarized Hodge structure $(V, Q, \rho \circ \varphi)$, is compact and is the centralizer $Z(\varphi(S^1))$ of the circle $\varphi(S^1)$ in $M(\mathbb{R})$.*

PROOF. For simplicity of notation, we shall drop reference to ρ. Then the closed subgroup of $M(\mathbb{R})$ fixing the Hodge structure is

$$H_\varphi = \{g \in M(\mathbb{R}) : g(V_\varphi^{p,q}) = V_\varphi^{p,q}\}$$

where, as usual, $V_\mathbb{C} = \oplus V_\varphi^{p,q}$ is the Hodge decomposition given by φ. Since the subgroup of $G(\mathbb{R})$ that preserves the polarized Hodge structure is compact (it is a product of unitary groups and, in even weight, an orthogonal group of a definite quadratic form), the closed subgroup H_φ of $M(\mathbb{R})$ is also compact.

To see that $H_\varphi = Z(\varphi(S^1))$, we note that since $\varphi(z) = e^{i(2p-n)\xi}$ on $V_\varphi^{p,n-p}$ it follows that for $g \in M(\mathbb{R})$

$$g\varphi(z) = \varphi(z)g \text{ if, and only if, } g(V_\varphi^{p,q}) \subseteq V_\varphi^{p,q}.$$

Let A be a maximal connected abelian subgroup of $M(\mathbb{R})$ containing $\varphi(S^1)$. Then

$$A \subseteq Z(\varphi(S^1)) = H_\varphi$$

so A is compact and connected, and hence a compact maximal torus. Because A is maximal abelian, $\dim(A) = \operatorname{rank}(M)$. □

COROLLARY: *There does not exist a Hodge representation of* $\mathrm{SL}_n(\mathbb{Q})$ *for* $n \geqq 3$.

We will see later in this section, and again in Section IV.E, that a converse to Proposition (IV.A.2) is true (cf. Theorem (IV.E.1)). The proof given there will make use of the theory of semi-simple Lie algebras. This result has the following consequence:

(IV.A.3) *Up to isogeny, the semi-simple Mumford-Tate groups are exactly those semi-simple $\mathbb{Q}$-algebraic groups M whose maximal torus is anisotropic.*

This condition is equivalent to the group $M(\mathbb{R})$ containing the compact maximal torus T. It is also equivalent to the existence of a maximal compact subgroup $M_c(\mathbb{R})$ of $M(\mathbb{C})$ that is isomorphic to the compact form of $M(\mathbb{R})$.

The caveat about up to isogeny, or equivalently, up to finite coverings is because of the following: In the lattice of semi-simple $\mathbb{Q}$-algebraic groups with the same Lie algebra, there may be some that admit faithful Hodge representations and some that do not.

Part II: Hodge representations over $\mathbb{R}$. Since the analysis is somewhat intricate, in the hopes that it will assist the reader in seeing the overall strategy, we begin with

Steps in the analysis of Hodge representations over $\mathbb{R}$.

Step one: Determine the irreducible representations over $\mathbb{R}$ of $\mathfrak{m}_{\mathbb{R}}$ that have an invariant bilinear form. By Schur's lemma, such a form will be non-degenerate. Moreover, for the Lie algebra $\mathfrak{m}_{\mathbb{R}}$, again by Schur's lemma $\operatorname{End}_{\mathfrak{m}_{\mathbb{R}}}(V_{\mathbb{R}})$ will necessarily be a division algebra over $\mathbb{R}$ and we will have the three possible cases

$$\operatorname{End}_{\mathfrak{m}_{\mathbb{R}}}(V_{\mathbb{R}}) = \begin{cases} \mathbb{R} & \text{(real case)} \\ \mathbb{C} & \text{(complex case)} \\ \mathbb{H} & \text{(quaternionic case),} \end{cases}$$

and then for $\operatorname{End}_{\mathfrak{m}_{\mathbb{R}}}(V_{\mathbb{R}}) \otimes \mathbb{C} \cong \operatorname{End}_{\mathfrak{m}_{\mathbb{R}}}(V_{\mathbb{C}})$, where $V_{\mathbb{C}} = V_{\mathbb{R}} \otimes_{\mathbb{R}} \mathbb{C}$, we have

$$\operatorname{End}_{\mathfrak{m}_{\mathbb{R}}}(V_{\mathbb{C}}) = \begin{cases} \mathbb{R} \otimes_{\mathbb{R}} \mathbb{C} = \mathbb{C} \\ \mathbb{C} \otimes_{\mathbb{R}} \mathbb{C} \cong \mathbb{C} \oplus \mathbb{C} \\ \mathbb{H} \otimes_{\mathbb{R}} \mathbb{C} \cong M_2(\mathbb{C}), \end{cases}$$

where as usual $\mathbb{H}$ are quaternions and $M_2(\mathbb{C})$ denotes the 2×2 matrices with complex entries. Only in the real case do we get a division algebra over $\mathbb{C}$, so $V_{\mathbb{C}}$ is reducible in the other two cases. The analysis of whether there are invariant forms and whether they are symmetric or alternating will necessitate considering the various cases arising from the three possibilities above.

Step two: We denote by $\operatorname{Res}_{\mathbb{C}/\mathbb{R}}$ the operation of *restriction of scalars* that considers a vector space over $\mathbb{C}$ as one over $\mathbb{R}$, and similarly for $\operatorname{Res}_{\mathbb{H}/\mathbb{R}}$,$\operatorname{Res}_{\mathbb{H}/\mathbb{C}}$.

We can associate to $V_{\mathbb{R}}$ an irreducible representation U of $\mathfrak{m}_{\mathbb{C}}$ over $\mathbb{C}$ such that as representations of $\mathfrak{m}_{\mathbb{C}}$

$$V_{\mathbb{C}} = \begin{cases} U & \text{(real case)} \\ U \oplus \check{U}, U \ncong \check{U} & \text{(complex case)} \\ U \oplus \check{U}, U \cong \check{U} & \text{(quaternionic case)} \end{cases}$$

and

$$V_{\mathbb{R}} \cong \operatorname{Res}_{\mathbb{C}/\mathbb{R}}(U) \cong \operatorname{Res}_{\mathbb{C}/\mathbb{R}}(\check{U}) \quad \text{(complex and quaternionic cases),}$$

while

$$V_{\mathbb{R}} \oplus V_{\mathbb{R}} \cong \operatorname{Res}_{\mathbb{C}/\mathbb{R}}(U) \quad \text{(real case).}$$

Furthermore, in the quaternionic case, there is an irreducible representation $\mathbb{U}$ of $\mathfrak{m}_{\mathbb{R}} \otimes_{\mathbb{R}} \mathbb{H}$ over $\mathbb{H}$ such that

$$V_{\mathbb{R}} \cong \mathrm{Res}_{\mathbb{H}/\mathbb{R}}(\mathbb{U}) \qquad \text{(quaternionic case)}$$

and then

$$U \cong \mathrm{Res}_{\mathbb{H}/\mathbb{C}}(\mathbb{U}) \qquad \text{(quaternionic case).}$$

Now

$$\mathrm{End}_{\mathfrak{m}_{\mathbb{R}}}(V_{\mathbb{R}}) \text{ acts on } V_{\mathbb{R}} \text{ as } \begin{cases} \mathbb{R} & \text{(real case)} \\ \mathbb{C} \text{ acting on } \mathrm{Res}_{\mathbb{C}/\mathbb{R}}(U) & \text{(complex case)} \\ \mathbb{H} \text{ acting on } \mathrm{Res}_{\mathbb{H}/\mathbb{R}}(\mathbb{U}) & \text{(quaternionic case).} \end{cases}$$

In the quaternionic case, if $\mathfrak{m}_{\mathbb{R}} \otimes_{\mathbb{R}} \mathbb{H}$ acts on the left, then $\mathrm{End}_{\mathfrak{m}_{\mathbb{R}}}(V_{\mathbb{R}}) \cong \mathbb{H}$ acts on the right.

Step three: We now let λ be the highest weight of U. There is a unique element w_0 of the Weyl group such that $w_0(\mathfrak{r}^+) = \mathfrak{r}^-$ [GW, page 133]. Further [GW, page 158]: U has an $\mathfrak{m}_{\mathbb{C}}$-invariant bilinear form if, and only if, $w_0(\lambda) = -\lambda$.

By Schur's Lemma, this is non-degenerate, unique up to a constant, and either alternating or symmetric.

If $\alpha_1, \dots \alpha_{\mathfrak{r}}$ are a choice of simple positive roots for $\mathfrak{m}_{\mathbb{C}}$ and h_{α_i} are the co-roots, let

$$h^0 = \sum_{\alpha \in \mathfrak{r}^+} h_\alpha .$$

Then [GW, page 160]:

> *The universal bilinear form is symmetric/alternating depending on whether $\langle \lambda, h^0 \rangle$ is even/odd.*

Further, [GW, page 160]:

$$\langle \alpha_i, h^0 \rangle = 2 \text{ for all } i.$$

Step four: If we write the decomposition into weight spaces

$$U = \bigoplus_{\omega} U_\omega$$

then

$$\check{U} = \bigoplus_{\omega} \check{U}_\omega, \qquad \check{U}_\omega \text{ has weight } -\omega.$$

On $V_{\mathbb{C}} \cong V_{\mathbb{R}} \otimes_{\mathbb{R}} \mathbb{C}$ conjugation gives an isomorphism $V_{\mathbb{C}} \xrightarrow{c} V_{\mathbb{C}}$ that gives a natural isomorphism of vector spaces $U_\omega \xrightarrow{c_\omega} \check{U}_\omega$. In the complex and quaternionic cases, $V_{\mathbb{C}} \cong U \oplus \check{U}$ and c is c_ω on U_ω and c_ω^{-1} on $\check{U}_\omega$. Now

$$\operatorname{Hom}_{\mathfrak{m}_{\mathbb{R}}}(S^2 V_{\mathbb{R}}, \mathbb{R}) \otimes_{\mathbb{R}} \mathbb{C} \cong \operatorname{Hom}_{\mathfrak{m}_{\mathbb{C}}}(S^2 V_{\mathbb{C}}, \mathbb{C})$$

and thus

$$\operatorname{Hom}_{\mathfrak{m}_{\mathbb{R}}}(S^2 V_{\mathbb{R}}, \mathbb{R}) \otimes_{\mathbb{R}} \mathbb{C}$$

$$= \begin{cases} \operatorname{Hom}_{\mathfrak{m}_{\mathbb{C}}}(S^2 U, \mathbb{C}) & \text{(real case)} \\ \operatorname{Hom}_{\mathfrak{m}_{\mathbb{C}}}(S^2 U, \mathbb{C}) \oplus \operatorname{Hom}_{\mathfrak{m}_{\mathbb{C}}}(U \otimes \check{U}, \mathbb{C}) \oplus \operatorname{Hom}_{\mathfrak{m}_{\mathbb{C}}}(S^2 \check{U}, \mathbb{C}) & \text{(complex and quaternionic cases).} \end{cases}$$

Similarly,

$$\operatorname{Hom}_{\mathfrak{m}_{\mathbb{R}}}(\Lambda^2 V_{\mathbb{R}}, \mathbb{R}) \otimes_{\mathbb{R}} \mathbb{C}$$

$$= \begin{cases} \operatorname{Hom}_{\mathfrak{m}_{\mathbb{C}}}(\Lambda^2 U, \mathbb{C}) & \text{(real case)} \\ \operatorname{Hom}_{\mathfrak{m}_{\mathbb{C}}}(\Lambda^2 U, \mathbb{C}) \oplus \operatorname{Hom}_{\mathfrak{m}_{\mathbb{C}}}(U \otimes \check{U}, \mathbb{C}) \oplus \operatorname{Hom}_{\mathfrak{m}_{\mathbb{C}}}(\Lambda^2 \check{U}, \mathbb{C}) & \text{(complex and quaternionic cases).} \end{cases}$$

Thus:

Real case: There is a unique (up to a constant) invariant bilinear form on $V_{\mathbb{R}}$, symmetric/alternating depending on the parity of $\langle \lambda, h^0 \rangle$.

Complex case: There are unique (up to constants) symmetric invariant bilinear and alternating invariant bilinear forms on $V_{\mathbb{R}}$.

Quaternionic case: There are unique (up to constants) symmetric invariant bilinear and alternating invariant bilinear forms Q on $V_{\mathbb{R}}$ that pair U and $\check{U}$, so that $Q(v, \overline{v})$ is non-degenerate on $V_{\mathbb{C}}$.

Step five: If R is the root lattice of $\mathfrak{m}_{\mathbb{C}}$, then the real form $\mathfrak{m}_{\mathbb{R}}$ has a Cartan involution $\mathfrak{m}_{\mathbb{R}} \xrightarrow{\theta} \mathfrak{m}_{\mathbb{R}}$ where

$$\theta = \begin{cases} 1 & \text{on } \mathfrak{k} \\ -1 & \text{on } \mathfrak{p}. \end{cases}$$

We recall the map

$$R \xrightarrow{\psi} \mathbb{Z}/2\mathbb{Z}$$

where

$$\psi(\alpha) = \begin{cases} 0 & \text{if } e_\alpha \in \mathfrak{k} \\ 1 & \text{if } e_\alpha \in \mathfrak{p}, \end{cases} \qquad \alpha \in \mathfrak{r}.$$

As noted above, because θ is a Lie algebra homomorphism ψ extends uniquely from $\mathfrak{r}$ to R as a group homomorphism. Now let

$$R \xrightarrow{\Psi} \mathbb{Z}/4\mathbb{Z}$$

be defined by

$$\begin{cases} \Psi(x) = 2 & \text{if, and only if, } \psi(x) = 1 \\ \Psi(x) = 0 & \text{if, and only if, } \psi(x) = 0. \end{cases}$$

Step six: Associated to $\mathfrak{m}_{\mathbb{R}}$ are connected Lie groups $M_{P'}$ for each lattice P' with

$$P \supseteq P' \supseteq R, \qquad R = \text{root lattice},\ P = \text{weight lattice}$$

where

$$\pi_1(M_{P'}) \cong P/P', \quad Z(M_{P'}) \cong P'/R.$$

Note

$$\begin{aligned} M_R &= M_a \text{ adjoint form, } M_a \overset{\text{Ad}}{\hookrightarrow} \text{Aut}(\mathfrak{m}_{\mathbb{R}}) \\ M_P &= M_s \text{ simply connected form, } \pi_1(M_s) = 0. \end{aligned}$$

The maximal torus T of $M_{P'}$ is

$$T = \mathfrak{t}/\Lambda, \quad \Lambda \cong \text{Hom}(P', \mathbb{Z}).$$

In order to have U defined on $M_{P'}$, we need $\lambda \in P'$.

Step seven: The weights that occur for U belong to $\lambda + R$, and

$$\text{span}_{\mathbb{Z}}(\text{weights of } U) = \mathbb{Z}\lambda + R.$$

Note that $\lambda \in R \otimes_{\mathbb{Z}} \mathbb{Q}$, so this is not a direct sum. Let

$$P' = \mathbb{Z}\lambda + R, \quad \Lambda = \text{Hom}(P', \mathbb{Z}).$$

Now let

$$\mathfrak{l}_{\varphi} \in \Lambda$$

be the lattice point such that the line $\mathbb{R}\mathfrak{l}_{\varphi}$ projects in $T \subset M(\mathbb{R})$ to give the circle $\varphi(S^1)$.

The key computation that must be done is:

Let $\mathbb{Z}\lambda + R \xrightarrow{\mathfrak{l}_\varphi} \mathbb{Z}$ *project to* $\mathbb{Z}\lambda + R \xrightarrow{\tilde{\mathfrak{l}}_\varphi} \mathbb{Z}/4\mathbb{Z}$*. Then* $\mathfrak{l}_\varphi$ *gives a polarized Hodge structure for* Q *or* $-Q$ *if, and only if,*

$$\begin{cases} \tilde{\mathfrak{l}}_\varphi|_R = \Psi \\ \tilde{\mathfrak{l}}_\varphi(\lambda) \ \textit{even/odd} \quad \textit{if, and only if, } Q \textit{ is symmetric/alternating.} \end{cases}$$

Step eight: In the complex and quaternionic cases, there exist both symmetric and alternating Q's, so the parity of $\tilde{\mathfrak{l}}_\varphi(\lambda)$ can always be matched.

To deal with the real case, we need an additional result. In the real case, $w_0(\lambda) = -\lambda$. Since for any element w of the Weyl group and any $\lambda \in P$, we have $w(\lambda) \equiv \lambda \mod R$, it follows that $\lambda - w_0(\lambda) \in R$, and then $2\lambda \in R$. Write

$$2\lambda = \sum_{i=1}^{r} m_i \alpha_i \qquad m_i \in \mathbb{Z}$$

where $\alpha_1, \ldots, \alpha_i$ are the simple positive roots. Then we will prove in (IV.E.4) that we are in the

- real case if, and only if, $\sum_{\psi(\alpha_i)=0} m_i$ is even;
- quaternionic case if, and only if, $\sum_{\psi(\alpha_i)=0} m_i$ is odd.

Now

$$\tilde{\mathfrak{l}}_\varphi(\lambda) = \frac{1}{2}\Sigma m_i \Psi(\alpha_i) = \sum_{\psi(\alpha_i)=1} m_i$$

$$\langle \lambda, h^0 \rangle = \left\langle \tfrac{1}{2} \sum_i m_i \alpha_i, h^0 \right\rangle = \sum_i m_i = \tilde{\mathfrak{l}}_\varphi(\lambda) + \sum_{\psi(\alpha_i)=0} m_i.$$

In the real case, this implies

$$\langle \lambda, h^0 \rangle \equiv \tilde{\mathfrak{l}}_\varphi(\lambda) \pmod 2,$$

and thus in the real case

Q is symmetric if, and only if, $\tilde{\mathfrak{l}}_\varphi(\lambda)$ is even;
Q is alternating if, and only if, $\tilde{\mathfrak{l}}_\varphi(\lambda)$ is odd.

We thus have:

In all cases — real, complex, quaternionic — for an appropriate choice of invariant Q,

$$\mathfrak{l}_\varphi \textit{ gives a polarized Hodge structure if, and only if,} \tilde{\mathfrak{l}}_\varphi|_R = \Psi.$$

Step nine: Since the weights of $V_{\mathbb{C}}$ belong to the $\lambda + R$, we have $\varphi(z)$ acts on V_ω as $z^{\mathfrak{l}_\varphi(\omega)}$.[2] Thus

$$V_\omega \subseteq V^{p,q}, \quad \text{where } p - q = \mathfrak{l}_\varphi(\omega).$$

The weight n must satisfy

$$\begin{cases} n & \geq \max \mathfrak{l}_\varphi(\omega), \quad \omega \text{ a weight of } V_{\mathbb{C}} \\ n & \equiv \tilde{\mathfrak{l}}(\lambda) \bmod 2. \end{cases}$$

Once such a weight n is chosen,

$$V_\omega \subseteq V^{p,q} \text{ where } p = \frac{n + \mathfrak{l}_\varphi(\omega)}{2}, \quad q = \frac{n - \mathfrak{l}_\varphi(\omega)}{2}.$$

At this stage the analysis proceeds by considering the action on V_ω of an sl_2 generated by $h_\alpha, e_\alpha, e_{-\alpha}$ (see the proof of Theorem (IV.B.6) for the detailed argument).

Step ten: It is possible to compute ψ, and hence Ψ, using the Vogan diagram [K]. For the compact form, $\psi = 0$. For other real forms, using $\alpha_1, \ldots, \alpha_r$ to denote the simple positive roots corresponding to the Dynkin diagram,

$$\psi(\alpha_i) = \begin{cases} 1 & \text{if node } i \text{ is "painted" in the Vogan diagram} \\ 0 & \text{if node } i \text{ is "unpainted" in the Vogan diagram.} \end{cases}$$

The existence of a compact maximal torus is equivalent to the Vogan diagram being "non-folded." A table of the non-folded Vogan diagrams is in [K] on page 414 (classical groups) and page 416 (exceptional groups).

This strategy will be carried out in detail in the proof of Theorem (IV.E.4).

(IV.A.4) **Example:** We shall illustrate the above in the simplest case when $V = \mathbb{Q}^2$ thought of as column vectors with bilinear form $Q(u, v) = {}^t v Q u$ where $Q = \left(\begin{smallmatrix} 0 & -1 \\ 1 & 0 \end{smallmatrix}\right)$. There are essentially two ways that we shall compute examples in this monograph. One is the "linear algebra" method, where one works directly with the matrix representations. The other is the "root-weight" method, as just outlined above.

---

[2]We use the notation $\mathfrak{l}_\varphi(\omega)$ for the pairing $\langle \omega, \mathfrak{l}_\varphi \rangle$ between $\check{\mathfrak{t}}$ and $\mathfrak{t}$.

In this case the group is SL_2 and the maximal torus $T = \mathrm{SO}(2)$ is given by $\left\{\left(\begin{smallmatrix} \cos 2\pi\theta & -\sin 2\pi\theta \\ \sin 2\pi\theta & \cos 2\pi\theta \end{smallmatrix}\right)\right\}$. Thus, identifying $\mathfrak{t} \cong \mathbb{R}$ with coordinate θ, we have $T \cong \mathbb{R}/\mathbb{Z}$. We set $H = \left(\begin{smallmatrix} 0 & -1 \\ 1 & 0 \end{smallmatrix}\right)$ and $H_l = lH$ and will check by linear algebra that

$\exp(-i \log z H_l)$ *gives a polarized Hodge structure if, and only if,*

$$l \equiv 1 \pmod 4.$$

Here we are thinking of $z = e^{2\pi i \xi} \in S^1 = \mathbb{R}/\mathbb{Z}$ so that for $l = 1$, $z \to \exp(-i \log z H)$ gives the circle S^1 in $\mathrm{SL}_2(\mathbb{R})$.

The eigenvectors and eigenvalues of H are given by setting

$$v_+ = \begin{pmatrix} 1 \\ -i \end{pmatrix}, \quad v_- = \begin{pmatrix} 1 \\ i \end{pmatrix} = \overline{v}_+,$$

and then

$$Hv_\pm = \pm i v_\pm.$$

We note that

$$Q(v_+, \overline{v}_+) = -2i$$

so that

$$\begin{cases} iQ(v_+, \overline{v}_+) > 0 \\ i^3 Q(v_-, \overline{v}_-) > 0. \end{cases}$$

Since Q is alternating, the weight n must be odd. The only possible Hodge decompositions are

$$V_\mathbb{C} = V^{n,0} \oplus V^{0,n}$$

where $V^{n,0} = V_\pm$. Thus $n = l$ and the bilinear relation

$$i^l Q(v, \overline{v}) > 0 \qquad v \in V^{n,0}$$

gives

$$\begin{cases} l \equiv 1 \pmod 4 & V^{n,0} = V_+ \\ -l \equiv 3 \pmod 4 & V^{n,0} = V_-. \end{cases}$$

The second is redundant, so that we have confirmed the italicized statement above.

Although very elementary the above contains the essential conceptual idea of the linear algebra computations that follow. Of course, the calculations in more general cases are more intricate.

For the root-weight approach, since the roots are purely imaginary it is more convenient notationally to set

$$h = -iH = \begin{pmatrix} 0 & i \\ -i & 0 \end{pmatrix}.$$

The root spaces are then the spans of

$$\begin{cases} e = \frac{1}{2}\left(\begin{smallmatrix} 1 & -i \\ -i & -1 \end{smallmatrix}\right) \\ f = \frac{1}{2}\left(\begin{smallmatrix} 1 & i \\ i & -1 \end{smallmatrix}\right) = \overline{e}.^{3} \end{cases}$$

Then

$$\begin{cases} [h,e] = 2e \\ [h,f] = -2f \\ [e,f] = h \end{cases}$$

and

$$\begin{cases} h \cdot v_{+} = v_{+} \\ e \cdot v_{+} = 0. \end{cases}$$

This gives us that, identifying it with $\mathbb{R}$ where $h \leftrightarrow 1$, the weight and root lattices are

$$\begin{array}{l} P \cong \mathbb{Z} \\ \cup \\ R \cong 2\mathbb{Z}. \end{array}$$

Moreoever, the standard representation of SL_2 on $\mathbb{Q}^2$ has highest weight 1. Thus, in the above notations we have:

- $\mathcal{U} = \mathbb{C}^2 = \mathbb{C}v_{+} \oplus \mathbb{C}v_{-}$;
- $\langle \lambda, h \rangle = 1$ and v_{+} is the highest weight vector;
- $\langle \alpha, h \rangle = 2$ where $[h, e] = \alpha e$;
- $\psi(\alpha) = 1$, $\Psi(\alpha) = 2$.

Setting $l_{\varphi} = lh$, $l_{\varphi}(\alpha) = 2l$. Thus the condition $\tilde{l}_{\varphi}|_{R} = \Psi$ on the map $\mathbb{Z}\lambda + R \to \mathbb{Z}/4\mathbb{Z}$ is

$$2l = \langle \alpha, l_{\varphi} \rangle \equiv 2 \pmod 4.$$

This is exactly the condition that l_{φ} give a polarized Hodge structure for $\pm Q$ ($+Q$ when $l \equiv 1 \pmod 4$, $-Q$ when $l \equiv 3 \pmod 4$).

---

[3]For the polarized Hodge structure on $\mathrm{sl}_{2,\varphi}$ where $\varphi = i \in \mathcal{H}$, we have $\mathrm{sl}_{2,\varphi}^{(0,0)} = \mathbb{C}h$, $\mathrm{sl}_{2,\varphi}^{-1,1} = \mathbb{C}f$ and $\mathrm{sl}_{2,\varphi}^{1,-1} = \mathbb{C}e$ (see example (IV.A.11)).

Example: Sp(4), I. Next, we shall determine the condition that an irreducible representation

$$\rho : \mathrm{Sp}(4) \to V^{\lambda}$$

with highest weight λ leads to a Hodge representation; i.e., there exists an invariant form $Q_\lambda : V^\lambda \otimes V^\lambda \to \mathbb{Q}$ and a circle $\varphi : S^1 \to \mathrm{Sp}(4, \mathbb{R})$ such that $(V^\lambda, Q_\lambda, \rho \circ \varphi)$ is a polarized Hodge structure. The result is

(IV.A.5) PROPOSITION: *Every such representation leads to a Hodge representation.*

PROOF. We shall use an alternative "weight" method to do this example. For this we shall use the notation and results from [GW], which are summarized in the appendix to this chapter. For the fundamental weights ω_1, ω_2 we have

$$\lambda = m_1\omega_1 + m_2\omega_2 \qquad\qquad m_1, m_2 \geqq 0.$$

We set $V^\lambda = V^{(m_1,m_2)}$. The notation is not to be confused with the $V_\varphi^{p,q}$ occurring in a Hodge structure on $V_\mathbb{C}$. Thus, $V^{(1,0)}$ is the standard representation.

Since in this case $w_0 = -I$, we have $w_0(\lambda) = -\lambda$ and there is always an invariant form Q_λ. One may verify that

$$\langle \lambda, h^0 \rangle = \sum_{i=1}^{2} i(4-i)m_i \equiv m_1 \pmod 2.$$

Thus

$$\begin{cases} m_1 \equiv 0 (\mathrm{mod}\, 2) & \text{if, and only if, } Q_\lambda \text{ is symmetric} \\ m_1 \equiv 1 (\mathrm{mod}\, 2) & \text{if, and only if, } Q_\lambda \text{ is alternating.} \end{cases}$$

The fundamental weights are $\omega_1 = \epsilon_1$, $\omega_2 = \epsilon_1 + \epsilon_2$; $V = V^{(1,0)}$ has weights $\pm\epsilon_1 \pm \epsilon_2$. Thus $\Lambda^2 V$ has dominant weight $\epsilon_1 + \epsilon_2 = \omega_2$. It follows that

$$V^{(m_1,m_2)} \subseteq S^{m_1} V \otimes S^{m_2}(\Lambda^2 V) \subseteq \otimes^{m_1+2m_2} V.$$

Consequently, if $\varphi : S^1 \to \mathrm{Sp}(4, \mathbb{R})$ gives a polarized Hodge structure on V, it gives one on $\otimes^n V$ for all n, and hence one on V^λ for all λ. □

Example: Sp(4), II. We shall now use the linear algebra method to determine explicitly the conditions that give Hodge representations for the standard representation, and we shall relate these conditions with the root-weight perspective in step seven to the above steps in the analysis of Hodge representations.

For the classical group $\mathrm{Sp}(2r)$, there are two common matrix realizations for the standard representation on V. The most common is that of matrices preserving the alternating form $\left(\begin{smallmatrix} 0 & -I_r \\ I_r & 0 \end{smallmatrix}\right)$. The other is to consider V as a direct sum of the standard representation space for $\mathrm{Sp}(2)$, and this is what we shall do here. Thus, for $\mathrm{Sp}(4)$ we set $H = \left(\begin{smallmatrix} 0 & -1 \\ 1 & 0 \end{smallmatrix}\right)$ and use the notations

$$Q = \begin{pmatrix} H & 0 \\ 0 & H \end{pmatrix}, \quad H_1 = \begin{pmatrix} H & 0 \\ 0 & 0 \end{pmatrix}, \quad H_2 = \begin{pmatrix} 0 & 0 \\ 0 & H \end{pmatrix}.$$

Then H_1 and H_2 give a basis for $\Lambda \subset \mathfrak{t}$ and the differential of a general co-character is

$$\mathfrak{l}_\varphi = l_1 H_1 + l_2 H_2, \qquad l_1, l_2 \in \mathbb{Z}.$$

We set

$$v_{1,\pm} = \begin{pmatrix} 1 \\ \pm i \\ 0 \\ 0 \end{pmatrix}, \quad v_{2,\pm} = \begin{pmatrix} 0 \\ 0 \\ 1 \\ \pm i \end{pmatrix}$$

and denote by $V_{1,\pm}$, $V_{2,\pm}$ the lines in $V_{\mathbb{C}}$ they span. Then $V_{1,\pm}, V_{2,\pm}$ are the $(\pm i)$-eigenspaces for H_1, H_2, and are thus the $\pm l_1 i$, $\pm l_2 i$ eigenspaces for $\mathfrak{l}_\varphi$.

Any polarized Hodge structure for $(V, Q, \mathfrak{l}_\varphi)$ must have odd weight and the $V^{p,q}$ are direct sums of the 1-dimensional eigenspaces $V_{1,\pm}$, $V_{2,\pm}$. Then the same argument as in example (IV.A.4) gives the result:

(IV.A.6) $\mathfrak{l}_\varphi = (l_1, l_2)$ *gives a polarized Hodge structure on* (V, Q) *if, and only if,*

$$l_1 \equiv 1 \pmod 4, \quad l_2 \equiv 1 \pmod 4)$$

and on $(V, -Q)$ *if, and only if,*

$$l_1 \equiv 3 \pmod 4, \quad l_2 \equiv 3 \pmod 4).$$

From a more abstract perspective, the Vogan diagram of $\mathrm{Sp}(4, \mathbb{R})$ is

1 ○═══● 2
α_1 α_2

and

$$\begin{aligned} 2\epsilon_1 &= 2\alpha_1 + \alpha_2, \\ 2\epsilon_2 &= \alpha_2. \end{aligned}$$

Since $\psi(\alpha_1) = 0$, $\psi(\alpha_2) = 1$, to have $\tilde{\mathfrak{l}}_\varphi|_R = \Psi$ is equivalent to

$$\begin{aligned} \tilde{\mathfrak{l}}_\varphi(\epsilon_1) &= \tfrac{1}{2}\Psi(\alpha_2) = 1 \text{ or } 3 \text{ in } \mathbb{Z}/4\mathbb{Z} \\ \tilde{\mathfrak{l}}_\varphi(\epsilon_2) &= \tilde{\mathfrak{l}}_\varphi(\epsilon_1) - \tilde{\mathfrak{l}}_\varphi(\alpha_1) = \tilde{\mathfrak{l}}_\varphi(\epsilon_1). \end{aligned}$$

Note that for ϵ_1, if $2\epsilon_1 = m_1\alpha_1 + m_2\alpha_2$, then $m_1 = 2$, $m_2 = 1$, so

$$\sum_{\psi(\alpha_i)=0} m_i = 2,$$

which is even, so, since $w_0 = -I$, $V^{(1,1)}$ is real, and similarly $V^{(m_1,m_2)}$ is real. From

$$\langle m_1\omega_1 + m_2\omega_2, h^0\rangle = m_1\langle \epsilon_1, h^0\rangle + m_2\langle \epsilon_1 + \epsilon_2, h^0\rangle = m_1 + 2m_2$$

we have $\langle m_1\omega_1 + m_2\omega_2, h^0\rangle \equiv m_1 \pmod 2$, as we found in our analysis of $\mathrm{Sp}(4)$.

(IV.A.7) **Remark:** The general method for determining when $\mathfrak{l}_\varphi \in \Lambda$ gives a polarized Hodge structure of weight n on (V, Q) is this: First, we recall the sign rule that

$$i^{p-q}Q(v, \overline{v}) > 0 \text{ for } 0 \neq v \in V^{p,q}.$$

Next, we will have direct sum decompositions $V_\mathbb{C} = V^+ \oplus V^-$ where $V^\pm$ is a direct sum of $V^{p,q}$'s and where for

$$n \equiv 0 \pmod 2 \text{ we have } V^+ = \{Q(v,\overline{v}) \geqq 0\},\ \ V^- = \{Q(v,\overline{v}) \leqq 0\},$$

and for

$$n = 1 \pmod 2 \text{ we have } V^+ = \{iQ(v,\overline{v}) \geqq 0\},\ \ V^- = \{iQ(v,\overline{v}) \leqq 0\}.$$

We denote by V_m the im-eigenspace of $\mathfrak{l}_\varphi = \Sigma l_j H_j$. Then the conditions are

$$\text{(IV.A.8)} \qquad \begin{cases} n \equiv 0 \pmod 2 & \begin{cases} V_m \subset V^+ & \text{if } m \equiv 0 \pmod 4 \\ V_m \subset V^- & \text{if } m \equiv 2 \pmod 4 \end{cases} \\ n \equiv 1 \pmod 2 & \begin{cases} V_m \subset V^+ & \text{if } m \equiv 1 \pmod 4 \\ V_m \subset V^- & \text{if } m \equiv 3 \pmod 4. \end{cases} \end{cases}$$

Once we choose a basis $H_1, \dots, H_r$ for Λ and determine the weight space decomposition of the H_j acting on $V_\mathbb{C}$, the conditions (IV.A.8) will be expressed in terms of congruences on the l_j's.

Remark: In the next Section IV.B we shall give a general method for determining when $\mathfrak{l}_\varphi$ gives a polarized Hodge structure for the adjoint representations (cf. Proposition (IV.B.3)). When applied to $\mathrm{Sp}(4)$ we shall recover (IV.A.6).

In Section IV.C this example will be extended to determine the conditions to have a Hodge representation for all the classical groups.

We now turn to the promised

(IV.A.9) PROPOSITION: *Let M be the adjoint group of a $\mathbb{Q}$-simple algebraic group and $\varphi : S^1 \to M(\mathbb{R})$ be a circle such that $(\mathfrak{m}, B, \operatorname{Ad}\varphi)$ gives a polarized Hodge structure. Then for a generic $m \in M(\mathbb{R})$, setting $\varphi_m = m^{-1}\varphi m$, M is the smallest $\mathbb{Q}$-algebraic subgroup of $M(\mathbb{R})$ such that $\varphi_m(S^1) \subseteq M(\mathbb{R})$.*

(IV.A.10) COROLLARY: *M_a is the Mumford-Tate group of the polarized Hodge structure on $(\mathfrak{m}, B, \operatorname{Ad}\varphi_m)$, where $\operatorname{Ad}\varphi_m$ is a generic conjugate of $\operatorname{Ad}\varphi$.*

PROOF OF PROPOSITION (IV.A.9). Let $m \in M(\mathbb{R})$ be generic, and take $M_0 \subset M$ to be the smallest $\mathbb{Q}$-algebraic subgroup such that $\operatorname{Ad}\varphi_m(S^1) \subseteq M_0(\mathbb{R})$. Then $\mathfrak{m}_0 \subset \mathfrak{m}$ is a sub-Lie algebra and an $\operatorname{Ad}\varphi_m$-sub-Hodge structure polarized by $-B$. Furthermore, referring to (VII.A.4)*ff.*, since φ_m is a *generic* conjugate of φ we have

$$(*) \qquad \mathfrak{m}_0^{-i,i} = \mathfrak{m}^{-i,i}, \qquad i < 0.$$

Here the Hodge decomposition is with respect to $\operatorname{Ad}\varphi_m$.

Let $\mathfrak{m}_1$ be the B-orthogonal complement of $\mathfrak{m}_0$; it is an $\operatorname{Ad}\varphi_m$-sub-Hodge structure of $\mathfrak{m}$. Since $\mathfrak{m}_1 \subset \mathfrak{m}^{0,0}$ by $(*)$, $\varphi_m(S^1)$ acts via Ad trivially on $\mathfrak{m}_1$. Therefore so does its $\mathbb{Q}$-closure M_0, which implies that $[\mathfrak{m}_0, \mathfrak{m}_1] = 0$.[4] We conclude that $[\mathfrak{m}_0, \mathfrak{m}] \subset \mathfrak{m}_0$. Hence $\mathfrak{m}_0 \subset \mathfrak{m}$ is an ideal defined over $\mathbb{Q}$, and therefore over all of $\mathfrak{m}$ by the simplicity assumption. □

(IV.A.11) **Example:** We conclude this section with an example of a different type of Hodge representation that plays a central role in the analysis of a degenerating Hodge structure; this is Schmid's SL_2-*orbit theorem* [Schm1].[5] Given a variation of polarized structure over the punctured disc $\Delta^* = \{0 < |z| < 1\}$, by a result of Borel (see [GS] for a proof) the monodromy transformation T is quasi-unipotent. By a base change we may assume that T is unipotent with logarithm

$$N = \log T = (T - I) - \frac{(T-I)^2}{2} + \cdots \in \mathfrak{g}.$$

Schmid's theorem states that the given variation of Hodge structure may, in a very precise sense, be approximated by an equivariant variation of the standard Hodge structure over the upper half plane arising from a Hodge representation of $\mathrm{SL}_2(\mathbb{R})$. The question of doing this over $\mathbb{Q}$ will be discussed in footnote 9 later in this Chapter.

[4]Here it is crucial that $\mathfrak{m}_1$ is defined over $\mathbb{Q}$.

[5]This example is an extension of the discussion in Section I.C, especially (I.C.1).

For $U = \mathbb{Q}^2$ thought of as column vectors $u = \left(\begin{smallmatrix} u_1 \\ u_2 \end{smallmatrix}\right)$ with alternating form

$$Q_0(u, v) = -u \wedge v,$$

we have the standard picture of the upper half plane $\mathcal{H} = \{\tau \in \mathbb{C} : \operatorname{Im} \tau > 0\}$ as the period domain for polarized weight one Hodge structures

$$U_{\mathbb{C}} = U_\tau^{1,0} \oplus U_\tau^{0,1}$$

where

$$\begin{cases} U_\tau^{1,0} = \mathbb{C} u_\tau \\ u_\tau \ = \frac{1}{\sqrt{2}} \left(\begin{smallmatrix} \tau \\ 1 \end{smallmatrix}\right) \text{ with } iQ_0(u_\tau, \overline{u}_\tau) = 1. \end{cases}$$

Here, the dual period domain is $\check{\mathcal{H}} = \mathbb{P}^1$ with the inclusion $\mathcal{H} \hookrightarrow \check{\mathcal{H}}$ given in terms of homogeneous coordinate column vectors in $\mathbb{P}^1$ by

$$\tau \to \begin{bmatrix} \tau \\ 1 \end{bmatrix}.$$

The action of $\mathrm{SL}_2(\mathbb{R})$ on $\mathcal{H}$ is the standard one: for $g = \left(\begin{smallmatrix} a & b \\ c & d \end{smallmatrix}\right)$,

$$g(\tau) = a\tau + b/c\tau + d.$$

We pick as reference Hodge structure $\tau_0 = i$. Then the polarized Hodge structure on sl_2 has the Hodge decomposition

$$\mathrm{sl}_{2,\mathbb{C}} = \mathrm{sl}_2^{-1,1} \oplus \mathrm{sl}_2^{0,0} \oplus \mathrm{sl}_2^{1,-1}$$

where

$$\begin{cases} \mathrm{sl}_2^{-1,1} = \mathbb{C} X_+, & X_+ = \frac{1}{2} \left(\begin{smallmatrix} -i & 1 \\ 1 & i \end{smallmatrix}\right) \\ \mathrm{sl}_2^{0,0} = \mathbb{C} Z, & Z = \left(\begin{smallmatrix} 0 & -i \\ i & 0 \end{smallmatrix}\right) \\ \mathrm{sl}_2^{1,-1} = \mathbb{C} X_- = \overline{\mathrm{sl}_2^{-1,1}}, & X_- = \frac{1}{2} \left(\begin{smallmatrix} i & 1 \\ 1 & -i \end{smallmatrix}\right). \end{cases}$$

Given $\varphi \in D$, the SL_2-orbit arises from a real Hodge representation[6]

$$\rho : \mathrm{SL}_2(\mathbb{R}) \to G(\mathbb{R})$$

with the following properties:

(i) $\rho_* : \mathrm{sl}_{2,\mathbb{R}} \to \mathfrak{g}_{\mathbb{R}}$ is a morphism of $\mathbb{R}$-Hodge structures, where $\mathfrak{g}_{\mathbb{R}} \subset \operatorname{Hom}(V_{\mathbb{R}}, V_{\mathbb{R}})$ has the Hodge structure induced by V_φ;

[6] See footnote 9 on page 104.

(ii) for $F = \begin{pmatrix} 0 & 1 \\ 0 & 0 \end{pmatrix}$,

$$\rho_*(F) = N.$$

Letting $o = \left[\begin{smallmatrix} 0 \\ 1 \end{smallmatrix}\right] \in \check{\mathcal{H}}$, $\mathcal{H}$ is identified with the nilpotent orbit

$$\tau \to \exp(\tau F) \cdot o, \qquad \tau \in \mathcal{H}.$$

The Hodge representation then gives an equivariant variation of Hodge structure

$$\begin{array}{ccc} \mathcal{H} & \xrightarrow{\tilde{\mathcal{O}}} & D \\ \downarrow & & \downarrow \\ \Delta^* & \xrightarrow{\mathcal{O}} & \Gamma_T \backslash D \end{array}$$

where $\Gamma_T = \{T^k\}_{k \in \mathbb{Z}}$ and

$$\begin{array}{ccc} \widetilde{\mathcal{O}}(\tau) & = & \exp(\tau \rho_*(F)) \cdot \varphi_0 \\ \| & & \| \\ \exp(\tau F) \cdot o & & \exp(\tau N) \cdot \varphi_0 \end{array}$$

for a point $\varphi_0 \in \check{D}$ (note that $\widetilde{\mathcal{O}}(i) = \varphi \in D$).[7]

Suppose now that

(IV.A.12) $$\Phi : \Delta^* \to \Gamma_T \backslash D$$

is a variation of Hodge structure whose Mumford-Tate group is M_Φ, defined as in Section III.A as the Mumford-Tate group of the lift $\tilde{\Phi}(\eta) \in D$ of $\Phi(\eta)$ at a generic point $\eta \in \Delta^*$. Setting, for simplicity of notation, $M = M_\Phi$ we have seen (Section I.C and (III.A)) that $T \in M$ and there is a Mumford-Tate domain $D_M \subset D$ such that (IV.A.12) is

(IV.A.13) $$\Phi : \Delta^* \to \Gamma_T \backslash D_M \subset \Gamma_T \backslash D.$$

(IV.A.14) PROPOSITION: *The real Hodge representation in Schmid's* SL_2-*orbit theorem has image in* $M(\mathbb{R})$*; i.e., with suitable choices we have*

$$\rho : \mathrm{SL}_2(\mathbb{R}) \to M(\mathbb{R}).$$

Proof. The argument consists of an analysis of Schmid's proof (cf. [Schm1] and [GS] for an outline). The basic observation is first that D_M has a compact

[7] We are using $\mathcal{O}$ and $\widetilde{\mathcal{O}}$ to denote the variation of Hodge structure associated to a nilpotent orbit. These approximate a variation of Hodge structure given by Φ as in (IV.A.12) and (IV.A.13).

dual $\check{D}_M = M(\mathbb{C})/P_M$ where P_M is a parabolic subgroup of $M(\mathbb{C})$, and we have

$$\begin{array}{ccc} D & \subset & \check{D} \\ \cup & & \cup \\ D_M & \subset & \check{D}_M. \end{array}$$

Schmid's analysis takes place in $D \subset \check{D}$, and we will sketch the steps to check that it remains in $D_M \subset \check{D}_M$.

Step one: From (IV.A.13) we have

$$\begin{array}{ccccccc} \tau & \in & \mathcal{H} & \stackrel{\tilde{\Phi}}{\to} & D_M & \subset & \check{D}_M \\ \downarrow & & \downarrow & & \downarrow & & \\ e^{2\pi i\tau} = z & \in & \Delta^* & \stackrel{\Phi}{\to} & \Gamma_T\backslash D_M. & & \end{array}$$

Noting that $\exp(\tau N) \in M(\mathbb{C})$, we may define

$$\widetilde{\Psi}(\tau) = \exp(-\tau N)\widetilde{\Phi}(\tau) \in \check{D}_M.$$

From $\widetilde{\Psi}(\tau+1) = \widetilde{\Psi}(\tau)$ we see that there is an induced map

$$\Psi : \Delta^* \to \check{D}_M.$$

This "unwinding the period map" is Schmid's first step. His argument gives that Ψ extends across the origin to define $\Psi(0) = \varphi_0 \in \check{D}_M$, and that the *nilpotent orbit*

$$\mathcal{N}(z) := \exp\left(\tfrac{\log z}{2\pi i} N\right) \cdot \varphi_0 \in \Gamma_T\backslash \check{D}_M$$

has the properties

(i) $\mathcal{N}(z) \in \Gamma_T\backslash D_M$ for $0 < |z| < \epsilon$, and it gives a variation of Hodge structure; i.e., the infinitesimal period relation is satisfied, which writing $\varphi_0 = \{F_0^p\}$ is equivalent to

$$N(F_0^p) \subseteq F_0^{p-1};$$

(ii) $\Phi(z)$ is "well approximated" by $\mathcal{N}(z)$.

Step two: By the Jacobson-Morosov theorem, $N \in \mathfrak{m}$ may be completed to an $\mathrm{sl}_2 \subset \mathfrak{m}$. Schmid selects the sl_2 over $\mathbb{R}$ in the following manner. Lifting $\mathcal{N}(z)$ to

$$\widetilde{\mathcal{N}}(\tau) = \exp(\tau N) \cdot \varphi_0$$

and setting $\tau = u + iv$, $v > 0$, for $\operatorname{Im}\tau > C$ the map

$$v \to \exp(ivN) \cdot \varphi_0 \tag{IV.A.15}$$

gives a real curve in $D_M = M(\mathbb{R})/H_M$. Choosing a reference point in D_M to define a Hodge structure on $\mathfrak{m}$, the $\operatorname{Ad} H_M$-invariant splitting

$$\mathfrak{m}_{\mathbb{R}} = \mathfrak{m}^{0,0} \oplus \left(\underset{i \neq 0}{\oplus} \mathfrak{m}^{-i,i} \right)_{\mathbb{R}}$$

defines an invariant connection in the principal bundle $H_M \to M(\mathbb{R}) \to D_M$, and there is an essentially unique lifting

$$v \to f(v) \in M(\mathbb{R})$$

of (IV.A.15) that is tangent to this curve.[8] Schmid defines three $\mathfrak{m}_{\mathbb{R}}$-valued functions

$$\begin{cases} A(v) = -2f(v)^{-1}f'(v) \\ F(v) = \operatorname{Ad} f(v)^{-1}N \\ E(v) = -C_0F(v), \end{cases}$$

where C_0 is the Weil operator for the reference Hodge structure. The infinitesimal period relation and the holomorphicity of $\widetilde{\mathcal{N}}(\tau)$ imply that

(i) $A(v) \in \left(\mathfrak{m}^{-1,1} \oplus \mathfrak{m}^{1,-1}\right)_{\mathbb{R}}$;

(ii) $F(v)$ represents the tangent to $\widetilde{\mathcal{N}}(u+iv)$ in the $\partial/\partial u$-direction along the imaginary axis; and

(iii) $-\frac{1}{2}A(v)$ represents the tangent to $\widetilde{\mathcal{N}}(u+iv)$ in the $\partial/\partial v$-direction, also along the imaginary axis.

Here, by translation back to the reference point in D_M we are identifying the holomorphic tangent spaces to $D_M \subset \check{D}_M$ with

$$\mathfrak{m}_{\mathbb{C}}/\mathfrak{p}_M \cong \underset{i>0}{\oplus} \mathfrak{m}^{-i,i}.$$

Then Schmid shows that, in addition to (i) above

$$\begin{cases} E(v), F(v) \in \left(\mathfrak{m}^{-1,1} \oplus \mathfrak{m}^{0,0} \oplus \mathfrak{m}^{1,-1}\right)_{\mathbb{R}} \\ A(v) + 2iF(v) \in \mathfrak{m}^{0,0} \oplus \mathfrak{m}^{1,-1}. \end{cases}$$

[8] It is unique once we specify the lift over one point.

The key point is that the conditions that $f(u)$ lift $\widetilde{\mathcal{N}}(iv)$ to be tangent to the horizontal spaces in the connection, together with the fact that $\widetilde{\mathcal{N}}(u+iv)$ is holomorphic, give the differential equations

$$\text{(IV.A.16)} \qquad \begin{cases} 2E'(v) &= -[A(v), E(v)] \\ 2F'(v) &= [E(v), F(v)] \\ A'(v) &= -[E(v), F(v)]. \end{cases}$$

Schmid proves that the functions $A(v)$, $E(v)$, $F(u)$ have Laurent series

$$\text{(IV.A.17)} \qquad \begin{cases} A(v) &= A_0 v^{-1} + A_1 v^{-1-1/2} + \cdots \\ E(v) &= E_0 v^{-1} + E_1 v^{-1-1/2} + \cdots \\ F(v) &= F_0 v^{-1} + F_1 v^{-1-1/2} + \cdots \end{cases}$$

around $v = 0$. This gives

$$\text{(IV.A.18)} \qquad \begin{cases} [A_0, E_0] &= 2E_0 \\ [A_0, F_0] &= -2F_0 \\ [E_0, F_0] &= A_0, \end{cases}$$

so that A_0, E_0, F_0 span on $\mathrm{sl}_{2,\mathbb{R}}$ in $\mathfrak{m}_{\mathbb{R}}$.[9] With the above notations the SL_2-orbit has essentially (cf. [Schm1] where the precise statement is given)

$$N = [A_0, F_0].$$

Remark: The following remarks may be useful as providing background for Schmid's argument. First, there are two general principles:

(a) A map $f : B \to G(\mathbb{R})$, where B is a connected manifold, is uniquely determined by its value at one point and the pullback $f^*(\omega)$ of the $\mathfrak{g}_{\mathbb{R}}$-valued Maurer-Cartan form $\omega = g^{-1}dg$. This form satisfies the Maurer-Cartan equations

$$\text{(IV.A.19)} \qquad d\omega = \tfrac{1}{2}[\omega, \omega].$$

(b) In case B is a Lie group, f is a homomorphism if, and only if, $f^*(\omega)$ is constant under the left action of B on itself. For $B = \mathbb{R}^{>0}$ with coordinate v, this is equivalent to

$$f^*(\omega) = C\frac{dv}{v}$$

where $C \in \mathfrak{g}_{\mathbb{R}}$.

[9]With some slight modifications, one may make choices so that the sl_2 is defined over $\mathbb{Q}$ [Schm1]. There is a subtlety in that if we choose a *general* point in $\mathcal{H}$ as reference point, this would imply that M is $\rho(\mathrm{SL}_2(\mathbb{Q}))$, which is not true.

Secondly, with the special points

$$i = \begin{bmatrix} i \\ 1 \end{bmatrix}, \quad o = \begin{bmatrix} 0 \\ 1 \end{bmatrix}, \quad \infty = \begin{bmatrix} 1 \\ 0 \end{bmatrix}$$

in $\mathbb{P}^1 = \check{\mathcal{H}}$, the standard nilpotent orbit is

$$\exp(\tau F)o = \begin{bmatrix} \tau \\ 1 \end{bmatrix}.$$

However, this is a 1-parameter subgroup of the complex group $\mathrm{SL}_2(\mathbb{C})$, and we cannot "go to infinity" using the real 1-parameter group $\exp(uF) \in \mathrm{SL}_2(\mathbb{R})$. For this we need to use the orbit of the 1-parameter subgroup

$$\exp(-\tfrac{1}{2}\log vA) = \begin{pmatrix} v^{1/2} & 0 \\ 0 & v^{-1/2} \end{pmatrix} := g(v)$$

in $\mathrm{SL}_2(\mathbb{R})$, where $A = \left(\begin{smallmatrix} -1 & 0 \\ 0 & 1 \end{smallmatrix}\right)$. Note that $g(v) \cdot i = iv$ and

$$\exp(iF) \cdot o = \exp(-\tfrac{1}{2}\log vA)\exp(iF) \cdot o.$$

The crucial observations are

(i) $g(v)$ is the horizontal lift to $\mathrm{SL}_2(\mathbb{R})$ of the curve $\exp(ivF) \cdot o$, $v > 0$, in $\mathcal{H}$;

(ii) $\omega = \Big((-\tfrac{1}{2})A\Big)\frac{dv}{v}$.

Schmid proves that in general around $v = \infty$ there is a convergent series expansion

$$f^*(\omega) = \Big(A_0 + A_1 v^{-\frac{1}{2}} + A_2 v^{-1} + \dots\Big)\frac{dv}{v}. \tag{IV.A.20}$$

In the Jacobson-Morosov theorem, knowing the N and the A with $[A, N] = -2N$ is enough to determine the sl_2. The expansion (IV.A.20) gives the candidate A_0 for A. Even though the arc $f(v)$ in $G_{\mathbb{R}}$ is 1-dimensional so that on the face of it the Maurer-Cartan equations (IV.A.19) give no information, using the Cauchy-Riemann equations it may be seen that N represents the left translate back to the identity in $G_{\mathbb{R}}$ of the horizontal lift of $\partial/\partial u$, so that in effect we have in $G_{\mathbb{R}}$ the arc $f(v)$ together with first order information normal to $f(v)$, and then the Maurer-Cartan equations together with the condition that $f^*(\omega)$ have values in $\bigoplus_{i \neq 0} \mathfrak{m}^{-i,i}$ give (IV.A.18).

Part III: Hodge representations over $\mathbb{Q}$. We shall summarize the basic results from the theory of representations of a $\mathbb{Q}$-algebraic group M that we shall need for this monograph. The basic reference we shall use is [KMRT].

Before giving the summary, we remark that a consequence will be that the basic results Theorem (IV.B.6) and Theorem (IV.E.2) hold under the assumption that M is a simple $\mathbb{Q}$-algebraic group. The results we shall use from [KMRT] are established there under the assumption that M is *absolutely simple*, meaning that $M(\mathbb{C})$ is simple. However, what we need for the use in Hodge representations is the existence of an invariant bilinear form Q defined over $\mathbb{Q}$. Proof analysis plus a further simple argument given below (the *general principle*) shows that it is sufficient for us to assume that M is simple.

Let M be a simple $\mathbb{Q}$-algebraic group and assume that

(IV.A.21) *M contains an anisotropic maximal torus.*

As we have noted, this assumption is satisfied by any Mumford-Tate group. It is equivalent to saying that $M(\mathbb{R})$ contains a compact maximal torus T. From this it follows that

(IV.A.22) *the roots are purely imaginary on* $\mathfrak{t}$.

This property will imply an important simplification in the Mumford-Tate group case over the general case.

It what follows we will use the following

Notation: *For any field $L \supset \mathbb{Q}$, M_L will be the corresponding L-algebraic group obtained by extending scalars from $\mathbb{Q}$ to L.*

We shall now summarize the needed results.

(i) There is a finite Galois extension L of $\mathbb{Q}$ and a system Π of simple roots defined for M_L. Set $\Gamma = \mathrm{Gal}(L/\mathbb{Q})$ and let W be the Weyl group. If $\gamma \in \Gamma$, then $\gamma \cdot \Pi$ is another system of simple roots, and thus there exists a unique $w_\gamma \in W$ such that

$$w_\gamma(\gamma \cdot \Pi) = \Pi.$$

We denote the map $w_\gamma \cdot \gamma$ by

$$\gamma_* : \Pi \to \Pi. \tag{IV.A.23}$$

It has the property that it is an isomorphism of the Dynkin diagram, and so in the case that M is absolutely simple we must have that γ_* acts on the Dynkin diagram. There are no isomorphisms of the Dynkin diagram for B_n, C_n, E_7, E_8, F_4, G_2. In the other cases we will analyze what is needed for the purposes of this monograph. As will be noted below, this will imply for M absolutely simple that any dominant weight gives rise to an irreducible representation defined over $\mathbb{Q}$. In case M is not absolutely simple γ_* may permute some of the components of the Dynkin diagram of M_L.

We note that there is a unique $w^* \in W$ such that $w^*(-\Pi) = \Pi$ (see the appendix to Chapter IV). It follows that $\gamma_*\lambda = -w^*(\lambda)$ when $\gamma_* = -\operatorname{id}$.

(ii) The mapping (IV.A.23) preserves the sets P and R of weights and roots; i.e., we have

$$\begin{array}{ccc} P & \xrightarrow{\gamma_*} & P \\ \cup & & \cup \\ R & \xrightarrow{\gamma_*} & R. \end{array}$$

We choose a set of positive roots with corresponding Weyl chamber $\mathbf{C}$. Let $\lambda \in \mathbf{C} \cap P$ be a dominant weight and set

$$\begin{cases} \Gamma_\lambda = \{\gamma \in \Gamma : \gamma_*\lambda = \lambda\} \\ K_\lambda = \text{fixed field of } \Gamma_\lambda. \end{cases}$$

Then the irreducible representation of $M(\mathbb{C})$ corresponding to λ can be defined over K_λ, and it cannot be defined over any proper subfield of K_λ. We denote it

$$\rho_\lambda : M_{K_\lambda} \to \mathrm{GL}(V^\lambda_{K_\lambda}).^{10}$$

Then

(a) $\operatorname{Res}_{K_\lambda/\mathbb{Q}}(V^\lambda_{K_\lambda})$ is an irreducible representation, defined over $\mathbb{Q}$, of M;

(b) $\operatorname{Res}_{K_\lambda/\mathbb{Q}}(V^\lambda_{K_\lambda}) \otimes_{\mathbb{Q}} K_\lambda \cong \bigoplus_{\gamma \in \Gamma/\Gamma_\lambda} V^{\gamma_*\lambda}_{K_{\gamma_*\lambda}}$

(c) all irreducible representations of M are of this form for M absolutely simple over $\mathbb{Q}$.

(iii) The simple Lie algebras where γ_* is not trivial, together with the automorphisms of their Dynkin diagrams, are

$$\begin{cases} A_n & \mathbb{Z}/2\mathbb{Z} \\ D_n & \mathbb{Z}/2\mathbb{Z} \qquad (n \neq 4) \\ E_6 & \mathbb{Z}/2\mathbb{Z} \\ D_4 & S_3. \end{cases}$$

We will treat the case of D_4 (triality) separately.

---

[10] In the discussion of the real case in Part II, this $V^\lambda_{K_\lambda}$ is what was denoted by $\mathcal{U}$ there.

For E_6, since $\gamma_* : P/R \to P/R$ takes $\mathbb{Z}/3\mathbb{Z} \to \mathbb{Z}/3\mathbb{Z}$, it must be of the form $n \to n^2$, which is of order two. This gives in the non-triality, i.e., not D_4, cases a homomorphism

$$\Gamma \to \mathbb{Z}/2\mathbb{Z}.$$

We let Γ_0 be the kernel and denote by $L_0 \subset L$ its fixed field. Then $[L_0 : \mathbb{Q}] = 2$ if γ acts non-trivially and we observe that by (IV.A.22) *if L_0 were real then $M(\mathbb{R})$ would not have a compact maximal torus.* Thus we conclude that

- L_0 must be a quadratic imaginary extension of $\mathbb{Q}$; i.e., $L_0 = \mathbb{Q}(\sqrt{-d})$ where $d \in \mathbb{Q}$ is positive;
- if $\gamma \in \mathrm{Gal}(L_0/\mathbb{Q})$ has $\gamma(\sqrt{-d}) = -\sqrt{-d}$, then $\gamma = -(\text{Identity})$ on P.[11]

The analysis now breaks into two cases:

Case 1: $\gamma_*\lambda \neq \lambda$.
Then the representation V^λ of M given by $\mathrm{Res}_{L_0/\mathbb{Q}} V^\lambda_{L_0}$ is irreducible. Moreover, since $\gamma\lambda = -\lambda$ the highest weight of $\check{V}^\lambda$ is $\gamma_*\lambda$. This is because $\gamma\lambda = -\lambda$ and the general fact that $w_\gamma(-\lambda)$ is the highest weight of $\check{V}^\lambda$.

Case 2: $\gamma_*\lambda = \lambda$.
Then the representation V^λ is defined over $\mathbb{Q}$; we have $\rho_\lambda : M \to \mathrm{GL}(V^\lambda)$.

(IV.A.24) PROPOSITION: *In each of cases 1 and 2, there is an invariant form Q defined over $\mathbb{Q}$. In case 1 the forms may be chosen to be either symmetric or alternating. In case 2 the parity of Q is given by $\langle \lambda, h^0 \rangle$.*

PROOF. In case 2 we have the isomorphisms over $\mathbb{Q}$ of M-modules

$$\check{V}^\lambda \cong V^{\gamma_*\lambda} \cong V^\lambda,$$

and the pairing $V^\lambda \otimes \check{V}^\lambda \to \mathbb{Q}$ is defined over $\mathbb{Q}$.

Case 1 is more interesting. We have the isomorphism over $\mathbb{Q}$ of M-modules

$$\begin{aligned}\mathrm{Res}_{L_0/\mathbb{Q}}(V^\lambda_{L_0}) \otimes_{\mathbb{Q}} L_0 &\cong V^\lambda \oplus V^{\gamma_*\lambda} \\ &\cong V^\lambda \oplus \check{V}^\lambda.\end{aligned}$$

[11] Up to this point, the discussion holds for any field k of characteristic zero replacing $\mathbb{Q}$. Taking k to be $\mathbb{R}$, the w_γ is the w_0 in the strategy for the analysis of real Hodge representations in Part II of Section IV.A. Although we could have given a summary of what is needed from [KMRT] for an arbitrary field k of characteristic zero and then applied this to the cases $k = \mathbb{Q}, \mathbb{R}$, we have chosen to treat the $k = \mathbb{R}$ separately as that is where the detailed root/weight analysis will be done.

This gives an invariant form on $V^\lambda = \mathrm{Res}_{L_0/\mathbb{Q}}(V^\lambda_{L_0})$ that is defined over L_0. The issue is how to get one that is defined over $\mathbb{Q}$. For this we shall use the following general principle:

> *Let $\mathcal{U}, W$ be representations of M over $\mathbb{Q}$. Denoting by $\mathbb{Q}[M]$ the group ring of M, the equations that*
>
> $$F \in \mathrm{End}_{\overline{\mathbb{Q}}}\left(\mathrm{Hom}_{\mathbb{Q}[M]}(\mathcal{U}, W)\right)$$
>
> *be fixed by M are linear over $\mathbb{Q}$. Hence the solutions over $\overline{\mathbb{Q}}$ are spanned by solutions defined over $\mathbb{Q}$.*

The reason for this principle is that $\mathrm{Hom}_{\mathbb{Q}[M]}(\mathcal{U}, W)$ is a representation space of M defined over $\mathbb{Q}$. Thus there is a Zariski dense set of $m_i \in M$ for which the $m_i \in \mathrm{End}_{\mathbb{Q}}\left(\mathrm{Hom}_{\mathbb{Q}[M]}(\mathcal{U}, W)\right)$ have entries in $\mathbb{Q}$.

For $F \in \mathrm{End}_{\mathbb{Q}}\left(\mathrm{Hom}_{\mathbb{Q}[M]}(\mathcal{U}, W)\right)$,

$$(m_i - Id)F = 0$$

is a set of linear equations over $\mathbb{Q}$. Taking the union of these equations over the m_i gives the above principle.

To complete the proof of Proposition (IV.A.24) in case 2, the principle says that if we can find an M-invariant form on $\mathrm{Res}_{L_0/\mathbb{Q}}(V^\lambda_{L_0})$ defined over $\overline{\mathbb{Q}}$, we can find one defined over $\mathbb{Q}$. However, over $\overline{\mathbb{Q}}$

$$V^{\gamma_*\lambda} \cong \check{V}^\lambda$$

and on $\mathrm{Res}_{L_0/\mathbb{Q}}(V^\lambda_{L_0}) \otimes_{\mathbb{Q}} L_0 \cong V^\lambda_{L_0} \oplus \check{V}^\lambda_{L_0}$ we may take

$$Q\left((u, w), (u', w')\right) = \langle u, v' \rangle \pm \langle u', w \rangle$$

to have either a symmetric or alternating invariant form.

In case 1 the existence of an M-invariant bilinear form is ensured by [GW], and it is symmetric or alternating depending on the parity of $\langle \lambda, h^0 \rangle$. The general principle says we can find one defined over $\mathbb{Q}$. □

IV.B THE ADJOINT REPRESENTATION AND CHARACTERIZATION OF WHICH WEIGHTS GIVE FAITHFUL HODGE REPRESENTATIONS

In this section we assume given a semi-simple $\mathbb{Q}$-algebraic group M such that $M(\mathbb{R})$ contains a compact maximal torus T. We denote by $K \subset M(\mathbb{R})$ the maximal compact subgroup with $T \subset K$. Then we have the Cartan decomposition

(IV.B.1) $$\mathfrak{m}_{\mathbb{R}} = \mathfrak{k} \oplus \mathfrak{p}$$

where $\mathfrak{t} \subset \mathfrak{k}$, and the standard bracket relations

(IV.B.2) $$\begin{cases} [\mathfrak{k}, \mathfrak{p}] \subseteq \mathfrak{p} \\ [\mathfrak{p}, \mathfrak{p}] \subseteq \mathfrak{k} \end{cases}$$

hold. We will denote by $\alpha_1, \ldots, \alpha_d$ the roots of T belonging to $\mathfrak{k}$ (the *compact roots*), by $\beta_1, \ldots, \beta_e$ the roots of T belonging to $\mathfrak{p}$ (the *non-compact roots*). Then the basic observations are (see the appendix to Chapter IV)

(i) the representation $\mathrm{Ad} : M \to \mathrm{Aut}(\mathfrak{m}, B)$ preserves the symmetric bilinear form B;

(ii) B is negative on the compact root spaces $\mathfrak{m}_{\alpha_j}$ and is positive on the non-compact root spaces $\mathfrak{m}_{\beta_k}$.

This means that $B < 0$ on $(\mathfrak{m}_{\alpha_j} \oplus \mathfrak{m}_{-\alpha_j}) \cap \mathfrak{k} = (\mathfrak{m}_{\alpha_j} \oplus \mathfrak{m}_{-\alpha_j})_{\mathbb{R}}$, and $B > 0$ on $(\mathfrak{m}_{\beta_k} \oplus \mathfrak{m}_{-\beta_k}) \cap \mathfrak{p} = (\mathfrak{m}_{\beta_t} \oplus \mathfrak{m}_{-\beta_k})_{\mathbb{R}}$. Thus, *both the issue of an invariant form and the signs of the form on eigenspaces are determined in this case.*

We consider a co-character

$$\varphi : S^1 \to T$$

given by

$$\varphi(z) = \left(z^{l_1}, \ldots, z^{l_r}\right)$$

where $\mathfrak{l}_\varphi = (l_1, \ldots, l_r) \in \mathrm{Hom}(\mathbb{Z}, \Lambda)$. As before, we identify $\mathfrak{l}_\varphi$ with $\mathfrak{l}_\varphi(1) \in \Lambda$.

(IV.B.3) PROPOSITION: *φ gives a polarized Hodge structure on $(\mathfrak{m}, B)$ if, and only if*

$$\begin{cases} \langle \alpha_j, \mathfrak{l}_\varphi \rangle \equiv 0 \pmod 4 \\ \langle \beta_k, \mathfrak{l}_\varphi \rangle \equiv 2 \pmod 4. \end{cases}$$

PROOF. Since B is symmetric, the weight $n = 2n'$ must be even in (IV.A.3). In fact, by tensoring with a Tate twist $\mathbb{Q}(n')$ we may assume that $n = 0$. Because, as previously noted

$$\mathfrak{k}_{\mathbb{C}} = \mathfrak{t}_{\mathbb{C}} \oplus \left(\bigoplus_j \mathfrak{m}_{\alpha_j}\right) = \bigoplus_i \mathfrak{m}^{-2i,2i}$$
$$\mathfrak{p}_{\mathbb{C}} = \bigoplus_j \mathfrak{m}_{\beta_j} = \bigoplus_i \mathfrak{m}^{-2i-1,2i+1},$$

the conditions in the proposition exactly mean that the form $Q = -B$ satisfies the second Hodge-Riemann bilinear relations (IV.A.8). □

Remarks: (i) The Lie algebra $\mathfrak{h}_\varphi$ of the isotropy group is given by

$$\mathfrak{h}_\varphi = \mathfrak{t} \oplus \bigoplus_{\langle \alpha_j, \mathfrak{l}_\varphi \rangle = 0} (\mathfrak{m}_{\alpha_j} \oplus \mathfrak{m}_{-\alpha_j})_\mathbb{R}. \tag{IV.B.4}$$

We note the inclusion $\mathfrak{h}_\varphi \subset \mathfrak{k}$, consistent with the fact that H_φ is compact.

(ii) We have a map

$$\Lambda/4\Lambda \xrightarrow{(\alpha_1,\dots,\alpha_d,\beta_1,\dots,\beta_e)} \Big(\bigoplus_{\frac{1}{2}(\dim \mathfrak{k}-r)} \mathbb{Z}/4\mathbb{Z}\Big) \oplus \Big(\bigoplus_{\frac{1}{2}(\dim \mathfrak{p})} \mathbb{Z}/4\mathbb{Z}\Big) \tag{IV.B.5}$$

where $r = \dim T$ is the rank, and the conditions in the proposition are conditions on this map. This map will be analyzed in detail in sections IV.E and IV.F.

The reason that all the congruences are "mod 4" is of course that $i^4 = 1$; more specifically

- the 2nd bilinear relations are $i^{p-q}Q(v, \bar{v}) > 0$ for $0 \neq v \in V^{p,q}$;
- the $V^{p,q}$ are eigenspaces V_m with eigenvalues mi for the action of the differential $\mathfrak{l}_\varphi = (l_1, \dots, l_r)$ of φ;
- thus on the one hand $p-q = m$, so that i^{p-q} depends only on $m \pmod 4$, while on the other hand for the adjoint representation the V_m are direct sums of root spaces $\mathfrak{m}_{\alpha_j}, \mathfrak{m}_{\beta_k}$ so that the m's above are given by $m = \langle \alpha_j, \mathfrak{l}_\varphi \rangle$, $m = \langle \beta_k, \mathfrak{l}_\varphi \rangle$.

We now come to the main theorem on Hodge representations, which deals with the following situation:

- $\rho : M \to \mathrm{Aut}(V)$;
- V is an irreducible representation over $\mathbb{Q}$ of M;
- $T \subset M(\mathbb{R})$ is a compact maximal torus, $T = \mathfrak{t}/\Lambda$;
- $\varphi : S^1 \to T$ is a co-character corresponding to $\mathfrak{l}_\varphi \in \Lambda$;
- $V_\mathbb{C} = \begin{cases} U & \text{real case} \\ U \oplus \check{U} & \text{complex or quaternionic case;} \end{cases}$
- U is an irreducible representation of $M(\mathbb{C})$ over $\mathbb{C}$ with highest weight λ.

(IV.B.6) THEOREM: *Assume that M is a simple $\mathbb{Q}$-algebraic group that contains an anisotropic maximal torus. Then $(V, \rho \circ \varphi)$ gives a polarized Hodge structure for some invariant bilinear form Q if, and only if, the map $\mathbb{Z}\lambda + R \xrightarrow{\mathfrak{l}_\varphi} \mathbb{Z}$ satisfies $\mathfrak{l}_\varphi|_R \equiv \Psi \pmod 4$.*[12]

PROOF. If $(V, \rho \circ \varphi, Q)$ gives a polarized Hodge structure, then we note that $\mathfrak{m} \to \mathrm{End}(V)$ is injective as the kernel is an ideal and M is simple. Thus

$$\mathrm{span}_{\mathbb{Z}}(\text{weights of } V_{\mathbb{C}}) = \mathbb{Z}\lambda + R$$

and $\mathbb{Z}\lambda + R \xrightarrow{\mathfrak{l}_\varphi} \mathbb{Z}$.

For each weight ω with weight space we observe that

(i) the map $\rho \circ \varphi : S^1 \to \mathrm{GL}(V_{\mathbb{C}})$ preserves V_ω and acts there by $z^{\langle \omega, l_\varphi \rangle}$;

(ii) each $V^{p,q}$ is a direct sum of weight spaces V_ω where

$$p - q = \langle \omega, \mathfrak{l}_\varphi \rangle ;$$

(iii) the weight

$$\omega = \lambda - \sum_{\alpha \in \mathfrak{r}^+} m_\alpha(\omega)\alpha$$

so that from (ii) we have

$$p - q = \langle \lambda, l_\varphi \rangle - \sum_{\alpha \in \mathfrak{r}^+} m_\alpha(\omega) \langle \alpha, l_\varphi \rangle .$$

We write this as

$$l_\varphi(\lambda) - \sum m_\alpha(\omega) l_\varphi(\alpha) = p - q.$$

Since the 2nd Hodge-Riemann bilinear relation depends only on $p - q \pmod 4$, this suggests the map

$$\mathbb{Z}\lambda + R \xrightarrow{\tilde{l}_\varphi} \mathbb{Z}/4\mathbb{Z}.$$

The idea now is to consider the action of the sl_2 generated by $e_\alpha, e_{-\alpha}, h_\alpha$ on $V_{\mathbb{C}}$. Let ω be a weight of ρ and $v_\omega \in V_\omega$, and let

$p =$ maximum positive integer such that $\omega + p\alpha$ is a weight
$q =$ maximum positive integer such that $\omega - q\alpha$ is a weight.

---

[12] The earlier argument in (IV.A.9) and (IV.A.10) shows that the generic conjugate of φ by $g \in M(\mathbb{R})$ has Mumford-Tate group M.

Then $e_\alpha, e_{-\alpha}, h_\alpha$ acting on V_ω generate an irreducible sl_2-module of dimension $p+q+1$. We may think of this module as homogeneous polynomials of degree $p+q+1$ in variables x, y where

$$\begin{cases} e_\alpha = y\partial/\partial x, e_{-\alpha} = x\partial/\partial y \\ v_\omega = x^p y^q. \end{cases}$$

Then, by direct calculation

$$e_\alpha e_{-\alpha} v_\omega = p(q+1)v_\omega.$$

We note that $p(q+1) > 0$ if $e_{-\alpha}v_\omega \neq 0$.

At this point we assume that Q is symmetric. The same argument works in the alternating case where we replace Q by iQ. By the invariance of Q

$$Q(e_{-\alpha}v_\omega, \overline{e_{-\alpha}v_\omega}) = -Q(\overline{e}_{-\alpha}e_{-\alpha}v_\omega, \overline{v}_\omega),$$

which using the definition of $\psi(\alpha)$ gives

$$Q(e_{-\alpha}v_\omega, \overline{e_{-\alpha}v_\omega}) = (-1)^{\psi(\alpha)}p(q+1)Q(v_\omega, \overline{v}_\omega).$$

Thus

> *The sign of* $Q(e_{-\alpha}v_\omega, \overline{e_{-\alpha}, v_\omega})$ *changes from that of* $Q(v_\omega, \overline{v}_\omega)$ *if* $\psi(\alpha) = 0$ *and stays the same if* $\psi(\alpha) = 1$.

Since $(-1)^{\psi(\alpha)} = i^{2\psi(\alpha)}$ it follows that

(IV.B.7) $\quad i^{2\psi(\alpha)}Q(e_{-\alpha}v_\omega, \overline{e_{-\alpha}v_\omega})$ and $Q(v_\omega, \overline{v}_\omega)$ have the same sign.

Because we have a polarized Hodge structure,

$$i^{\mathfrak{l}_\varphi(\omega-\alpha)}Q(e_{-\alpha}v_\omega, \overline{e_{-\alpha}v_\omega}) > 0, \qquad i^{\mathfrak{l}_\varphi(\omega)}Q(v_\omega, \overline{v}_\omega) > 0.$$

Putting this together with (IV.B.7) we obtain

$$\begin{cases} \dfrac{i^{2\psi(\alpha)}Q(e_{-\alpha}v_w, \overline{e_{-\alpha}v_w})}{Q(v_w, \overline{v}_w)} > 0 \\ \dfrac{i^{\mathfrak{l}_\varphi(w-\alpha)}Q(e_{-\alpha}v_w, \overline{e_{-\alpha}v_w})}{i^{\mathfrak{l}_\varphi(w)}Q(e_{-\alpha}v_w, \overline{e_{-\alpha}v_w})} > 0. \end{cases}$$

It follows that

$$i^{2\psi(\alpha)+\mathfrak{l}_\varphi(\alpha)} > 0$$

and thus

$$2\psi(\alpha) + \mathfrak{l}_\varphi(\alpha) \equiv 0 \pmod 4.$$

Then since $-2 \equiv 2 \pmod 4$

$$\begin{aligned}\mathfrak{l}_\varphi(\alpha) \equiv -2\psi(\alpha) &\equiv 2\psi(\alpha) \pmod 4\\ &\equiv \Psi(\alpha) \pmod 4,\end{aligned}$$

which gives

$$\mathfrak{l}_\varphi \mid_R \equiv \Psi \pmod 4$$

as desired.

Conversely, if we have a φ with $\mathfrak{l}_\varphi|_R \equiv \Psi \pmod 4$ then for any invariant bilinear form Q on $V_\mathbb{R}$, if we can arrange that $i^{\mathfrak{l}_\varphi(\lambda)}Q(v_\lambda, \overline{v}_\lambda) > 0$, it will follow that this holds for all weights ω by (IV.B.7).

We now must consider the case where $V_\mathbb{R}$ is real, complex, or quaternionic.

Real Case: We have

$$V_\mathbb{C} \cong U \text{ irreducible and therefore } \mathfrak{m}_\mathbb{C} v_\lambda = U.$$

We know that $w_0(\lambda) = -\lambda$, and by a standard result quoted in the appendix to this chapter we have that Q is symmetric/antisymmetric according to whether $\langle \lambda, h^0 \rangle$ is even/odd. By the representation-theoretic result given by Theorem (IV.E.4), if $2\lambda = \sum_{i=1}^n m_i\alpha_i$ where $\alpha_1, \dots, \alpha_r$ are the simple positive roots and where $m_i \in \mathbb{Z}$, then $\sum_{\psi(\alpha_i)=0} m_i$ is even while

$$\langle \lambda, h^0 \rangle = \tfrac{1}{2}\sum_{i=1}^n m_i \langle \alpha_i, h^0 \rangle = \sum_{i=1}^n m_i.$$

Now

$$\begin{aligned}\mathfrak{l}_\varphi(\lambda) = \tfrac{1}{2}\Sigma m_i \mathfrak{l}_\varphi(\alpha_i) &\equiv \tfrac{1}{2}\Sigma m_i \Psi(\alpha_i) \pmod 2\\ &\equiv \Sigma m_i \psi(\alpha_i) \pmod 2\\ &\equiv \langle \lambda, h^0 \rangle \pmod 2.\end{aligned}$$

It follows that

$$\begin{cases}\mathfrak{l}_\varphi(\lambda) \equiv 1 \text{ or } 3 \pmod 4 & \text{if } Q \text{ is anti-symmetric}\\ \mathfrak{l}_\varphi(\lambda) \equiv 2 \text{ or } 4 \pmod 4 & \text{if } Q \text{ is symmetric}\end{cases}$$

and thus we have

$$i^{\mathfrak{l}_\varphi(\lambda)}Q(v_\lambda, \overline{v}_\lambda) \in \mathbb{R}$$

and hence, possibly replacing Q by $-Q$,

$$i^{\mathfrak{l}_\varphi(\psi)}Q(v_\lambda, \overline{v}_\lambda) > 0.$$

This completes the real case.

Complex Case: We have $V_{\mathbb{C}} \cong U \oplus \check{U}$, where $U \not\cong \check{U}$. We note that Q is a multiple of the contraction operation $U \otimes \check{U} \xrightarrow{Q_0} \mathbb{C}$. Since U is irreducible, $i^{\mathfrak{l}_\varphi(\omega)} Q(v_\omega, \overline{v}_\omega)$ has the same sign as $i^{\mathfrak{l}_\varphi(\lambda)} Q(v_\lambda, \overline{v}_\lambda)$ for $v_\omega, v_\lambda \in U$.

We now extend Q to $V_{\mathbb{C}} \otimes V_{\mathbb{C}} \to \mathbb{C}$ by $V_{\mathbb{C}} \cong U \oplus \check{U}$; i.e.,

$$Q = \begin{pmatrix} 0 & Q_0 \\ \pm Q_0 & 0 \end{pmatrix}$$

where we choose the $+$ sign if $\mathfrak{l}_\varphi(\lambda)$ is even and the $-$ sign if $\mathfrak{l}_\varphi(\lambda)$ is odd. Now $i^{\mathfrak{l}_\varphi(\lambda)} Q(v_\lambda, \overline{v}_\lambda) \in \mathbb{R}$ and we proceed as before. One may verify that $Q|_{V_{\mathbb{R}} \otimes V_{\mathbb{R}}}$ is real.

Quaternionic Case:

$$V_{\mathbb{C}} \cong U \oplus \check{U}, \qquad U \cong \check{U}.$$

Since

$$\overline{(v_1 \oplus v_2)} = \overline{v}_2 \oplus \overline{v}_1$$

we see that it is only the part of Q on $U \otimes \check{U} \to \mathbb{C}$ that matters in evaluating $Q(v, \overline{v})$. As in the complex case, we take

$$Q = \begin{pmatrix} 0 & Q_0 \\ \pm Q & 0 \end{pmatrix}.$$

To complete the proof of the theorem, it remains only to note that from Part III in Section IV.A we may assume that the bilinear form Q is defined over $\mathbb{Q}$. □

(IV.B.8) COROLLARY: *If* $(V_{\mathbb{R}}, \rho \circ \varphi)$ *gives a Hodge structure, then* $(V, Q, \rho \circ \varphi)$ *gives a polarized Hodge structure for some invariant bilinear form* Q *if, and only if,* $(\mathfrak{m}, B, \mathrm{Ad}\,\varphi)$ *gives a polarized Hodge structure.*

PROOF OF THE COROLLARY. The two conditions, by the main theorem, are both equivalent to $\mathfrak{l}_\varphi|_R \equiv \Psi \pmod 4$. The notation $\pm Q$ means that either Q or $-Q$ gives the polarizing form. □

One way the corollary will be used in practice is the following: First, from the root structure of the Lie algebra, determine the conditions that $(\mathfrak{m}, B, \mathrm{Ad}\,\varphi)$ give a polarized Hodge structure. Then for every representation $\rho : M \to \mathrm{Aut}(V, Q)$, replacing Q by $-Q$ if necessary, we will obtain a Hodge representation $(V, Q, \rho \circ \varphi)$. This avoids having to get into the weight structure of the individual representation.

The other way it will be used is that in many cases, e.g., for the classical groups, it is easier to determine the polarized Hodge structures for the natural representation than for the adjoint representation.

We conclude this section by revisiting, for the adjoint representation, the question:

What is the minimal degree in which the algebra of Hodge tensors is effectively generated?

Recall that effective generation means that adding additional generators does not *set-theoretically* decrease the locus in the period domain containing the given algebra of Hodge tensors. We will prove the

(IV.B.9) PROPOSITION: *Let* $(\mathfrak{m}, B, \mathrm{Ad}\,\varphi)$ *be a polarized Hodge structure with* $D_{\mathfrak{m}_\varphi} \subset D$ *the Mumford-Tate domain embedded in the corresponding period domain. Assume that* $\mathfrak{m}_\mathbb{R}$ *is a simple real Lie algebra. Then we have:*

(i) *The bracket* $[\ ,\ ]$ *is a Hodge tensor;*

(ii) *In* $T^{\bullet,\bullet}$ *we denote by* $T^{\bullet,\bullet}([\ ,\ ])$ *the sub-algebra generated by the polarizing form* B *and bracket* $[\ ,\ ]$ *and by* $D_{[\ ,\ ]} \subset D$ *the polarized Hodge structures* $(\mathfrak{m}, B, \psi)$ *such that* $\mathrm{Hg}_\psi^{\bullet,\bullet} \supseteq T^{\bullet,\bullet}([\ ,\ ])$.[13] *Then the components of* $D_{\mathfrak{m}_\varphi}$ *and* $D_{[\ ,\ ]}$ *through* $\mathrm{Ad}\,\varphi$ *coincide.*

Thus, although B and $[\ ,\ ]$ may not be enough to generate $\mathrm{Hg}_{\mathrm{Ad}\,\varphi}^{\bullet,\bullet}$, they are enough to set-theoretically cut out the component through $\mathrm{Ad}\,\varphi \in D_{\mathfrak{m}_\varphi}$ of the Mumford-Tate domain $D_{\mathfrak{m}_\varphi}$.

PROOF. We first note that

$$[\ ,\ ] \in \mathrm{Hom}(\Lambda^2\mathfrak{m}, \mathfrak{m}) \subset T^{1,2}$$

is a rational tensor. Writing

$$\mathfrak{m}_\mathbb{C} = \bigoplus_i \mathfrak{m}^{-i,i}$$

and using the bracket relation

$$[\mathfrak{m}^{-i,i}, \mathfrak{m}^{-j,j}] \subseteq \mathfrak{m}^{-(i+j),i+j}$$

gives that $[\ ,\ ] \in \mathrm{Hg}_{\mathrm{Ad}\,\varphi}^{1,2}$ is a Hodge tensor.

[13] *Here,* $\psi : S^1 \to \mathrm{Aut}(\mathfrak{m}_\mathbb{R}, B)$ *need not be of the form* $\mathrm{Ad}\,\varphi'$ *for a circle* $\varphi : S^1 \to M(\mathbb{R})$.

Next, we recall that $F \in \operatorname{Aut}(\mathfrak{m})$ is a Lie algebra automorphism, written $F \in \operatorname{Aut}(\mathfrak{m}, [\ ,\ ])$, if

$$[F(X), F(Y)] = F([X, Y]), \qquad X, Y \in \mathfrak{m}.$$

This is equivalent to saying that the diagram

$$\begin{array}{ccc} \mathfrak{m} & \xrightarrow{F} & \mathfrak{m} \\ \uparrow & & \uparrow \\ \mathfrak{m} \otimes \mathfrak{m} & \xrightarrow{F \otimes F} & \mathfrak{m} \otimes \mathfrak{m} \end{array}$$

is commutative. Denoting by $M_a(\mathbb{C})$ the adjoint group of $M(\mathbb{C})$ acting on $\mathfrak{m}_{\mathbb{C}}$ since $\mathfrak{m}_{\mathbb{C}}$ is simple, from [FH], page 498 we have that

$$M_a(\mathbb{C}) = \operatorname{Aut}^0(\mathfrak{m}_{\mathbb{C}}, [\ ,\ ]), \tag{IV.B.10}$$

and that $\operatorname{Aut}(\mathfrak{m}_{\mathbb{C}}, [\ ,\ ]) / \operatorname{Aut}^0(\mathfrak{m}_{\mathbb{C}}, [\ ,\ ])$ is identified with the automorphisms of the Dynkin diagram. Since the Mumford-Tate group of $(\mathfrak{m}, B, \operatorname{Ad}\varphi)$ is defined by preserving all Hodge tensors, we may infer from (IV.B.10) the second statement in the proposition. □

IV.C EXAMPLES: THE CLASSICAL GROUPS

In this section we shall use Theorem IV.B.6 and (IV.B.3) to determine which co-characters of the classical groups give polarized Hodge structures for which representations. These examples will illustrate both the linear algebra and Vogan diagram methods. We shall use the notations from [GW], which are summarized in the appendix. We begin by considering the standard, non-compact real forms of the classical simple complex Lie groups.

$\mathbf{SO(p, q)}$. We first note that in general the issue of whether $\varphi : S^1 \to M(\mathbb{R})$ gives a polarized Hodge structure on $(\mathfrak{m}, B, \operatorname{Ad}\varphi)$ depends only on the real Lie group $M(\mathbb{R})$ and not on the $\mathbb{Q}$-algebraic group M whose real points are $M(\mathbb{R})$. Thus we may take

$$\begin{pmatrix} I_p & 0 \\ 0 & -I_q \end{pmatrix}$$

to be the matrix of the quadratic form that defines $\mathrm{SO}(p, q)$. Also, in the discussion of these examples we shall use A, B, C to denote matrices and shall use $\mathbf{B}$ to denote the Cartan-Killing form. Then

$$\mathrm{so}(p, q) = \left\{ \begin{pmatrix} A & B \\ {}^tB & C \end{pmatrix} : A = -{}^tA \text{ and } C = -{}^tC \right\}.$$

We note that $B(M_1, M_2) = \mathrm{Tr}(M_1 M_2)$ and thus

$$\begin{cases} \mathbf{B} \text{ is negative on } & \left\{\begin{pmatrix} A & 0 \\ 0 & C \end{pmatrix}\right\} \\ \mathbf{B} \text{ is positive on } & \left\{\begin{pmatrix} 0 & B \\ {}^tB & 0 \end{pmatrix}\right\}. \end{cases}$$

We shall denote these negative and positive spaces by $\mathrm{so}^-(p,q)$ and $\mathrm{so}^+(p,q)$. Taking the case where $p = 2a$ and $q = 2b$ are both even, so that the rank $r = a + b$, we consider the co-character

$$\mathfrak{l}_\varphi = \begin{pmatrix} F & 0 \\ 0 & G \end{pmatrix}$$

where for $E = \left(\begin{smallmatrix} 0 & -1 \\ 1 & 0 \end{smallmatrix}\right)$

$$F = \begin{pmatrix} k_1 E & & 0 \\ & \ddots & \\ 0 & & k_a E \end{pmatrix}, \quad G = \begin{pmatrix} l_1 E & & 0 \\ & \ddots & \\ 0 & & l_b E \end{pmatrix}.$$

(IV.C.1) PROPOSITION: *The conditions* (IV.B.3) *that* $\mathfrak{l}_\varphi$ *give a polarized Hodge structure on* $(\mathfrak{m}, \mathbf{B})$ *are*

$$\text{(IV.C.2)} \qquad \begin{cases} \textit{all } k_m \equiv 0 \pmod 4 \textit{ and all } l_n \equiv 2 \pmod 4, \textit{ or} \\ \textit{all } k_m \equiv 2 \pmod 4 \textit{ and all } l_m \equiv 0 \pmod 4. \end{cases}$$

PROOF. Under the action of $\mathrm{ad}\,\mathfrak{l}_\varphi$ we have

$$\begin{cases} A \to [F, A] \\ B \to FB - BG \\ C \to [G, C]. \end{cases}$$

On the set of A's the eigenvalues are

$$\begin{cases} 0 & a\text{-times} \\ \pm k_m i \pm k_n i, & 1 \leqq m < n \leqq a. \end{cases}$$

On the C's the eigenvalues are

$$\begin{cases} 0 & b\text{-times} \\ \pm l_m i \pm l_n i & 1 \leqq m < n \leqq b. \end{cases}$$

The conditions in Proposition (IV.B.3) give that all k_m are even and all $k_m - k_n \equiv 0 \pmod 4$, which is equivalent to all $k_m \equiv 0 \pmod 4$ or all $k_m \equiv 2 \pmod 4$. Together with a similar argument of the l_m's on $\mathrm{so}^-(p,q)$ we obtain the conditions

$$\begin{cases} \text{all } k_m \equiv 0 \pmod 4 \ \text{ or all } \ k_m \equiv 2 \pmod 4 \\ \text{all } l_n \equiv 0 \pmod 4 \ \text{ or all } \ l_n \equiv 2 \pmod 4. \end{cases}$$

On B the eigenvalues are $\pm k_m i \pm l_n i$, $1 \leqq m \leqq a$ and $1 \leqq n \leqq b$. Thus, on $\mathrm{so}^+(p,q)$ the conditions are

$$k_m \pm l_n \equiv 2 \pmod 4 \qquad\qquad 1 \leqq m \leqq a, 1 \leqq n \leqq b.$$

Combining this with the conditions on $\mathrm{so}^-(p,q)$ gives the proposition. □

If $p = 2a$ is even and $q = 2b+1$ is odd, then on C's we get eigenvalues

$$\begin{cases} 0 \ b\text{-times}, & \pm l_n i \pm l_n i \quad \text{for } 0 \leqq n < n \leqq b \\ \quad 0, & \qquad \pm l_m i \quad \text{for } 0 \leqq m \leqq b, \end{cases}$$

and on the B's we get

$$\begin{cases} \pm k_m i \pm k_n i & \text{for } 0 \leqq m < n \leqq a \\ \qquad \pm k_m & \text{for } 0 \leqq m \leqq a. \end{cases}$$

This forces all $l_m \equiv 0 \pmod 4$, and then all $k_m \equiv 2 \pmod 4$. If p and q are both odd, then all $k_m \equiv 0 \pmod 4$ as well and the conditions are incompatible. Thus

> *In case p is even and q is odd, the conditions are* (IV.C.2). *The case p and q both odd has no solutions.*

(IV.C.3) COROLLARY: *The conditions in the proposition are necessary and sufficient that any representation of* $\mathrm{SO}(p,q)$ *lead to a Hodge representation.*

This means that given any non-trivial and irreducible representation ρ : $\mathrm{SO}(p,q) \to \mathrm{GL}(V)$, there is a Q and a φ such that $(V, Q, \rho \circ \varphi)$ gives a polarized Hodge structure. This follows because every representation of $\mathrm{SO}(p,q)$ is a sub-representation of $\Lambda^k V$ for some k, where V is the standard representation.

Example 1: This is the standard case, where V is the standard representation of $\mathrm{SO}(2a,b)$ and $\mathfrak{m} = \Lambda^2 V$. In the above notation, a of the $k_m = 2$, a of the $k_n = -2$ and all $l_m = 0$. This corresponds to the Hodge structure of $\mathrm{so}(2a,b)$ at a point of the period domain $\mathrm{SO}(2a,b)/\mathcal{U}(a) \times \mathrm{SO}(b)$ parametrizing polarized Hodge structures of weight two with $h^{2,0} = a$, $h^{1,1} = b$.

Example 2: $\mathrm{SO}(2,2)$: We may take $k = k_1 \equiv 2 \pmod 4$, $l = l_1 \equiv 0 \pmod 4$. The eigenvalues are $\pm ki, \pm li$ for the standard representation and $0, 0, \pm(k \pm l)i$ for the adjoint representation. The isotropy group $H_\varphi = T$ is a compact maximal torus. For a few choices the corresponding Hodge numbers for the standard representation are

$$\text{(IV.C.4)} \qquad \begin{cases} k = 2, l = 0 & \text{weight 2 and Hodge numbers } 1, 2, 1 \\ k = 6, l = 4 & \text{weight 6 and Hodge numbers } 1, 1, 0, 0, 0, 1, 1 \\ k = 6, l = 0 & \text{weight 6 and Hodge numbers } 1, 0, 0, 2, 0, 0, 1. \end{cases}$$

We shall return to this example.

Example 3: $\mathrm{SO}(2,3)$: Here we must take $k = k_1 \equiv 2 \pmod 4$, $l = l_1 \equiv 0 \pmod 4$. For the adjoint representation the eigenvalues are $0, 0, \pm li, \pm ki$, $(\pm k \pm l)i$. Some illustrations are

$$\text{(IV.C.5)} \qquad \begin{cases} k = 2, l = 0 & \text{weight 2 and Hodge numbers } 3, 4, 3 \\ k = 2, l = 4 & \text{weight 6 and Hodge numbers } 1, 1, 2, 2, 2, 1, 1 \\ k = 6, l = 0 & \text{weight 6 and Hodge numbers } 3, 0, 0, 4, 0, 0, 3. \end{cases}$$

For a general $\mathrm{so}(p, q)$, from a more abstract perspective, the Vogan diagram of $\mathrm{so}(p, q)$ is

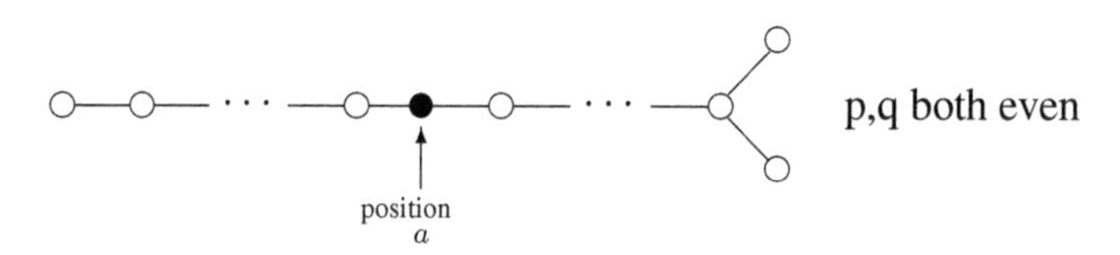

p even, q odd.

position a

In the first case the simple roots are

$$\alpha_1 = \epsilon_1 - \epsilon_2, \alpha_2 = \epsilon_2 - \epsilon_3, \ldots, \alpha_{a+b-1} = \epsilon_{a+b-1} - \epsilon_{a+b}, \alpha_{a+b} = \epsilon_{a+b-1} + \epsilon_{a+b}.$$

Now

$$\psi(\alpha_i) = \begin{cases} 0 & i \neq a \\ 1 & i = a. \end{cases}$$

Thus

$$\iota_\varphi(\epsilon_1) \equiv \cdots \equiv \iota_\varphi(\epsilon_a) \equiv 2, \quad \iota_\varphi(\epsilon_{a+1}) \equiv \cdots \equiv \iota_\varphi(\epsilon_{a+b}) \equiv 0 \pmod 4)$$

or

$$\mathfrak{t}_\varphi(\epsilon_1) \equiv \cdots \equiv \mathfrak{t}_\varphi(\epsilon_a) \equiv 0, \quad \mathfrak{t}_\varphi(\epsilon_{a+1}) \equiv \cdots \equiv \mathfrak{t}_\varphi(\epsilon_{a+b}) \equiv 2 \pmod 4.$$

For the p even, q odd case,

$$\alpha_1 = \epsilon_1 - \epsilon_2, \ldots, \alpha_{a+b-1} = \epsilon_{a+b-1} - \epsilon_{a+b}, \alpha_{a+b} = \epsilon_{a+b}.$$

Now this fixes

$$\mathfrak{t}_\varphi(\epsilon_1) \equiv \cdots \equiv \mathfrak{t}_\varphi(\epsilon_a) \equiv 2, \qquad \mathfrak{t}_\varphi(\epsilon_{a+1}) \equiv \cdots \equiv \mathfrak{t}_\varphi(\epsilon_{a+b}) \equiv 0 \pmod 4.$$

Since

$$\mathfrak{t}_\varphi = \sum_{i=1}^{a} k_i \breve{e}_i + \sum_{j=1}^{b} l_j \breve{e}_{a+j}$$

the congruences above translate into the congruences on the k_i, l_j.

Note that there are other representations, but these are only defined on the 2-sheeted covers of $\mathrm{SO}(p,q)$.

Sp(2r,$\mathbb{R}$): We first consider the adjoint representation. Again, $B(M_1, M_2) = \mathrm{Tr}(M_1 M_2)$. Using analogous notations to those for $\mathrm{so}(p,q)$ above, we have

$$\begin{aligned} \mathrm{sp}^-(2g) &= \left\{ \begin{pmatrix} A & B \\ -B & A \end{pmatrix} : A = -{}^tA \text{ and } B = {}^tB \right\} \\ \mathrm{sp}^+(2g) &= \left\{ \begin{pmatrix} A & B \\ B & -A \end{pmatrix} : A = {}^tA \text{ and } B = {}^tB \right\}. \end{aligned}$$

We first assume that $g = 2h$ is even and, setting $E = \left(\begin{smallmatrix} 0 & -1 \\ 1 & 0 \end{smallmatrix}\right)$ and $I = \left(\begin{smallmatrix} 1 & 0 \\ 0 & 1 \end{smallmatrix}\right)$ we take

$$\mathfrak{t}_\varphi = \left(\begin{array}{ccc|ccc} k_1 E & & 0 & l_1 I & & 0 \\ & \ddots & & & \ddots & \\ 0 & & k_h E & 0 & & l_h I \\ \hline -l_1 I & & 0 & k_1 E & & 0 \\ & \ddots & & & \ddots & \\ 0 & & -l_h I & 0 & & k_h E \end{array} \right),$$

which is

$$A = \begin{pmatrix} k_1 E & & 0 \\ & \ddots & \\ 0 & & k_h E \end{pmatrix}, B = \begin{pmatrix} l_1 I & & 0 \\ & \ddots & \\ 0 & & l_h I \end{pmatrix}.$$

For $g = 2h+1$ odd we take

$$A = \begin{pmatrix} k_1 E & & & 0 \\ & \ddots & & \\ & & k_h E & \\ 0 & & & 0 \end{pmatrix}, \quad B = \begin{pmatrix} l_1 E & & & 0 \\ & \ddots & & \\ & & l_h E & \\ 0 & & & 0 \end{pmatrix}.$$

The case $g = 2$ already exhibits the ranges of possibilities. Then for $k = k_1$ and $l = l_1$,

$$\mathfrak{l}_\varphi = \begin{pmatrix} kE & lI \\ -lI & kE \end{pmatrix}.$$

On $\mathrm{sp}^-(4)$ the action of $\mathrm{ad}\,\mathfrak{l}_\varphi$ is given by

$$\begin{cases} A \to k[E, A] \\ B \to k[E, B] \end{cases}$$

with eigenvalues

$$0, 0, \pm 2ki.$$

On $\mathrm{sp}^+(4)$ the action is

$$\begin{cases} A \to k[E, A] + 2lB \\ B \to k[E, B] - 2lA \end{cases}$$

where A and B are symmetric. The eigenvalues are

$$\pm 2li, \quad \pm 2ki \pm 2li.$$

Thus we obtain:

> $\mathfrak{l}_\varphi$ *gives a polarized Hodge structure if, and only if,* $k \equiv 0 \pmod 2$ *and* $l \equiv 1 \pmod 2$,

confirming (IV.A.6).

For the adjoint representation, some illustrations are

$$\text{(IV.C.6)} \quad \begin{cases} k = 0, \; l = 1 & \text{weight 2 and Hodge numbers } 3, 4, 3 \\ k = 2, \; l = 1 & \text{weight 6 and Hodge numbers } 1, 2, 1, 2, 1, 2. \end{cases}$$

For the standard representation, we extend in the obvious way the example of $\mathrm{Sp}(4)$ to have $Q = \oplus E$ and as the differential of a co-character take

$$H = l_1 H_1 + \cdots + l_r H_r$$

where

$$H_j = \begin{pmatrix} 0 & & & & 0 \\ & \ddots & & & \\ & & E & & \\ & & & \ddots & \\ 0 & & & & 0 \end{pmatrix} j.$$
$$j$$

Then the conditions are (cf. (IV.A.6))

(IV.C.7) $$l_j \equiv 1 \pmod 4 \qquad j = 1, \dots, r.$$

In terms of Vogan diagrams, $\mathrm{sp}(2r)$ has the diagram

$$\alpha_1 = \epsilon_1 - \epsilon_2, \dots, \alpha_{r-1} = \epsilon_{r=1} - \epsilon_r, \alpha_r = \epsilon_r$$

so

$$\psi(\alpha_i) = \begin{cases} 0 & i \neq r \\ 1 & i = r. \end{cases}$$

It follows that $\mathfrak{l}_\varphi(\epsilon_i) \equiv 1 \pmod 4)$ for all i. The roots are $\pm 2\epsilon_i$, $\pm\epsilon_i \pm \epsilon_j$, so if $\lambda = \sum_i m_i \epsilon_i$ then $\mathfrak{l}_\varphi(\lambda) \equiv \sum_i m_i \pmod 4$ and Q is symmetric/antisymmetric according to whether Σm_i is even/odd.

Example: $\mathbf{SU(r+1)}$. This is an example of a compact real form. For the compact maximal torus we take the usual diagonal matrices

$$\begin{pmatrix} e^{i\theta_1} & & \\ & \ddots & \\ & & e^{i\theta_{r+1}} \end{pmatrix}, \qquad \sum_j \theta_j \equiv 0 \pmod{2\pi}.$$

The roots are $\epsilon_i - \epsilon_j$ for $i \neq j$ and where $\epsilon_1 + \cdots + \epsilon_{r+1} = 0$. Then by (IV.B.3) a co-character given by $\mathfrak{l}_\varphi = (l_1, \dots, l_{r+1})$ where

$$\theta_j = l_j \xi$$

gives a polarized Hodge structure on $(\mathrm{su}(r+1), \mathbf{B})$ if, and only if,

$$l_i - l_j \equiv 0 \pmod 4, \qquad i \neq j.$$

If we write $l_j = l_1 + 4m_j$ for $j \geqq 2$, then $l_1 + \cdots + l_{r+1} = 0$ gives $(r+1)l_1 \equiv 0 \pmod 4$. Thus we have

$$\text{(IV.C.8)} \quad \begin{cases} r+1 \equiv 1 \text{ or } 3 \pmod 4 \Rightarrow & \text{all } l_j \equiv 0 \pmod 4 \\ r+1 \equiv 0 \pmod 4 \Rightarrow & l_1 \text{ arbitrary and} \\ & l_j \equiv l_1 \pmod 4 \text{ for } j \geqq 2 \\ r+1 \equiv 2 \pmod 4 \Rightarrow & l_1 \equiv 0 \pmod 2 \text{ and} \\ & l_j \equiv l_1 \pmod 4 \text{ for } j \geqq 2. \end{cases}$$

This is a new phenomenon, where the arithmetic of r comes into play.

Example: $\mathbf{SU(p,q)}$. where $\mathbf{p > 0, q > 0}$. The Cartan-Killing form is negative on $\epsilon_i - \epsilon_j$ if $1 \leqq i, j \leqq p$ or $p+1 \leqq i, j \leqq p+q$, and is positive on $\epsilon_i - \epsilon_j$ if $i \in \{1, \ldots, p\}$, $j \in \{p+1, \ldots, p+q\}$ or vice-versa. Then (IV.B.3) leads to:

The conditions that the co-character given by the exponential of $\mathfrak{l}_\varphi = (l_1, \ldots, l_{r+1})$ *where* $l_1 + \cdots + l_{r+1} = 0$, *give a polarized Hodge structure on* $(\mathrm{su}(p,q), B)$ *are*

$$\begin{cases} l_j \equiv l_1 \pmod 4, & 2 \leqq j \leqq p \\ l_j \equiv l_1 + 2 \pmod 4, & p+1 \leqq j \leqq p+q \\ (p+q)l_1 \equiv 2q \pmod 4. \end{cases}$$

We will discuss the case of $\mathrm{SU}(2,1)$ in detail in (IV.F) and (IV.G).

(IV.C.9) **Example: AII** (notation from [K]). In this case there is a compact maximal torus only when $r+1 = 2$, i.e., for sl_2.

Example: $\mathbf{SO^*(2r)}$ (also known as DIII). The Lie algebra is given by

$$\left\{ \begin{pmatrix} A & B \\ -\overline{B} & \overline{A} \end{pmatrix} : A = -{}^tA \text{ and } B = {}^t\overline{B} \right\}.$$

Here the matrices A, B are of size $r \times r$. If

$$D = \begin{pmatrix} d_1 & & 0 \\ & \ddots & \\ 0 & & d_r \end{pmatrix}, \qquad d_i \in \mathbb{R}$$

then the matrices

$$\begin{pmatrix} 0 & D \\ -D & 0 \end{pmatrix}$$

give the Lie algebra of a compact maximal torus. For the Cartan-Killing form $\mathbf{B}$ we have

$$\mathrm{Tr}\left(\begin{pmatrix} A & B \\ -\overline{B} & \overline{A} \end{pmatrix}^2\right) = -2\mathrm{Tr}(\mathrm{Re}\, A\, {}^t\mathrm{Re}\, A - \mathrm{Im}\, A\, {}^t\mathrm{Im}\, A) \\ - 2\mathrm{Tr}(\mathrm{Re}\, B\, {}^t\mathrm{Re}\, B - \mathrm{Im}\, B\, {}^t\mathrm{Im}\, B).$$

Thus, $\mathbf{B}$ is negative when A and B are real and positive when A and B are purely imaginary. Letting H_i correspond to D with 1 in the ith spot, zeroes elsewhere, straightforward calculations lead to

(IV.C.10) *$\sum_j l_j H_j$ gives a polarized Hodge structure if, and only if, the conditions*

$$\begin{cases} l_j = 2m_j + 1, & m_j \in \mathbb{Z} \\ m_i - m_j \equiv 0 \pmod 2 \end{cases}$$

are satisfied.

Example: $\mathrm{Sp}(\mathrm{p}, \mathrm{q})$ (sometimes denoted by CII). Setting $I_{p,q} = \begin{pmatrix} I_p & 0 \\ 0 & -I_q \end{pmatrix}$, $K_{p,q} = \begin{pmatrix} I_{p,q} & 0 \\ 0 & I_{p,q} \end{pmatrix}$, $J = \begin{pmatrix} 0 & I_{p+q} \\ -I_{p+q} & 0 \end{pmatrix}$ and $\Omega = \begin{pmatrix} 0 & I_{p,q} \\ -I_{p,q} & 0 \end{pmatrix}$

$$\mathrm{Sp}(p,q) = \left\{ g \in \mathrm{Sp}(\mathbb{C}^{2r}, \Omega) : g = K_{p,q} \left({}^t\overline{g}\right)^{-1} K_{p,q} \right\}$$

where $\mathrm{Sp}(\mathbb{C}^{2r}, \Omega)$ is the complex symplectic group corresponding to the alternating form Ω. The Lie algebra of a maximal torus is

$$\begin{pmatrix} id_1 & & & & & \\ & \ddots & & & & \\ & & id_r & & & \\ & & & -id_1 & & \\ & & & & \ddots & \\ & & & & & -id_r \end{pmatrix}.$$

Letting H_j correspond to $d_j = 1$, and all other $d_k = 0$, calculations similar to the above lead to:

(IV.C.11) *The conditions that $\sum_j l_j H_j$ give a polarized Hodge structure on $(\mathrm{sp}(p,q), B)$ are*

$$\begin{cases} l_j = 2m_j \\ m_i - m_j \equiv 0 \pmod 2 \text{ if } i, j \in \{1, \dots, p\} \text{ or } \{p+1, \dots, p+q\} \\ m_i - m_j \equiv 1 \pmod 2 \text{ if } i \in \{q, \dots, p\}, j \in \{p+1, \dots, p+q\}. \end{cases}$$

IV.D EXAMPLES: THE EXCEPTIONAL GROUPS

Example: $\mathbf{G_2}$. Because this is the first time, to our knowledge, that G_2 as a $\mathbb{Q}$-algebraic group has appeared explicitly as a Mumford-Tate group, and because of the interesting geometry associated to the corresponding Mumford-Tate domains (see IV.F), we shall go into some detail regarding this example.

We will begin by establishing some notation. In $\mathbb{R}^7$ with standard basis $e_1, \ldots, e_7$, we set

$$\omega = (e_1 \wedge e_4 + e_2 \wedge e_3 + e_3 \wedge e_6) \wedge e_7 - 2e_1 \wedge e_2 \wedge e_3 + 2e_4 \wedge e_5 \wedge e_6.$$

Then (see [Ag], whose notations we shall generally follow)

$$G_2(\mathbb{R}) = \{g \in \mathrm{GL}(7, \mathbb{R}) : g(\omega) = \omega\}.$$

There is an obvious $\mathbb{Q}$-algebraic group G_2 whose real points are $G_2(\mathbb{R})$.

Step 1: Make a change of basis

$$\begin{aligned} u_1 &= e_1 - e_4, & v_1 &= e_1 + e_4 \\ u_2 &= e_2 - e_5, & v_2 &= e_2 + e_5 \\ u_3 &= e_3 - e_6, & v_3 &= e_3 + e_6 \\ & & v_4 &= e_7. \end{aligned}$$

Then

$$\begin{aligned} 4\omega = &- u_1 \wedge u_2 \wedge u_3 + u_1 \wedge (v_1 \wedge v_4 - v_2 \wedge v_3) + u_2 \wedge (v_2 \wedge v_4 - v_3 \wedge v_1) \\ &+ u_3 \wedge (v_3 \wedge v_4 - v_1 \wedge v_2). \end{aligned}$$

Step 2: Define $Q(X, Y) = (X \rfloor \beta) \wedge (Y \rfloor \beta) \wedge \beta$ where $\beta = -4\omega$. In terms of the basis $u_1, u_2, u_3, v_1, \ldots, v_4$,

$$Q = \begin{pmatrix} -I_3 & 0 \\ 0 & I_4 \end{pmatrix}.$$

Step 3: $\mathfrak{g}_{2,\mathbb{R}} \subset \mathrm{so}(4, 3)$ is defined by infinitesimally preserving β. If

$$\begin{matrix} 3\{ \\ 4\{ \end{matrix} \begin{pmatrix} \overbrace{A}^{3} & \overbrace{B}^{4} \\ {}^tB & C \end{pmatrix}, \qquad \text{where } A = -{}^tA, C = -{}^tC \text{ is an element of } \mathrm{so}(4, 3),$$

then the equations to preserve β are

$$\begin{aligned} a_{12} &= c_{12} + c_{43} & b_{14} &= b_{32} - b_{23} \\ a_{23} &= c_{23} + c_{41} & b_{24} &= b_{13} - b_{31} \\ a_{31} &= c_{31} + c_{42} & b_{34} &= b_{21} - b_{12} \\ & & b_{11} &+ b_{22} + b_{33} = 0. \end{aligned}$$

Step 4: Note that if $E = \left(\begin{smallmatrix} 0 & -1 \\ 1 & 0 \end{smallmatrix}\right)$ and

$$H_1 = \left(\begin{array}{cc|cc} E & 0 & 0 & 0 \\ 0 & 0 & 0 & 0 \\ \hline 0 & 0 & E & 0 \\ 0 & 0 & 0 & 0 \end{array}\right), \quad H_2 = \left(\begin{array}{cc|cc} 0 & 0 & 0 & 0 \\ 0 & 0 & 0 & 0 \\ \hline 0 & 0 & E & 0 \\ 0 & 0 & 0 & E \end{array}\right)$$

satisfy the equations of Step 3, mutually commute, and $\exp(tH_1)$, $\exp(tH_2)$ are circles in $G_2(\mathbb{R})$ with period $2\pi i$. They commute and span a maximal torus T. The exponentials of $2\pi i$ times their real linear combinations give a torus T in $G_2(\mathbb{R})$, which must then be a maximal torus since G_2 has rank two.

(IV.D.1) THEOREM: *The co-character φ whose differential is $\mathfrak{l}_\varphi = l_1 H_1 + l_2 H_2$ gives a polarized Hodge structure for* **every** *representation of G_2 if, and only if, the conditions*

$$\text{(IV.D.2)} \qquad \begin{cases} l_1 \equiv 0 \pmod 4 \\ l_2 \equiv 2 \pmod 4 \end{cases}$$

are satisfied.

PROOF. We first show that the standard representation of G_2 on $V \cong \mathbb{Q}^7$, with Q as in Step 2, has a polarized Hodge structure. For this we think of $V_\mathbb{R}$ as column vectors and let

$$\begin{cases} V^- = \text{column vectors} \begin{pmatrix} * \\ * \\ * \\ 0 \\ 0 \\ 0 \\ 0 \end{pmatrix} & \text{where } Q < 0 \\ V^+ = \text{column vectors} \begin{pmatrix} 0 \\ 0 \\ 0 \\ * \\ * \\ * \\ * \end{pmatrix} & \text{where } Q > 0. \end{cases}$$

Then $l_1 H_1 + l_2 H_2$ has

$$\begin{cases} \text{eigenvalues } \pm l_1 i, 0 & \text{on } V^- \\ \text{eigenvalues } \pm (l_1 + l_2) i, \pm l_2 i & \text{on } V^+. \end{cases}$$

This gives a polarized Hodge structure if, and only if, $l_1 \equiv 0 \pmod 4$, $l_2 \equiv 2 \pmod 4$ and $l_1 + l_2 \equiv 2 \pmod 4$. The third condition is a consequence of the first two, which are just (IV.D.2).

At this point we recall the root diagram of $\mathfrak{g}_2$ with positive roots

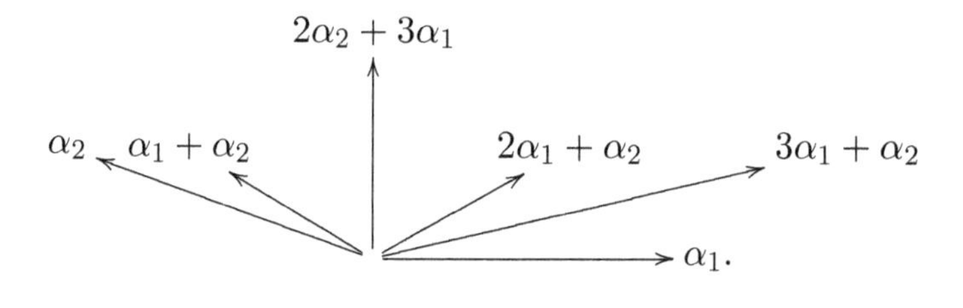

For this choice the co-roots are

$$\begin{cases} h_{\alpha_1} &= H_1 \\ h_{\alpha_2} &= H_2 - H_1. \end{cases}$$

From pages 351–355 in [FH] we have that the dominant weights of the irreducible $\mathfrak{g}_{2,\mathbb{C}}$-modules are linear combinations

$$\lambda = m_1\lambda_1 + m_2\lambda_2,$$

where m_1, m_2 are non-negative integers, and where

$$\begin{cases} \lambda_1 &= 2\alpha_1 + \alpha_2 \\ \lambda_2 &= 3\alpha_1 + 2\alpha_2. \end{cases}$$

The standard representation has highest weight λ_1, corresponding to the co-weight H_1+H_2. The adjoint representation has highest weight $\lambda_2 = 3\alpha_1+2\alpha_2$. It follows that the representation with highest weight $\lambda = m_1\lambda_1+m_2\lambda_2$ occurs in $S^{m_1}V \otimes S^{m_2}\mathfrak{g}_2$, and hence has a polarized Hodge structure when (IV.D.2) is satisfied. □

We shall now examine some special cases, and for this we shall use the following

(IV.D.3) PROPOSITION: *The eigenvalues of the adjoint action of $l_1H_2 + l_2H_2$ on the non-compact root spaces $\mathfrak{g}_2^-$ and compact root spaces $\mathfrak{g}_2^+$ are*

$$\begin{cases} 0, 0, \pm l_1 i, \pm(l_1 + 2l_2)i \ \text{ on } \ \mathfrak{g}_2^+ \\ \pm(2l_1 + l_2)i, \pm l_2 i, \pm(l_1 + l_2)i, \pm(l_1 - l_2)i \ \text{ on } \ \mathfrak{g}_2^-. \end{cases}$$

PROOF. The proof is by explicit computation using, in the above notation,

$$\mathfrak{g}_2^- = \{ \left(\begin{smallmatrix} A & 0 \\ 0 & C \end{smallmatrix}\right) : A = -{}^tA, C = -{}^tC \text{ and the equations of Step 3 are satisfied} \}$$

$$\mathfrak{g}_2^+ = \{ \left(\begin{smallmatrix} 0 & B \\ {}^tB & 0 \end{smallmatrix}\right) : B \text{ satisfies the equations of Step 3.} \}$$ □

For later reference we note that

$$l_1 H_1 + l_2 H_2 = \begin{pmatrix} l_1 E & 0 & 0 & 0 \\ 0 & 0 & 0 & 0 \\ 0 & 0 & (l_1+l_2)E & 0 \\ 0 & 0 & 0 & l_2 E \end{pmatrix}.$$

We also note that: *The equations of Step 3 determine A from C.*

(IV.D.4) COROLLARY: *The dimensions of the Lie algebra $\mathfrak{h}_\varphi$ of the isotropy group of the polarized Hodge structures are given by*

$$\begin{cases} \dim \mathfrak{h}_\varphi = 2 \text{ if, and only if, } l_1 \neq 0, l_1 + 2l_2 \neq 0 \\ \dim \mathfrak{h}_\varphi = 4 \text{ if, and only if, } l_1 = 0 \text{ or } l_1 \neq 0, l_1 = -2l_2. \end{cases}$$

We now consider the individual cases.

(i) $l_1 \neq 0, l_2 + 2l_2 \neq 0$.

Recalling that A is determined by C, we see that $C = \begin{pmatrix} 0 & a & 0 & 0 \\ -a & 0 & 0 & 0 \\ 0 & 0 & 0 & b \\ 0 & 0 & -b & 0 \end{pmatrix}$ where a, b are arbitrary, and thus $\dim \mathfrak{h}_\varphi = 2$ and

$$\mathfrak{h}_\varphi = \mathfrak{t};$$

i.e., H_φ is a maximal torus T.

(ii) $l_1 = 0$.

Then $\mathfrak{h}_\varphi$ is determined by the C's in the span of $\left(\begin{smallmatrix} E & 0 \\ 0 & 0 \end{smallmatrix}\right)$, $\left(\begin{smallmatrix} 0 & 0 \\ 0 & E \end{smallmatrix}\right)$, $\left(\begin{smallmatrix} 0 & E \\ E & 0 \end{smallmatrix}\right)$, $\left(\begin{smallmatrix} 0 & I \\ -I & 0 \end{smallmatrix}\right)$. Thus $\dim \mathfrak{h}_\varphi = 4$, and in fact

$$\mathfrak{h}_\varphi \cong \mathrm{sl}_2 \oplus \mathfrak{a}$$

where $\mathfrak{a}$ is a 1-dimensional abelian sub-algebra.

(iii) $l_1 \neq 0, l_1 = -2l_2$.

Then $\mathfrak{h}_\varphi$ is determined by the C's in the span of

$$\begin{pmatrix} E & 0 \\ 0 & 0 \end{pmatrix}, \begin{pmatrix} 0 & 0 \\ E & 0 \end{pmatrix}, \left(\begin{array}{cc|cc} 0 & 0 & 0 & 1 \\ 0 & 0 & 1 & 0 \\ \hline 0 & -1 & 0 & 0 \\ -1 & 0 & 0 & 0 \end{array}\right), \left(\begin{array}{cc|cc} 0 & 0 & 1 & 0 \\ 0 & 0 & 0 & -1 \\ \hline -1 & 0 & 0 & 0 \\ 0 & 1 & 0 & 0 \end{array}\right).$$

Thus $\dim \mathfrak{h}_\varphi = 4$, and again

$$\mathfrak{h}_\varphi \cong \mathrm{sl}_2 \oplus \mathfrak{a}$$

as above, but where the sl_2 and $\mathfrak{a}$ are *different* sub-algebras of $\mathfrak{g}_2$ than in case (ii).

Finally, we consider a few specific cases of Hodge representations for the standard 7-dimensional G_2-module.

(a) $l_1 = 0, l_2 = 2$.

The eigenvalues of $2H_2$ are $0, 0, 0, \pm 2i, \pm 2i$. Thus we obtain a polarized Hodge structure of *weight two* and *Hodge numbers* $h^{2,0} = 2$, $h^{1,1} = 3$.[14]

(b) $l_1 = 4, l_2 = -2$.

This gives a polarized Hodge structure of *weight four* and with *Hodge numbers* $h^{4,0} = 1$, $h^{3,1} = 2$, $h^{2,2} = 1$.

(c) $l_1 = 4, l_2 = 6$.

This gives a polarized Hodge structure of *weight six* and with all *Hodge numbers* $h^{p,6-p} = 1$.

We shall revisit (a) and (c) in Section IV.F when we discuss the various G_2-Mumford-Tate domains.

From a more abstract perspective, the Vogan diagram for the non-compact real form of G_2 is

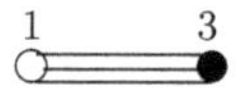

thus,

$$\psi(\alpha_1) = 0, \quad \psi(\alpha_2) = 1.$$

The weight lattice equals the root lattice; i.e., $P = R$ so there is only the adjoint form. We have

$$\mathfrak{l}_\varphi(\alpha_1) \equiv 0, \quad \mathfrak{l}_\varphi(\alpha_2) \equiv 2 \pmod 4$$

and this is equivalent to the computation above.

[14]This leads to a 5-dimensional Mumford-Tate domain in the 8-dimensional period domain — see Section IV.F, especially (IV.F.13).

Example: $\mathbf{E_6}$. There are four real forms for E_6, of which two (EI and EIV) have folded Vogan diagrams and hence do not contain a compact maximal torus, which rules them out as Mumford-Tate groups. The other two have Vogan diagrams

EII

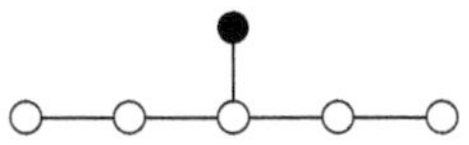

EIII

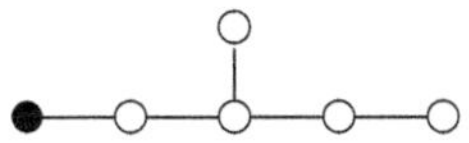

and we have that $P/R \cong \mathbb{Z}/3\mathbb{Z}$. Each real form of E_6 has two connected Lie groups associated with it — the adjoint form and the simply-connected form. Because 3 is invertible in $\mathbb{Z}/4\mathbb{Z}$, Ψ extends uniquely to a map

$$P \xrightarrow{\tilde{\Psi}} \mathbb{Z}/4\mathbb{Z}.$$

For $\lambda \in P$, $\lambda \notin R$ we have that $\mathbb{Z}\lambda + R = P$ and any $\mathfrak{l}_\varphi$ in the lattice for the simply connected form gives a polarized Hodge structure if, and only if,

$$\mathfrak{l}_\varphi \equiv \tilde{\Psi} \pmod 4.$$

The map $-w_0$ acts on the simple roots by the one non-trivial automorphism of the Dynkin diagram, flipping it horizontally. Thus if the highest weight λ of $V_\mathbb{C}$ is not in R, the representation is complex, while if $\lambda \in R$, the representation is real. A table of the 72 roots as linear combinations of $\alpha_1, \ldots, \alpha_6$ is on page 268 of [FH].

Example: $\mathbf{F_4}$. There are two real forms:

1 1 2 2

FI

1 1 2 2

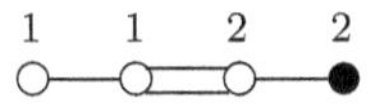

FII

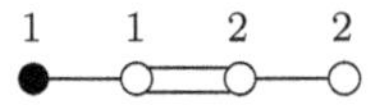

(see [K, p. 416]). We have $P = R$ so there is only the adjoint form in each case. For FI,

$$\mathfrak{l}_\varphi(\alpha_i) \equiv \begin{cases} 0 & i = 1, 2, 3 \\ 2 & i = 4 \end{cases} \pmod 4.$$

For FII,

$$\mathfrak{l}_\varphi(\alpha_i) \equiv \begin{cases} 2 & i = 1 \\ 0 & i = 2, 3, 4 \end{cases} \pmod 4.$$

There is a table of the roots as linear combinations of $\alpha_1, \alpha_2, \alpha_3, \alpha_4$ on page 274 of [FH]. There are 48 roots.

IV.E CHARACTERIZATION OF MUMFORD-TATE GROUPS

In this section we will prove the following

(IV.E.1) THEOREM: *The adjoint group of a simple $\mathbb{Q}$-algebraic group M is a Mumford-Tate group if, and only if, M contains an anisotropic maximal torus.*

From the theory of linear algebraic groups [Bor], it follows that M contains an anisotropic maximal torus if, and only if, the associated real Lie group $M(\mathbb{R})$ contains a compact maximal torus T.

PROOF. The implication one way is given in Proposition (IV.A.2). To prove the converse we shall assume that $M(\mathbb{R})$ is simple. In the general case, $\mathfrak{m}_\mathbb{R}$ is a direct sum of simple Lie algebras. Analysis of the proof below shows that the argument extends to this setting.

Under the hypothesis, a compact maximal torus T in the adjoint group $M(\mathbb{R})$ has

$$T \cong \mathfrak{t}/\Lambda, \qquad \Lambda \cong \mathrm{Hom}(R, \mathbb{Z}).$$

Because the Cartan involution θ for $\mathfrak{m}_\mathbb{R}$ is a Lie algebra homomorphism, the map $\mathfrak{r} \xrightarrow{\psi} \mathbb{Z}/2\mathbb{Z}$ defined by

$$\psi(\alpha) = \begin{cases} 0 & \text{if } \alpha \text{ is a compact root} \\ 1 & \text{if } \alpha \text{ is a non-compact root} \end{cases}$$

extends linearly to a group homomorphism $R \xrightarrow{\psi} \mathbb{Z}/2\mathbb{Z}$, and we take

$$\Psi(\alpha) = \begin{cases} 0 & \text{if } \psi(\alpha) = 0 \\ 2 & \text{if } \psi(\alpha) = 1. \end{cases}$$

We saw in Section IV.B that

$$\begin{cases} B(e_\alpha, \overline{e}_\alpha) > 0 & \text{if } \alpha \text{ is a non-compact root; i.e., } \psi(\alpha) = 1 \\ B(e_\alpha, \overline{e}_\alpha) < 0 & \text{if } \alpha \text{ is a compact root; i.e., } \psi(\alpha) = 0. \end{cases}$$

Thus $\mathfrak{l}_\varphi \in \Lambda$ gives a polarized Hodge structure for $(\mathfrak{m}_\mathbb{R}, B)$ if, and only if,

$$\begin{cases} \mathfrak{l}_\varphi(\alpha) \equiv 0 \pmod 4 & \alpha \text{ compact root} \\ \mathfrak{l}_\varphi(\alpha) \equiv 2 \pmod 4 & \alpha \text{ non-compact root.} \end{cases}$$

The latter condition is $\mathfrak{l}_\varphi \equiv \Psi \pmod 4$. Thus any lifting

$$\begin{array}{ccc} & & \mathbb{Z} \\ & {\scriptstyle \mathfrak{l}_\varphi} \nearrow & \downarrow \\ R & \xrightarrow{\Psi} & \mathbb{Z}/4\mathbb{Z} \end{array}$$

gives a polarized Hodge structure for $(\mathfrak{m}_\mathbb{R}, B)$, and these are the only co-characters that do so. Once we have a Hodge representation then by (IV.A.9), (IV.A.10) we obtain a realization of the adjoint group M_a as a Mumford-Tate group. □

Finally, as noted in Section IV.D, each of the exceptional Lie algebras arises from a $\mathbb{Q}$-algebraic group having a real form containing a compact maximal torus.

We next ask:

> *If M is of an absolutely simple $\mathbb{Q}$-algebraic group having an anisotropic maximal torus, when does an irreducible representation $M \to \mathrm{GL}(V)$ lead to a Hodge representation?*

In part III of Section IV.A we have discussed the aspects of the theory of representations of $\mathbb{Q}$-algebraic groups M that are relevant to the study of Mumford-Tate groups. In particular we discussed there how a dominant weight gives rise to an irreducible representation defined over $\mathbb{Q}$, and that these representations of M have invariant bilinear forms defined over $\mathbb{Q}$. This allows us to concentrate on the irreducible representations of $M(\mathbb{R})$, as that is where the circle giving Hodge structures lives. We thus pass to the real representation and let

$$V_\mathbb{R} \otimes_\mathbb{R} \mathbb{C} = \begin{cases} U & \text{real case} \\ U \oplus \check{U} & \text{complex or quaternionic case} \end{cases}$$

and denote by λ the highest weight of U. Let $\delta \in \mathbb{Z}^+$ be the minimal positive integer with the property $\delta\lambda \in R$. Define

$$P' = \mathbb{Z}\lambda + R;$$

We note that

$$P \supseteq P' \supseteq R, \quad \Lambda = \mathrm{Hom}(P', \mathbb{Z}).$$

(IV.E.2) THEOREM: *There exists* $\mathfrak{l}_\varphi \in \Lambda$ *and an invariant form* Q *giving polarized Hodge structure if, and only if,* $\Psi(\delta\lambda) \equiv \delta m \pmod 4$ *for some integer* m.

PROOF. We have already seen that $\mathfrak{l}_\varphi \in \Lambda$ gives a polarized Hodge structure for V for some invariant Q if, and only if, $\mathfrak{l}_\varphi|_R \equiv \Psi \pmod 4$. In order to extend Ψ from R to P' we need to define $\mathfrak{l}_\varphi(\lambda)$, and for this extension we must have

$$\delta\mathfrak{l}_\varphi(\lambda) \equiv \Psi(\delta\lambda) \pmod 4.$$

This completes the proof. □

(IV.E.3) COROLLARY: *If* δ *is odd, we can always do this. If* $2 \mid \delta$ *but* $4 \nmid \delta$, *we can do this because* $2 \mid \Psi(\delta\lambda)$ *by the definition of* Ψ. *If* $4 \mid \delta$, *we exactly need that* $\Psi(\delta\lambda) \equiv 0 \pmod 4$.

The integer δ defined above divides the determinant D of the Cartan matrix. The values of D for the simple Lie algebras are:

	D	P/R
A_r	$r+1$	$\mathbb{Z}/(r+1)\mathbb{Z}$
B_r	2	$\mathbb{Z}/2\mathbb{Z}$
C_r	2	$\mathbb{Z}/2\mathbb{Z}$
D_r	4	$\mathbb{Z}/4\mathbb{Z}$ (r odd) or $\mathbb{Z}/2\mathbb{Z} \oplus \mathbb{Z}/2\mathbb{Z}$ (r even)
E_6	3	$\mathbb{Z}/3\mathbb{Z}$
E_7	2	$\mathbb{Z}/2\mathbb{Z}$
E_8	1	$\{e\}$
F_4	1	$\{e\}$
G_2	1	$\{e\}$

Thus the only possible cases where a representation can fail to have a φ that induces a polarized Hodge structure are A_{4k-1}, D_{2k-1}. It is not difficult to show that the condition

$$\Psi(\delta\lambda) \equiv \delta m \pmod 4 \quad \text{for some } \ m \in \mathbb{Z}$$

is not vacuous for some real form in either case.

Example: $\mathbf{A_3}$. Here

$$\omega_1 = \tfrac{1}{4}(3\alpha_1 + 2\alpha_2 + \alpha_3).$$

Now

$$\Psi(4\omega_1) = 3\Psi(\alpha_1) + 2\Psi(\alpha_2) + \Psi(\alpha_3).$$

For $\mathrm{su}(1,3)$ we have that $\Psi(4\omega_1) = 6$, which is not $\equiv 0 \pmod 4$. Therefore we do not get a polarized Hodge structure in this case. For $\mathrm{su}(3,1)$, $\Psi(4\omega_1) = 2$, so again no polarized Hodge structure exists. For $\mathrm{su}(2,2)$, $\Psi(4\omega_1) = 4$; consequently polarized Hodge structures do exist, and in this case $\mathfrak{l}_\varphi(\omega_1)$ can assume any value.

Example: $\mathbf{D_3}$. Here

$$\omega_3 = \tfrac{1}{2}\alpha_1 + \tfrac{1}{4}\alpha_2 + \tfrac{3}{4}\alpha_3.$$

So

$$\Psi(4\omega_3) = 2\Psi(\alpha_1) + \Psi(\alpha_2) + 3\Psi(\alpha_3).$$

For $\mathrm{so}(2,4)$, $\Psi(4\omega_3) = 4 \equiv 0 \pmod 4$, so here a polarized Hodge structure is possible.

For $\mathrm{so}^*(6)$, $\Psi(4\omega_3) = 6 \not\equiv 0 \pmod 4$, so again there is no polarized Hodge structure.

Indeed, for D_{2k-1}, and dominant weight any $\lambda \in P$ of order 4 in P/R, it is only $\mathrm{so}^*(4k-2)$ for which there is no polarized Hodge structure.

A complete list of non-compact real forms of simple Lie algebras that have Hodge representations is the following:

$$\begin{array}{ll}
A_r & \mathrm{su}(p,q),\ p+q=r+1,\ 0 \le p,q \le r+1;\ \mathrm{sl}(2,\mathbb{R}) \\
B_r & \mathrm{so}(2p,2q+1),\ p+q=r,\ 0 \le p,q \le r \\
C_r & \mathrm{sp}(p,q),\ p+q=r,\ 0 \le p,q \le r;\ \mathrm{sp}(r,\mathbb{R}) \\
D_r & \mathrm{so}(2p,2q),\ p+q=r,\ 0 \le p,q \le r;\ \mathrm{so}^*(2r) \\
E_6 & \text{EII, EIII} \\
E_7 & \text{EV, EVI, EVII} \\
E_8 & \text{EVIII, EIX} \\
F_4 & \text{FI, FII} \\
G_2 & \text{G.}
\end{array}$$

Of these, the only ones that have representations that cannot be Hodge representations are $\mathrm{su}(p,4k-p)$, $k \geqq 1$ and p odd and also $\mathrm{so}^*(4k-2)$, $k \geqq 2$.

The exceptional groups are from Cartan's classification. A complete list of the real forms of simple Lie algebras that do *not* admit Hodge representations is:

$$\mathrm{sl}(m,\mathbb{R}) \text{ for } n \ge 3$$

$$\mathrm{sl}(m,\mathbb{H}) \text{ for } n \ge 2$$

$$\text{EI}$$

$$\text{EIV.}$$

Since Mumford-Tate groups are $\mathbb{Q}$-algebraic groups, a natural question is which of the real forms that have Hodge representations are in fact the real forms of the Lie algebras of $\mathbb{Q}$-algebraic groups. To partially answer this we have the following

Observation: *For $M_{a,\mathbb{R}}$ the connected adjoint group of any real form of a simple Lie algebra, there is a simple $\mathbb{Q}$-algebraic group M whose associated real Lie group is $M_{a,\mathbb{R}}$.*

Of course there will in general be many distinct $\mathbb{Q}$-algebraic groups having the same associated real Lie group. But the observation is that, up to isogeny, the map of sets

$$\left\{ \begin{array}{c} \text{simple} \\ \mathbb{Q}\text{-algebraic} \\ \text{groups} \end{array} \right\} \longrightarrow \left\{ \begin{array}{c} \text{simple} \\ \text{real Lie} \\ \text{groups} \end{array} \right\}$$

is surjective.

PROOF OF THE OBSERVATION. Each simple, real Lie algebra $\mathfrak{m}_{\mathbb{R}}$ is the $\mathbb{R}$-form of a simple Lie algebra $\mathfrak{m}$ over $\mathbb{Q}$. Choose a basis $X_1, \dots, X_d$ for $\mathfrak{m}$; then the brackets $[X_i, X_j]$ are $\mathbb{Q}$-linear combinations of the X_k's. Let

$$\begin{aligned} \mathrm{Aut}_{\mathrm{LA}}(\mathfrak{m}_{\mathbb{R}}) &= \{A \in \mathrm{GL}(\mathfrak{m}_{\mathbb{R}}) : [AX, AY] = [X, Y] \text{ for all } X, Y \in \mathfrak{m}_{\mathbb{R}}\} \\ &= \{A \in \mathrm{GL}(\mathfrak{m}_{\mathbb{R}}) : [AX_i, AX_j] = [X_i, X_j]\}\,. \end{aligned}$$

These are quadratic equations over $\mathbb{Q}$ so that $\mathrm{Aut}_{\mathrm{LA}}(\mathfrak{m}_{\mathbb{R}})$ are the real points of a $\mathbb{Q}$-algebraic group.

From [K], page 102, for any Lie algebra $\mathfrak{m}_{\mathbb{R}}$

- the Lie algebra of $\mathrm{Aut}_{\mathrm{LA}}(\mathfrak{m}_{\mathbb{R}})$ is

$$\mathrm{Der}_{\mathbb{R}}(\mathfrak{m}_{\mathbb{R}}) := \left\{ \begin{array}{l} A \in \mathrm{End}(\mathfrak{m}_{\mathbb{R}}) : \\ A[X, Y] = [AX, Y] + [X, AY] \end{array} \right\}$$

- for $\mathfrak{m}_{\mathbb{R}}$ semi-simple,

$$\mathrm{Der}_{\mathbb{R}}(\mathfrak{m}_{\mathbb{R}}) = \mathrm{Int}_{\mathbb{R}}(\mathfrak{m}_{\mathbb{R}}) := \{\mathrm{ad}\, X : X \in \mathfrak{m}_{\mathbb{R}}\}.$$

It follows that $\mathrm{Aut}_{\mathrm{LA}}(\mathfrak{m}_{\mathbb{R}})$ has $M_{a,\mathbb{R}}$ as identity component. □

We now turn to a result purely in representation theory that underlies the above proof. As it may be of independent interest we shall isolate it here.

(IV.E.4) THEOREM: *For the irreducible $\mathfrak{m}_{\mathbb{R}}$-module V^λ associated to a dominant weight λ that satisfies*

$$\begin{cases} w_0(\lambda) = -\lambda \\ \quad 2\lambda = \sum_i m_i \alpha_i \end{cases}$$

where $\alpha_1, \dots, \alpha_r$ are the simple roots and $m_i \in \mathbb{Z}$, we have

$$\begin{cases} V^\lambda \quad \text{is real if, and only if,} \sum_{\alpha_i \text{ compact}} m_i \equiv 0 \pmod 2 \\ V^\lambda \quad \text{is quaternionic if, and only if,} \sum_{\alpha_i \text{ compact}} m_i \equiv 1 \pmod 2. \end{cases}$$

PROOF OF (IV.E.4). We have

$$\langle \lambda, h^0 \rangle = \tfrac{1}{2} \sum_i m_i \langle \alpha_i, h^0 \rangle = \sum_i m_i$$

where the sum is over the simple roots and we recall that all $\langle \alpha_i, h^0 \rangle = 2$. In case $\mathfrak{m}_{\mathbb{R}}$ is compact, it is known (cf. [Ad], 3.50 on page 61) that the statement in the theorem is valid. Thus we have to extend that result to a general real simple Lie algebra. This will be done in three steps.

Step one: We let $\mathfrak{m}_c$ be a compact form of $\mathfrak{m}_{\mathbb{C}}$. We may assume that $\mathfrak{t} \subset \mathfrak{m}_c$. The idea is to compare the root structure of $\mathfrak{m}_{\mathbb{R}}$ and $\mathfrak{m}_c$. For this we denote by

$$\begin{cases} - : \mathfrak{m}_{\mathbb{C}} \to \mathfrak{m}_{\mathbb{C}}, & \overline{\mathfrak{m}}_{\mathbb{R}} = \mathfrak{m}_{\mathbb{R}} \\ \# : \mathfrak{m}_{\mathbb{C}} \to \mathfrak{m}_{\mathbb{C}}, & \mathfrak{m}_c^{\#} = \mathfrak{m}_c \end{cases}$$

the conjugations of $\mathfrak{m}_{\mathbb{C}}$ relative to the respective real forms $\mathfrak{m}_{\mathbb{R}}$ and $\mathfrak{m}_c$.

Now we let $e_\alpha, f_\alpha, h_\alpha \in \mathfrak{m}_{\mathbb{C}}$ be a standard triple spanning an sl_2. We set

$$D_\alpha = e_\alpha + \overline{e}_\alpha, \quad D'_\alpha = \left(\frac{1}{i}\right)(e_\alpha - \overline{e}_\alpha).$$

Together with $\mathfrak{t}$ these span $\mathfrak{m}_{\mathbb{R}}$ as α varies over the set $\mathfrak{r}$ of roots. Similarly, we set

$$D_\alpha^{\#} = e_\alpha + e_\alpha^{\#}, \quad D_\alpha'^{\#} = \left(\frac{1}{i}\right)(e_\alpha - e_\alpha^{\#}).$$

From

$$\begin{cases} B(e_\alpha, f_\alpha) & = 1 \\ B(e_\alpha, \overline{e}_\alpha) & < 0 \ \text{ for } \alpha \text{ a compact root} \\ B(e_\alpha, \overline{e}_\alpha) & > 0 \ \text{ for } \alpha \text{ a non-compact root} \end{cases}$$

we infer that

$$\overline{e}_\alpha = \begin{cases} -f_\alpha & \text{for } \alpha \text{ compact} \\ f_\alpha & \text{for } \alpha \text{ non-compact,} \end{cases}$$

while

$$e_\alpha^\# = -f_\alpha \qquad \text{for all } \alpha.$$

This gives

$$\begin{cases} D_\alpha = D_\alpha^\# \text{ and } D'_\alpha = D'^\#_\alpha & \text{for } \alpha \text{ compact} \\ D_\alpha = iD'^\#_\alpha \text{ and } D'_\alpha = -iD_\alpha^\# & \text{for } \alpha \text{ non-compact.} \end{cases}$$

Thus

$$D_\alpha, D'_\alpha \in i^{\psi(\alpha)}\mathfrak{m}_c.$$

Denote by $\mathfrak{U}(\mathfrak{m}_\mathbb{R}), \mathfrak{U}(\mathfrak{m}_c)$ the (real) universal enveloping algebras of $\mathfrak{m}_\mathbb{R}, \mathfrak{m}_c$. Then by the previous relation

$$D_{\beta_1} D_{\beta_2} \dots D_{\beta_k} \in i^{\Sigma\psi(\beta_j)}\mathfrak{U}(\mathfrak{m}_c)$$

for any roots $\beta_1, \dots, \beta_k$. This equation also holds if some of the D_{β_j} are replaced by D'_{β_j}.

Let

$$V_\mathbb{R} \otimes_\mathbb{R} \mathbb{C} = \begin{cases} W^\lambda & \text{(real case)} \\ W^\lambda \oplus \breve{W}^\lambda & \text{(quaternionic case)} \end{cases}$$

where W^λ is irreducible over $\mathbb{C}$. Using the assumption $w_0(\lambda) = -\lambda$ we infer that $-\lambda$ is the lowest weight. Then by the irreducibility of W^λ as an $\mathfrak{m}_\mathbb{C}$-module, we have

$$2\lambda = \beta_1 + \cdots + \beta_k \tag{IV.E.5}$$

for roots $\beta_1, \dots, \beta_k$. For $v_{\pm\lambda}$ a non-zero $\pm\lambda$-weight vector we have, after scaling, for some choice of roots β_i satisfying (IV.E.5),

$$D_{\beta_1} \dots D_{\beta_k} v_{-\lambda} = v_\lambda.$$

Step two: Let

$$W^\lambda = \bigoplus_\omega W_\omega$$

be the decomposition of W^λ into weight spaces W_ω where $\omega \in P$. We have $V^\lambda = \operatorname{Res}_{\mathbb{C}/\mathbb{R}} W^\lambda$ and we set

$$V_\omega = \operatorname{Res}_{\mathbb{C}/\mathbb{R}} W_\omega.$$

In the diagram

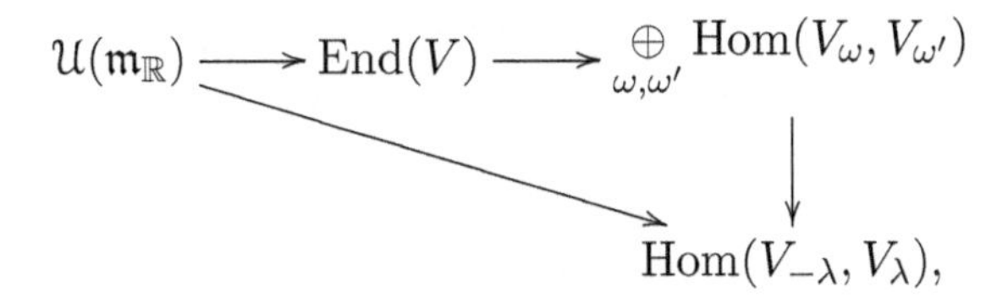

where the vertical map is the projection, we have from $V_\lambda = \operatorname{Res}_{\mathbb{C}/\mathbb{R}} W_\lambda$ that multiplication by i gives maps

$$V_\lambda \xrightarrow{i} V_\lambda, \quad V_{-\lambda} \xrightarrow{i} V_{-\lambda}.$$

In step one we have shown that

$$\text{(IV.E.6)} \qquad \begin{cases} \mathcal{U}(\mathfrak{m}_{\mathbb{R}}) \to \operatorname{Hom}(V_{-\lambda}, V_\lambda) \oplus \operatorname{Hom}(V_\lambda, V_{-\lambda}) \text{ is non-trivial} \\ \operatorname{Im}(\mathcal{U}(\mathfrak{m}_{\mathbb{R}})) = i^{\psi(2\lambda)} \operatorname{Im}(\mathcal{U}(\mathfrak{m}_c)). \end{cases}$$

We now analyze what happens in the two cases.

Real case. We will show that

(IV.E.7) *The image of* $\mathcal{U}(\mathfrak{m}_{\mathbb{R}})$ *in* (IV.E.6) *lies in*

$$\{(A, B) : A \in \operatorname{Hom}(V_{-\lambda}, V_\lambda),\ B \in \operatorname{Hom}(V_\lambda, V_{-\lambda}) : BA = cI, c > 0\}.$$

PROOF. Dropping the upper λ index for convenience, we have

$$W = V \otimes_{\mathbb{R}} \mathbb{C}$$

so

$$\operatorname{Res}_{\mathbb{C}/\mathbb{R}} W \cong V \oplus V$$

where the action of i is by $\left(\begin{smallmatrix} 0 & I \\ -I & 0 \end{smallmatrix}\right)$. The action of $\mathcal{U}(\mathfrak{m}_{\mathbb{R}})$ preserves each factor in this direct sum decomposition. From

$$V_\lambda \otimes_{\mathbb{R}} \mathbb{C} = W_\lambda \oplus W_\lambda,$$

if S^1 acts on $V_\lambda \cong \mathbb{R}^2$ by $\varphi(\theta) := \left(\begin{smallmatrix} \cos\theta & -\sin\theta \\ \sin\theta & \cos\theta \end{smallmatrix}\right)$, then (cf. example (IV.A.4))

$$W_\lambda = \mathbb{C}\begin{pmatrix} 1/2 \\ -i/2 \end{pmatrix}, \quad W_{-\lambda} = \mathbb{C}\begin{pmatrix} 1/2 \\ i/2 \end{pmatrix}$$

since $\varphi(\theta)$ acts in W_λ by $e^{i\theta}$ and on $W_{-\lambda}$ by $e^{-i\theta}$. Now $\mathcal{U}(\mathfrak{m}_{\mathbb{R}})$ acts on V_λ by 2×2 real matrices A:

$$A\begin{pmatrix} 1/2 \\ -i/2 \end{pmatrix} = u\begin{pmatrix} 1/2 \\ -i/2 \end{pmatrix} + v\begin{pmatrix} 1/2 \\ i/2 \end{pmatrix} \qquad u, v \in \mathbb{C}$$

and by conjugating

$$A\begin{pmatrix}1/2\\ i/2\end{pmatrix} = \overline{v}\begin{pmatrix}1/2\\ -i/2\end{pmatrix} + \overline{u}\begin{pmatrix}1/2\\ i/2\end{pmatrix},$$

which implies that the corresponding elements in $\mathrm{Hom}(W_\lambda, W_{-\lambda})$ and $\mathrm{Hom}(W_{-\lambda}, W_\lambda)$ are v and $\overline{v}$ respectively. Their composition is $|v|^2\mathrm{Id}$. This establishes (IV.E.7).

Quaternionic Case. We will show that

(IV.E.8) *The image of* $\mathfrak{U}(\mathfrak{m}_\mathbb{R})$ *in* (IV.E.6) *lies in*

$$\{(A,B) : A \in \mathrm{Hom}(V_{-\lambda}, V_\lambda),\ B \in \mathrm{Hom}(V_\lambda, V_{-\lambda}) : BA : cI, c < 0\}.$$

Proof of (IV.E.8). We have

$$W_\lambda \oplus W_{-\lambda} = \mathrm{Res}_{\mathbb{H}/\mathbb{C}}\, \mathbb{W}_\lambda$$

where $\mathbb{W}_\lambda$ is a 1-dimensional $\mathbb{H}$-vector space. The action of $\mathfrak{U}(\mathfrak{m}_R)$ is contained in the left action of the quaternions

$$W_\lambda \cong 1\cdot\mathbb{C}, \quad W_{-\lambda} \cong j\cdot\mathbb{C}$$

and

$$V_\lambda = \mathbb{R} \oplus \mathbb{R}i, \qquad V_{-\lambda} = \mathbb{R}j \oplus \mathbb{R}k.$$

The action on $\mathbb{H}$ on the left is

$$V_{-\lambda} \xrightarrow{aj+bk} V_\lambda, \qquad V_\lambda \xrightarrow{aj+bk} V_{-\lambda}.$$

Then

$$(aj+bk)^2 = a^2j^2 + ab(jk+kj) + b^2k^2 = (-a^2-b^2)\cdot 1,$$

which establishes (IV.E.8).

Step three: We apply the results of step two to $\mathfrak{m}_\mathbb{R}$ and $\mathfrak{m}_c$ and compare. Note that

$$BA = cI \Rightarrow (iB)\cdot(iA) = -cI.$$

The $D_{\beta_1}\dots D_{\beta_k} \in \mathfrak{U}(\mathfrak{m}_\mathbb{R})$ as constructed in step one gives a non-zero element $(A,B) \in \mathrm{Hom}(V_{-\lambda}, V_\lambda) \oplus \mathrm{Hom}(V_\lambda, V_{-\lambda})$, and

$$BA = cI \ \text{ where } \begin{cases} c > 0 & \text{in the real case} \\ c < 0 & \text{in the quaternionic case.} \end{cases}$$

By step one

$$i^{\Sigma\psi(\beta_j)} D_{\beta_1} \dots D_{\beta_k} \in \mathfrak{U}(\mathfrak{m}_c).$$

Thus

$$\begin{cases} V^\lambda \text{ is real for } \mathfrak{m}_c \Leftrightarrow (-1)^{\Sigma\psi(\beta_j)} c > 0 \\ V^\lambda \text{ is quaternionic for } \mathfrak{m}_c \Leftrightarrow (-1)^{\Sigma\psi(\beta_j)} c < 0. \end{cases}$$

By the result referenced above from [Ad], together with [GW], page 160, we have

$$\begin{cases} V^\lambda \text{ real for } \mathfrak{m}_c \text{ if, and only if, } \langle \lambda, h^0 \rangle \equiv 0(2) \Leftrightarrow (-1)^{\langle \lambda, h^0 \rangle} > 0 \\ V^\lambda \text{ quaternionic for } \mathfrak{m}_c \text{ if, and only if, } \langle \lambda, h^0 \rangle \equiv 1(2) \Leftrightarrow (-1)^{\langle \lambda, h^0 \rangle} < 0. \end{cases}$$

We now may complete the proof of Theorem (IV.E.4). From

$$\beta_1 + \dots + \beta_k = 2\lambda = m_1\alpha_1 + \dots + m_r\alpha_r$$

we have

$$(-1)^{\Sigma\psi(\beta_j)} = (-1)^{\Sigma m_j \psi(\alpha_j)},$$

while as above

$$\langle \lambda, h^0 \rangle = \tfrac{1}{2} \sum_j m_j \langle \alpha_j, h^0 \rangle = \sum_j m_j.$$

Thus

$$(-1)^{\Sigma m_j \psi(\alpha_j)} = (-1)^{\langle \lambda, h^0 \rangle - \Sigma_{\alpha_j\text{-compact}} m_j},$$

and it follows that

$$V^\lambda \text{ is real for } \mathfrak{m}_\mathbb{R} \Leftrightarrow c > 0$$

while

$$\begin{cases} V^\lambda \text{ is real for } \mathfrak{m}_c \Leftrightarrow (-1)^{\langle \lambda, h^0 \rangle} > 0, \text{ and equivalently} \\ V^\lambda \text{ is real for } \mathfrak{m}_c \Leftrightarrow (-1)^{\langle \lambda, h^0 \rangle - \Sigma_{\alpha_j\text{-compact}} m_j} c > 0. \end{cases}$$

In summary,

$$\begin{cases} c > 0 \Leftrightarrow \Sigma_{\alpha_j\text{-compact}} m_j \equiv 0(2) \\ c < 0 \Leftrightarrow \Sigma_{\alpha_j\text{-compact}} m_j \equiv 1(2), \end{cases}$$

which was to be proved. This completes the proof of (IV.E.4). □

The following is a simple application of (IV.E.8).

The Hermitian symmetric case. We will analyze the possible realizations of an Hermitian symmetric domain as a Hodge domain as defined in Section IV.F. In general the circle $\varphi : S^1 \to M(\mathbb{R})$ is given by

$$\mathfrak{l}_\varphi \in \Lambda = \mathrm{Hom}(R, \mathbb{Z})$$

lifting $\Psi : R \to \mathbb{Z}/4\mathbb{Z}$. In the Hermitian symmetric case we have, except when $\mathfrak{m}_\mathbb{R} = \mathrm{so}(2,2)$,

$$\begin{cases} \mathfrak{m}_\mathbb{R} &= \mathfrak{k} \oplus \mathfrak{p} \\ \mathfrak{k} &= \mathbb{R} \oplus \mathfrak{k}_{\mathrm{ss}} \end{cases}$$

where $\mathfrak{k}_{\mathrm{ss}}$ is the semi-simple part of $\mathfrak{k}$. Since $K = Z(\varphi(S^1))$ we must have that $\mathfrak{l}_\varphi(\alpha) = 0$ for all compact roots α, and thus

$$\mathfrak{l}_\varphi\Big((\text{root lattice of } \mathfrak{k}_{\mathrm{ss}}) \otimes \mathbb{Q} \cap R\Big) = 0.$$

There is still one generator α of R where we will have $\mathfrak{l}_\varphi(\alpha) \neq 0$, and the only constraint is

$$\mathfrak{l}_\varphi(\alpha) \equiv 2 \ (\mathrm{mod}\, 4).$$

The possibilities for the IPR distribution W are

$$\begin{cases} \mathfrak{l}_\varphi(\alpha) = \pm 2 \Rightarrow W = TD_{M_\varphi} \\ \mathfrak{l}_\varphi(\alpha) \neq \pm 2 \Rightarrow W = 0. \end{cases}$$

The first is the unconstrained case; the second is the rigid case.

The case of $\mathrm{so}(2,2)$ where $\mathfrak{k} = \mathbb{R}^2$ is more interesting. Then $R \cong \mathbb{Z}^2 = \mathrm{span}_\mathbb{Z}\{\alpha_1, \alpha_2\}$, and the constraints are $\mathfrak{l}_\varphi(\alpha_j) \equiv 2 \ (\mathrm{mod}\, 4)$ for $j = 1, 2$. We then have the possibilities for $k_j = \mathfrak{l}_\varphi(H_j)$ where $H_1 = \left(\begin{smallmatrix} E & 0 \\ 0 & 0 \end{smallmatrix}\right)$, $H_2 = \left(\begin{smallmatrix} 0 & 0 \\ 0 & E \end{smallmatrix}\right)$ and $E = \left(\begin{smallmatrix} 0 & -1 \\ 1 & 0 \end{smallmatrix}\right)$ and for the standard representation

$$\mathfrak{l}_\varphi = \begin{pmatrix} k_1 E & 0 \\ 0 & k_2 E \end{pmatrix} \in \Lambda$$

has eigenvalues $\pm k_1 i, \pm k_2 i$ where $k_1 \equiv 0 \ (\mathrm{mod}\, 4)$, and $k_2 \equiv 2 \ (\mathrm{mod}\, 4)$. For example

$$\begin{cases} k_1 = 0, k_2 = 2 & W = TD_{M_\varphi} \\ k_1 = 4, k_2 = 2 & \mathrm{rank}\, W = 1 \\ k_1 = 8, k_2 = 2 & W = (0). \end{cases}$$

In the first case have have a Mumford-Tate domain where

$$n = 2 \text{ and } h^{2,0} = 1, \ \ h^{1,1} = 2$$

(think of K3's with Picard number $\rho = 18$). In the second case we have a Mumford-Tate domain where

$$n = 4 \text{ and } h^{4,0} = h^{3,1} = 1, \ \ h^{2,2} = 0.$$

The exceptional Hermitian symmetric domains EIII **and** EVII. It can be shown that for each of these and for the adjoint representation, there is a *unique* polarized Hodge structure $(\mathfrak{m}, B, \varphi)$. In both cases the weight is two and the respective Hodge numbers are

$$\begin{cases} \text{EIII}: & h^{2,0} = 16, \ h^{1,1} = 46 \\ \text{EVII}: & h^{2,0} = 27, \ h^{1,1} = 79. \end{cases}$$

Also, in both cases the IPR is trivial.

We conclude this section with the following

Discussion. From theorem (IV.E.1) we have that, up to isogeny, Mumford-Tate groups M are exactly in those having the properties:

(i) M is a reductive, $\mathbb{Q}$-algebraic linear group; and

(ii) $M(\mathbb{R})$ contains a compact maximal torus T.

Moreover, when M is semi-simple and has the property (ii), the adjoint group M_a is a Mumford-Tate group.

In the 1950's, H. C. Wang [W] studied homogeneous complex manifolds and introduced the class of *Kähler C-spaces* $X = M_c(\mathbb{R})/H_c$ where $M_c(\mathbb{R})$ is a compact, semi-simple Lie group and $H_c \subset M_c(\mathbb{R})$ is the centralizer of a non-trivial torus. These are exactly rational homogeneous varieties of the form $X = M(\mathbb{C})/P$ where $P \subset M(\mathbb{C})$ is a parabolic subgroup with $P \cap M_c(\mathbb{R}) = H_c$. In [GS] the class of homogeneous complex manifolds of the form $D = M(\mathbb{R})/H$ where $M(\mathbb{R})$ is a real, semi-simple Lie group and $H \subset M(\mathbb{R})$ is a compact centralizer of a non-trivial torus was studied. There it was noted that each such D had a compact "dual" X as above in which D is embedded as an open $M(\mathbb{R})$-orbit.[15] From the above result, it follows that the spaces studied in [GS] are exactly the class of Hodge domains, the latter having the particular complex structure and canonical exterior differential system specified.

In the works of Harish-Chandra [HC1], [HC2], it is proved that the above Lie groups $M(\mathbb{R})$ are exactly those that have discrete series representations

[15]See [W1] for a general discussion of the open orbits in compact homogeneous algebraic varieties of real forms of a complex semi-simple Lie algebra.

occurring in $L^2(M(\mathbb{R}))$. In [Schm2], [Schm3] these representations are realized as L^2-cohomology arising from homogeneous vector bundles over D. The groups $M(\mathbb{R})$ are also those for which one may hope that the discrete series occur in $L^2(\Gamma\backslash M(\mathbb{R}))$ where $\Gamma \subset M(\mathbb{R})$ is a discrete subgroup with $\mathrm{vol}(\Gamma\backslash D) < \infty$ [WW]. Finally, because M is a $\mathbb{Q}$-algebraic group, the space $L^2(M(\mathbb{Q})\backslash M(\mathbb{A}))$ may be defined, and those automorphic representations that have as infinite component a discrete series, or a limit of such, representation of $M(\mathbb{R})$ are of particular arithmetic interest.

Reprise: In this subsection M will be a simple $\mathbb{Q}$-algebraic group and λ will be a dominant weight. A Hodge representation (M, ρ, φ) with associated dominant weight λ will be said to be *faithful* if $\rho : M \to \mathrm{GL}(V)$ is injective. In Part III of Section IV.A, we have discussed how one may associate to the pair (M, λ) an irreducible representation, defined over $\mathbb{Q}$, of M. Moreover, any irreducible representation of M arises in this way. We then say that λ is the dominant weight associated to the representation. Here we will use the results above, specifically theorems (IV.B.6) and (IV.E.2), to address the questions:

(i) *Which combinations (M, λ) give faithful Hodge representations?*

(ii) *Which λ can give Hodge representations of odd weight?*

We note that (i) will give an answer, not just one up to isogeny, to the question:

Which absolutely simple $\mathbb{Q}$-algebraic groups can be Mumford-Tate groups?

Question (ii) is interesting because we shall see that to have an odd weight Hodge representation is much more restrictive than just to have a Hodge representation.[16]

We recall that the connected simple real Lie groups correspond to the lattices between P and R. We shall denote by M a $\mathbb{Q}$-form of the corresponding real Lie group, and by Λ_M the corresponding lattice. Thus the maximal torus of $M(\mathbb{R})$ is

$$T = \mathfrak{t}/\Lambda_M.$$

When $\Lambda_M = \mathrm{Hom}_{\mathbb{Z}}(R, \mathbb{Z})$ we have that $M = M_a$ is the adjoint group, and when $\Lambda_M = \mathrm{Hom}_{\mathbb{Z}}(P, \mathbb{Z})$ $M = M_s$ is the simply connected group. A basic observation is that:

(IV.E.9) *The necessary and sufficient condition that the representation that has associated dominant weight λ be faithful on M is that*

$$\Lambda_M = \mathrm{Hom}(R + \mathbb{Z}\lambda, \mathbb{Z}).$$

[16] It is much easier to have odd weight Hodge representations in the reductive case.

We recall that in any case to have the representation defined on M is equivalent to

$$\Lambda_M \supseteq \mathrm{Hom}(R+\mathbb{Z}\lambda,\mathbb{Z}).$$

In the case when we have a reducible representation whose irreducible factors have associated dominant weights λ_i, the condition in (IV.E.9) becomes

(IV.E.10) $$\Lambda_M = \mathrm{Hom}(R+\bigoplus_i \mathbb{Z}\lambda_i,\mathbb{Z}).$$

If λ gives a faithful representation, then the representation restricted to T_M must be faithful, and this is equivalent to the pairing

$$\Lambda_M \times (R+\mathbb{Z}\lambda) \to \mathbb{Z}$$

having no left kernel, which is equivalent to

$$\Lambda_M \subseteq \mathrm{Hom}(R+\mathbb{Z}\lambda,\mathbb{Z}).$$

We next recall from theorem (IV.B.6) and its proof that the condition to have a Hodge representation with associated dominant weight λ is:

(IV.E.11) *Let δ be the smallest positive integer such that $\delta\lambda \in R$. Then*

$$\Psi(\delta\lambda) \equiv \delta m \pmod 4$$

for some $m \in \mathbb{Z}$. The parity of the corresponding polarized Hodge structure is then the parity of m.

We note that:

- δ odd implies that δ^{-1} exists (mod 4), which gives

  $$\delta^{-1}\Psi(\delta\lambda) \equiv m \pmod 4$$

  and forces m to be even;

- $\delta \equiv 2 \pmod 4$ implies that m always exists, and then m even is equivalent to $\Psi(\delta\lambda) \equiv 0 \pmod 4$ and m odd is equivalent to $\Psi(\delta\lambda) \equiv 2 \pmod 4$;

- $\delta \equiv 0 \pmod 4$ implies that m exists if, and only if, $\Psi(\delta\lambda) \equiv 0 \pmod 4$, in which case m can be even or odd.

With these general facts in hand, we now proceed to a case-by-case analysis, in order of difficulty.

Initial case: $\lambda \in R$. Then since $\operatorname{Hom}(R, \mathbb{Z}) = \Lambda_{M_s}$, $M = M_s$ is simply connected and

- *there is always a faithful Hodge representation with associated dominant weight λ;*
- *all of these representations have even Hodge weight.*

We note that this case happens whenever

$$P = R,$$

or equivalently $M_s = M_a$. The list of the simple Lie algebras for which this situation occurs is

$$E_8, F_4, G_2.$$

Remaining cases.

$P/R \cong \mathbb{Z}/2\mathbb{Z}$	A_1, B_l, C_l, E_7
$P/R \cong \mathbb{Z}/3\mathbb{Z}$	E_6, A_2
$P/R \cong \mathbb{Z}/(l+1)\mathbb{Z}$	A_l
$P/R \cong \mathbb{Z}/2\mathbb{Z} \oplus \mathbb{Z}/2\mathbb{Z}$	D_{2k}
$P/R \cong \mathbb{Z}/4\mathbb{Z}$	D_{2k+1}.

In the analysis of these cases, in several instances we shall show that Hodge representations exist for a given λ, and then give a prescription for finding the M between M_s and M_a on which the representation is faithful. Since we have already dealt with the case $\lambda \in R$, we need only consider the situation when λ is non-zero in P/R.

$\mathbf{C}_{\ell}$. A generator for P/R is

$$\lambda = \alpha_1 + 2\alpha_2 + \cdots + (l-1)\alpha_{l-1} + \left(\tfrac{1}{2}\right)\alpha_l \equiv \left(\tfrac{1}{2}\right)\alpha_l \ (\operatorname{mod} R).$$

Then $\delta = 2$, and since $\Psi(\alpha_i) = 0$ or 2 we have that

$$\begin{aligned} \Psi(2\lambda) &\equiv \Psi(\alpha_l) \ (\operatorname{mod} 4) \\ &\equiv \begin{cases} 2 & \text{for } \operatorname{sp}(2l) \\ 0 & \text{otherwise.} \end{cases} \end{aligned}$$

We conclude that λ gives a faithful Hodge representation for M_s, and this is an odd weight polarized Hodge structure for $\operatorname{Sp}(2l, \mathbb{R})$ and an even weight polarized Hodge structure for the other forms.

$\mathbf{B}_\ell$. A generator for P/R is

$$\lambda = \tfrac{1}{2}(\alpha_1 + 2\alpha_2 + \cdots + l\alpha_l).$$

Thus $\delta = 2$ and

$$\Psi(\delta\lambda) \equiv \begin{cases} 2 & \text{if the painted node in the} \\ & \text{Vogan diagram is at } 1, 3, 5, \ldots \\ 0 & \text{if the painted node is at} \\ & 2, 4, 6, \ldots, \text{ or if we} \\ & \text{have the compact form.} \end{cases}$$

Thus λ gives a faithful Hodge representation for M_s, and this is an odd weight polarized Hodge structure for $\mathrm{so}(4p+2, 2q+1)$ and an even weight polarized Hodge structure for $\mathrm{so}(4p, 2q+1)$ or for the compact form.

$\mathbf{E}_6$. Then $\delta = 3$ and $m \equiv \delta^{-1}\Psi(3\lambda) \equiv -\Psi(3\lambda) \equiv \Psi(3\lambda) \pmod 4$, from which we conclude that every λ gives a faithful Hodge representation of M_s of even weight.

$\mathbf{E}_7$. Then $\delta = 2$ and a generator for P/R is

$$\lambda = \tfrac{1}{2}(2\alpha_1 + 3\alpha_2 + 4\alpha_3 + 6\alpha_4 + 5\alpha_5 + 4\alpha_6 + 3\alpha_7).$$

It follows that every λ gives a faithful Hodge representation of M_s, which is of odd weight only in the cases EV, EVII, and is of even weight for EVI and the compact form.

$\mathbf{A}_\ell, \ell$ even. Then $l+1$ is odd and consequently all elements of $P/R \cong \mathbb{Z}/(l+1)\mathbb{Z}$ are of odd order, which implies that δ is odd. Thus a faithful Hodge representation always exists and the weight must be even. For a given λ the representation is faithful on the unique M such that $M \to M_a$ has degree equal to the order of $[\lambda]$ in $\mathbb{Z}/(l+1)\mathbb{Z}$.

$\mathbf{A}_\ell, \ell$ odd. Then $l+1$ is even and there are the two cases, $l+1 \equiv 0 \pmod 4$ and $l+1 \equiv 2 \pmod 4$. In the latter case $l+1 = 2d$ where d is odd, and then $\delta =$ order of $[\lambda]$ in $\mathbb{Z}/(l+1)\mathbb{Z}$ is either odd or equal to $2d'$ where d' is odd. If δ is odd a Hodge representation always exists and must be of even weight. If δ is even the issue is whether $\Psi(\delta\lambda) \equiv 0$ or $2 \pmod 4$.

From

$$\Psi(\delta\lambda) = \Psi(\alpha_1 + 2\alpha_2 + \cdots + l\alpha_l) \equiv \Psi(\alpha_1) + \Psi(\alpha_3) + \cdots \pmod 4$$

it follows that $\Psi(2\lambda) \equiv 2 \pmod 4$ if, and only if, the painted node in the Vogan diagram is at the $1^{\rm st}, 3^{\rm rd}, \ldots$ spot. These are the cases $\mathrm{su}(2p+1, l-2p)$, or equivalently su(odd,odd).

If $\Psi(\delta\lambda) \equiv 0 \pmod 4$, we have either the compact form or su(even,even) but $4 \nmid (l+1)$. So the only odd weight faithful Hodge representations when $2 \mid l+1$ but $4 \nmid (l+1)$ are su(odd,odd). These come from M where the degree of $M \to M_a$ is $2d'$.

If $l+1 \equiv 0 \pmod 4$, then $l+1 = 4d$ and the possibilities are the cases δ odd, $\delta = 2d'$, d' odd, and $\delta = 4d'$. The first case is as before. For the second case, a Hodge representation exists if, and only if,

$$\Psi(\delta\lambda) \equiv 0 \pmod 4 \equiv \Psi(\alpha_1 + 2\alpha_2 + 3\alpha_3 + \cdots) \equiv 0 \pmod 4,$$

which is equivalent to the compact form or to su(even,even). The Hodge weight can be either even or odd. The M for which the Hodge representation is faithful is determined as above. We note that:

> *This produces some M's for which no faithful Hodge representations exist*; e.g. $\mathrm{su}(1,3)$.

$\mathbf{D_{2k+1}}$. Then $P/R \cong \mathbb{Z}/4\mathbb{Z}$ and we have the two cases $\delta = 2$ or $\delta = 4$. In the first case

$$\begin{aligned} \Psi(\delta\lambda) &\equiv \Psi\big(2\alpha_1 + 4\alpha_2 + \cdots + 2(2k-1)\alpha_{2k-1} \\ &\qquad + (2k-1)\alpha_{2k} + (2k+1)\alpha_{2k+1}\big) \pmod 4 \\ &\equiv \Psi(\alpha_{2k} + \alpha_{2k+1}) \pmod 4 \\ &\equiv 0 \pmod 4 \text{ except for } \mathrm{so}^*(4k+2). \end{aligned}$$

This gives odd weight faithful Hodge representations on the compact form and on all $\mathrm{so}(p, 4k+1-q)$, where $M \to M_a$ is a double cover.

In the second case, we still have

$$\begin{aligned} \Psi(\delta\lambda) &\equiv \Psi(\alpha_{2k} + \alpha_{2k+1}) \pmod 4 \\ &\equiv 0 \pmod 4 \text{ except for } \mathrm{so}^*(4k+2). \end{aligned}$$

It follows that for M_s we get both odd and even weight faithful Hodge representations for the compact form and for all $\mathrm{so}(p, 4k+2-q)$, but no Hodge representation for $\mathrm{so}^*(4k+2)$.

$\mathbf{D_{2k}}$. This is the case when $P/R = \mathbb{Z}/2\mathbb{Z} \oplus \mathbb{Z}/2\mathbb{Z}$. Then $\delta = 2$ and for any λ there is a Hodge representation. There are three possibilities for $\Psi(\delta\lambda)$.

$$\begin{aligned} &\Psi\left(\alpha_1 + 2\alpha_2 + \cdots + (k-2)\alpha_{k-2} + \alpha_{k-1}\right) \\ &\qquad\qquad \equiv \Psi\left(\alpha_1 + \alpha_3 + \cdots + \alpha_{k-1}\right) \pmod 4. \end{aligned}$$

Since $\Psi(\alpha) \equiv 0, 2 \pmod 4$, this expression is $\equiv 0 \pmod 4$ in the $\mathrm{so}(4p, 4q)$ case, including the compact case, and in one of the two $\mathrm{so}(4k)^*$ cases.

The next possibility is

$$\Psi(\alpha_1 + 2\alpha_2 + \ldots + (k-1)\alpha_{k-1} + \alpha_k) \equiv \Psi(\alpha_1 + \alpha_3 + \cdots + \alpha_k) \pmod 4.$$

This is congruent to zero (mod 4) in the $\mathrm{so}(4p, 4q)$ case, including the compact case, and the other $\mathrm{so}(4k)^*$ case. Here there are two painted nodes in the Dynkin diagram corresponding to α_{k-1} and α_k.

The third case is

$$\Psi(\alpha_k + \alpha_k) \equiv 0 \pmod 4 \text{ except for } \mathrm{so}(4k)^*.$$

Conclusion: We can get an odd weight Hodge representation for $\mathrm{so}(4p+2, 4q+2)$ and $\mathrm{so}(4k)^*$, but not for $\mathrm{so}(4p, 4q)$.

The M on which the Hodge representation corresponding to a weight λ is faithful is determined by the method used above. For M_a we cannot get a faithful Hodge representation unless we can take a direct sum of different V^λ's.

We conclude this section by commenting on the general structure of semi-simple $\mathbb{Q}$-algebraic groups M.

- M is a product of factors $\mathrm{Res}_{L_i/\mathbb{Q}}(M_i)$ where M_i is absolutely simple over $\mathbb{Q}$; here $[L_i : \mathbb{Q}]$ is finite but not necessarily Galois.

- If M is a simple $\mathbb{Q}$-algebraic group, then $M = \mathrm{Res}_{L/\mathbb{Q}}(M')$ where M' is absolutely simple over L.

- In this case, denoting by $\mathrm{Dy}(*)$ the Dynkin diagram of $*$ we have $\mathrm{Dy}(M) \cong [L : \mathbb{Q}]$ disjoint copies of $\mathrm{Dy}(M')$; thus, $\mathrm{Aut}\,\mathrm{Dy}(M)$ incorporates $\mathrm{Aut}(\mathrm{Dy}(M'))$ and the symmetry group of $[L : \mathbb{Q}]$.

IV.F HODGE DOMAINS

In this section we will assume that M is simple. The essential aspects of the general conclusions we shall draw can, in a straightforward manner, be extended to the case where M is semi-simple.

We denote by D the period domain for all polarized Hodge structures on (V, Q) with a given set of Hodge numbers. Let (M, ρ, φ) be a Hodge representation with $\rho : M \to \mathrm{Aut}(V, Q)$ such that the polarized Hodge structure $(V, Q, \rho \circ \varphi)$ is a point in D. Recall that the Mumford-Tate domain $D_M \subset D$ is the orbit under $\rho(M(\mathbb{R})) \subset G(\mathbb{R})$ of this point. In this discussion we shall only be concerned with the component of D_M passing through $\varphi \in D$.

As noted in Section II.A, given a Hodge representation, $(\mathfrak{m}, B, \operatorname{Ad}\varphi)$ also gives a Hodge representation of the adjoint group M_a. Denote by $D_{\mathfrak{m},\varphi}$ the period domain for polarized Hodge structures of this type on $(\mathfrak{m}, B)$.

(IV.F.1) THEOREM: (i) *The subgroup $H_\varphi \subset M(\mathbb{R})$ that stabilizes the polarized Hodge structure (M, ρ, φ) is equal to the subgroup that stabilizes the polarized Hodge structure $(M_a, \operatorname{Ad}, \varphi)$. Consequently, as complex manifolds*

$$D_M = D_{\mathfrak{m},\varphi}.$$

(ii) *Under this identification the Pfaffian systems defining the infinitesimal period relations coincide.*

PROOF. We begin by noting that the Hodge structure on $\mathfrak{m}$ given by $\operatorname{Ad}\varphi$ is the same as the induced Hodge structure on $\rho_*(\mathfrak{m}) \subset \operatorname{End}_Q(V)$. In addition, as previously noted the polarizing form induced on the Hodge structure $\rho_*(\mathfrak{m})$ by that on $\operatorname{End}_Q(V)$ agrees with that given by the Cartan-Killing form. Finally, again as previously noted the stability subgroup in $M(\mathbb{R})$ of each of the above polarized Hodge structures is given by the centralizer $H_\varphi = Z(\varphi(S^1))$ of $\varphi(S^1)$ in $M(\mathbb{R})$. Noting that

$$\ker(M(\mathbb{R}) \to M_a(\mathbb{R})) \subseteq Z(M(\mathbb{R})) \subseteq Z(\varphi(S^1))$$

thus if temporarily we label $Z(\varphi(S^1))$ as $H_{\varphi,a}$ and H_φ is as above, we have

$$M_a(\mathbb{R})/H_{\varphi,a} \cong M(\mathbb{R})/H_\varphi.$$

This completes the proof of (i). The argument for (ii) is essentially just tracing through what needs to be shown, the essential point being that the adjoint representation of $M_a(\mathbb{R})$ on $\mathfrak{m}_\mathbb{R}$ is faithful.

First, we have

$$\rho_*(\mathfrak{m}_\mathbb{C}) \subset \operatorname{Hom}(V_\mathbb{C}, V_\mathbb{C})$$

and by definition

$$\rho_*(\mathfrak{m}^{-i,i}) = \rho_*(\mathfrak{m}_\mathbb{C}) \cap \Big(\bigoplus_p \operatorname{Hom}\big(V_\mathbb{C}^{p,n-p}, V_\mathbb{C}^{p-i,n-p+i}\big)\Big).$$

The infinitesimal period relation for D_M is given by $\mathfrak{m}^{-1,1}$.

Next, we have

$$\operatorname{ad}(\mathfrak{m}_\mathbb{C}) \subset \operatorname{Hom}(\mathfrak{m}_\mathbb{C}, \mathfrak{m}_\mathbb{C})$$

and the infinitesimal period relation for $D_{\mathfrak{m},\varphi}$ is given by

$$\operatorname{ad}(\mathfrak{m}_\mathbb{C}) \cap \Big(\bigoplus_i \operatorname{Hom}\big(\mathfrak{m}^{-i,i}, \mathfrak{m}^{-i-1,i+1}\big)\Big).$$

Since $\mathfrak{m}_{\mathbb{C}} = \underset{i}{\oplus}\, \mathfrak{m}^{-i,i}$ and

$$\left[\mathfrak{m}^{-1,1}, \mathfrak{m}^{-i,i}\right] \subseteq \mathfrak{m}^{-i-1,i+1},$$

it follows that under the isomorphism

$$\mathfrak{m}_{\mathbb{C}} \xrightarrow{\sim} \operatorname{ad} \mathfrak{m}_{\mathbb{C}} \subset \operatorname{Hom}(\mathfrak{m}_{\mathbb{C}}, \mathfrak{m}_{\mathbb{C}})$$

the above intersection is exactly equal to $\mathfrak{m}^{-1,1}$. □

An alternative interpretation is that if $\varphi(t) = \exp(t\mathfrak{l}_{\varphi})$ then

$$\text{(IV.F.2)} \qquad \begin{aligned} \mathfrak{h}_{\varphi} &= 0\text{-eigenspace } \mathfrak{l}_{\varphi} \\ \mathfrak{m}_{\mathbb{R}}^{-1,1} &= -2i\text{-eigenspace of } \mathfrak{l}_{\varphi} \end{aligned}$$

and since $M(\mathbb{R})$ and $M_a(\mathbb{R})$ have the same Lie algebra, the infinitesimal period relations coincide.

An obvious interpretation of the above result is:

> *The same complex manifold and exterior differential system may appear in many different ways as a Mumford-Tate domain parametrizing a family of Hodge representations and with the associated infinitesimal period relation.*[17]

In fact, there will be multiple realizations of the *same* $D_{\mathfrak{m}}$ as Mumford-Tate domains associated to polarized Hodge structures *of distinctly different weights and for the same* (V, Q).[18] In order to discuss this in examples, in the remainder of this section we will use the following

Definition: A *Hodge structure of minimal positive weight* n on a $\mathbb{Q}$-vector space is given by a Hodge decomposition

$$\begin{cases} V_{\mathbb{C}} = \displaystyle\bigoplus_{\begin{cases} p+q=n \\ p,q \geqq 0 \end{cases}} V^{p,q} \\ \\ V^{p,q} = \overline{V}^{q,p} \ \text{ and } \ V^{n,0} \neq 0. \end{cases}$$

A Hodge structure of any positive weight may, by a Tate twist, be converted to one of minimal positive weight.

---

[17]As will be discussed and illustrated in Section IV.G, the same complex manifold and EDS may appear in several different ways as a *homogeneous* complex manifold, and these should be considered as distinct Mumford-Tate domains because the Mumford-Tate groups at a generic point will be different.

[18]"Distinctly different" means not differing by a Tate twist.

Example: A Hodge representation (M, ρ, φ) induces a Hodge structure of weight zero on $\mathfrak{m}$. In the notation of Section III.B, in this case the minimal positive weight is $n = 2l(\mathfrak{m}_\varphi)$ where

(IV.F.3) $$l(\mathfrak{m}_\varphi) = \max\left\{i : \mathfrak{m}^{-i,i} \neq 0\right\}.$$

Note that $l(\mathfrak{m}_\varphi) = \max((\frac{1}{2i}) \cdot (\text{eigenvalues of } \mathfrak{l}_\varphi))$. Before turning to some examples, and in light of the above observations, we give the

Definition: A *Hodge group* is a semi-simple $\mathbb{Q}$-algebraic group M, in which there is a circle $\varphi : S^1 \to M_a(\mathbb{R})$ such that $(\mathfrak{m}, B, \operatorname{Ad}\varphi)$ gives a polarized Hodge structure. A *Hodge domain* is the homogeneous complex manifold $D_\mathfrak{m} = M_a(\mathbb{R})/H_\varphi$.[19]

A Hodge domain is thus determined by the pair (M, φ) consisting of the Hodge group M and the circle $\varphi : S^1 \to M(\mathbb{R})$. We think of it as a Mumford-Tate domain for the adjoint representation, a generic point of which has Mumford-Tate group M_a.

The data (M, φ) determines the invariant Pfaffian exterior differential system $I \subset T^* D_{M_\varphi}$ given by the infinitesimal period relation. We write

$$\mathfrak{m} = \oplus\, \mathfrak{m}^{-k,k}$$

where

(IV.F.4) $\mathfrak{m}^{-k,k}$ *is* $(-2ki)$*-eigenspace of the action of the differential of* $\operatorname{Ad}\varphi : S^1 \to \operatorname{Aut} \mathfrak{m}_\mathbb{C}$.

We note that, using the notations from the appendix to this chapter,

(IV.F.5a) $\mathfrak{m}^{-(2k+1,2k+1)}$ *is a direct sum of non-compact root spaces* $\mathfrak{m}_{\beta_j}$ *where* $\langle \beta_j, \mathfrak{l}_\varphi \rangle = -2(2k+1)i$,

(IV.F.5b) $\mathfrak{m}^{-2k,2k}$ *is a direct sum of compact root spaces* $\mathfrak{m}_{\alpha_j}$ *where* $\langle \alpha_j, \mathfrak{l}_\varphi \rangle = -2(2k)i$.

Before turning to examples, we recall a few aspects about the exterior differential system on a manifold X generated by a sub-bundle $I \subset T^*X$, everything here being in the holomorphic setting. We use the notations:

- $\mathcal{I}$ is the differential ideal in $\Omega^\bullet_X$ generated over $\Omega^\bullet_X$ by the sections θ of $\mathcal{O}_X(I)$, together with their exterior derivatives;

[19] A more precise notation would be $D_{\mathfrak{m},\varphi}$, but we shall omit reference to the particular φ.

- $I^\perp = W \subset TX$ will denote the distribution dual to I;
- *integral elements* of $\mathcal{I}$ are linear subspaces $E \subset T_xX$ such that $\psi|_E = 0$ for all $\psi \in \mathcal{I}$. This is equivalent to

$$\begin{cases} \theta|_E = 0 \\ d\theta|_E = 0 \end{cases}$$

where θ runs over a set of local generators of $\mathcal{O}_X(I)_x$.

Finally, since we shall be working in a homogeneous space setting, all calculations may be done at the identity coset, and there we shall denote the fibres of these bundles also by I and W.

For the exterior differential system given by the infinitesimal period relation, using the above notations and identifying the tangent space at the identity coset with $\mathfrak{m}$, we have

$$\begin{cases} W = \mathfrak{m}^{-1,1} \\ I \;=\; \bigoplus\limits_{k\geq 2} \check{\mathfrak{m}}^{-k,k}. \end{cases}$$

The exterior derivative induces a map

$$\delta : I \to \Lambda^2(\check{\mathfrak{m}}^-/I)$$

with dual

$$\delta^* : \Lambda^2 W \to \mathfrak{m}^-/W. \tag{IV.F.6}$$

Integral elements are linear subspaces $E \subset W$ such that

$$\delta(I)|_E = 0.$$

The map (IV.F.6) is

$$\delta^* : \Lambda^2\mathfrak{m}^{-1,1} \to \bigoplus_{k\geqq 2} \mathfrak{m}^{-k,k},$$

and by the Maurer-Cartan equation this map may, up to scaling, be identified with the bracket

$$[\ ,\ ] : \Lambda^2\mathfrak{m}^{-1,1} \to \mathfrak{m}^{-2,2}. \tag{IV.F.7}$$

This confirms the well known fact [CGG] that *integral elements are given by abelian Lie sub-algebras of* $\mathfrak{m}^{-1,1}$. More importantly for computational purposes, using (IV.F.5a), (IV.F.5b) we may identify (IV.F.7) with

$$[\ ,\ ] : \bigoplus_{\begin{cases} \langle\beta_j, l_\varphi\rangle = -2i \\ \langle\beta_k, l_\varphi\rangle = -2i \\ j<k \end{cases}} \mathfrak{m}_{\beta_j} \otimes \mathfrak{m}_{\beta_k} \to \bigoplus_{\begin{cases} \langle\beta_j, l_\varphi\rangle = -2i \\ \langle\beta_k, l_\varphi\rangle = -2i \\ j<k \end{cases}} \mathfrak{m}_{\beta_j+\beta_k}. \tag{IV.F.8}$$

The individual terms in (IV.F.8) *are non-zero exactly when* $\beta_j + \beta_k$ *is a root,* and thus, at least in principle, in this way the structure theory of the exterior differential system may be reduced to the root structure of the Lie algebra.

Illustration: Suppose that $\dim \mathfrak{m}^{-2,2} = 1$; say $\mathfrak{m}^{-2,2} = \mathbb{C}e_\alpha$ where $\langle \alpha, \mathfrak{l}_\varphi \rangle = -4i$, and $\mathfrak{m}^{-k,k} = 0$ for $k \geqq 3$. Then I is locally generated by a single 1-form θ and (IV.F.7) is

$$[\ ,\ ] : \Lambda^2 \mathfrak{m}^{-1,1} \to \mathbb{C}e_\alpha.$$

The system $\mathfrak{I}$ is a contact structure exactly when this skew-symmetric form is non-degenerate.

Before turning to examples we want to make one further observation from an exterior differential system perspective. We recall that a distribution W on a connected manifold X is said to be *accessible* if any two points of X may be joined by a chain of integral curves of W. By a well-known theorem of Chow, this is equivalent to W being *bracket generating*; i.e.,

$$W = [W, W] + [W, [W, W]] + \cdots = TX.$$

For the canonical distribution on a period domain, it is known [CGG] that the bracket generation condition is equivalent to the condition that *the sequence* $h^{n,0}, h^{n-1,0}, \ldots, h^{0,n}$ *of Hodge numbers has no gaps*; i.e., all $h^{p,n-p} \neq 0$. The observation is the following:

> *For a Hodge domain $D_{\mathfrak{m}}$, if we have accessibility for the canonical distribution, then for* **any** *Hodge representation (M, ρ, φ) the sequence of Hodge numbers has no gaps. The converse is* **false**.

Example: Let D be the period domain of weight six polarized Hodge structures on (V, Q) with Hodge numbers $h^{6,0} = h^{5,1} = h^{4,2} = 1$, $h^{3,3} = 0$. Then we do not have accessibility. Here D is the Mumford-Tate domain for $M = \mathrm{SO}(4,2)$. For the polarized Hodge structures on $(\Lambda^2 V, \Lambda^2 Q)$ the Hodge numbers are $h^{11,1} = \cdots = h^{8,4} = 1$, $h^{7,5} = 2$, $h^{6,6} = 3$; thus there are no gaps. However, the Mumford-Tate domain, together with its canonical distribution, for a generic Hodge structure $\Lambda^2(V, Q, \varphi)$ is biholomorphic to D (the notation means Λ^2 of the polarized Hodge structure (V, Q, φ)). Hence, for this Mumford-Tate domain, the sequence of Hodge numbers has no gaps but we do not have accessibility.

If we want to understand the compact dual of $D_{\mathfrak{m}}$, we let

$$\mathfrak{p} = \mathrm{span}\{e_\alpha | \langle \alpha, \mathfrak{l}_\varphi \rangle \geqq 0\}$$ [20]

and $P \subset M(\mathbb{C})$ the corresponding parabolic subgroup of the complex Lie group M. Then $D_{\mathfrak{m}} \hookrightarrow M(\mathbb{C})/P$ embeds holomorphically.

Note that $\mathfrak{h}_\varphi = \{\alpha \in r | \langle \alpha, \mathfrak{l}_\varphi \rangle = 0\}$ and thus $\dim(D_{\mathfrak{m}})$ is determined by $\{\alpha | \langle \alpha, \mathfrak{l}_\varphi \rangle = 0\}$. Further, the complex structure on $D_{\mathfrak{m}}$ is determined by $\{\alpha \in \mathfrak{r} \mid \langle \alpha, \mathfrak{l}_\varphi \rangle \geq 0\}$. This in turn has a Weyl chamber interpretation.

However, the differential system on $D_{\mathfrak{m}}$ is determined by the additional knowledge of $\{\alpha \in \mathfrak{r} | \langle \alpha, \mathfrak{l}_\varphi \rangle = -2i\}$.

For the computation of examples we need a notational structure that relates the $\mathfrak{l}_\varphi = \Sigma l_i H_i$ giving a co-character to the root diagram. We shall explain this for G_2; from this the general pattern will hopefully be clear.

First, the standard root diagram of $\mathfrak{g}_2$ is:

(IV.F.9)

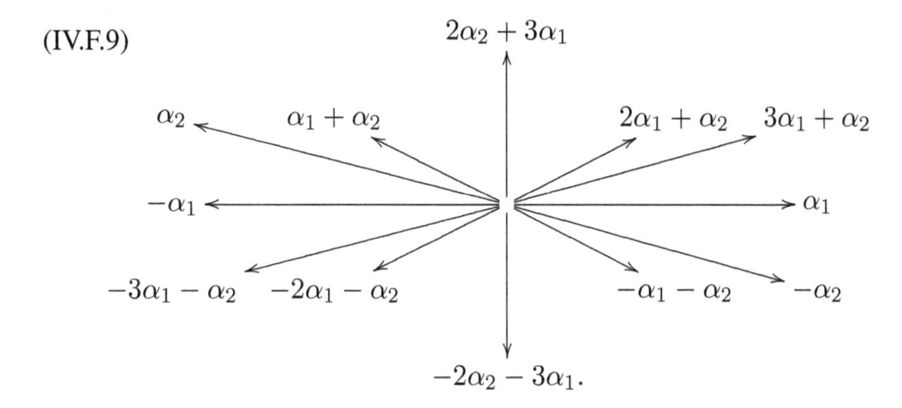

Recalling our notations $E = \left(\begin{smallmatrix} 0 & -1 \\ 1 & 0 \end{smallmatrix}\right)$ and

$$H_1 = \underbrace{\left(\begin{array}{cc|cc} E & 0 & 0 & 0 \\ 0 & 0 & 0 & 0 \\ \hline 0 & 0 & E & 0 \\ 0 & 0 & 0 & 0 \end{array}\right)}_{3 \quad\quad 4} \begin{array}{l} \}3 \\ \}4 \end{array}, \quad H_2 = \underbrace{\left(\begin{array}{cc|cc} 0 & 0 & 0 & 0 \\ 0 & 0 & 0 & 0 \\ \hline 0 & 0 & E & 0 \\ 0 & 0 & 0 & E \end{array}\right)}_{3 \quad\quad 4} \begin{array}{l} \}3 \\ \}4 \end{array}$$

for the basis of a Cartan sub-algebra of $\mathfrak{g}_2$, $H_{\mathfrak{l}_\varphi} = l_1 H_1 + l_2 H_2$ acts on each root space $\mathfrak{g}_2$ by the differential of a character. We want to know what that action is; i.e., how $H_{\mathfrak{l}_\varphi}$ acts on e_α. We shall use the notation

$$\begin{cases} \alpha_1 \leftrightarrow l_1 \\ \alpha_2 \leftrightarrow l_2 - l_1 \end{cases}$$

[20]Here, we follow the customary notations in the Lie algebra theory and omit the i.

to signify that $H_{\mathfrak{l}_\varphi}$ acts on e_{α_1} by $l_1 i$ and on e_{α_2} by $(l_2 - l_1)i$. Thus we have the table:

compact roots	non-compact roots
$l_1 \leftrightarrow \alpha_1$	$2l_1 + l_2 \leftrightarrow 3\alpha_1 + \alpha_2$
$l_1 + 2l_2 \leftrightarrow 2\alpha_2 + 3\alpha_1$	$l_2 \leftrightarrow \alpha_1 + \alpha_2$
	$l_2 - l_1 \leftrightarrow \alpha_2$
	$l_1 + l_2 \leftrightarrow 2\alpha_1 + \alpha_2;$

and with this notation the above diagram becomes:

(IV.F.10)

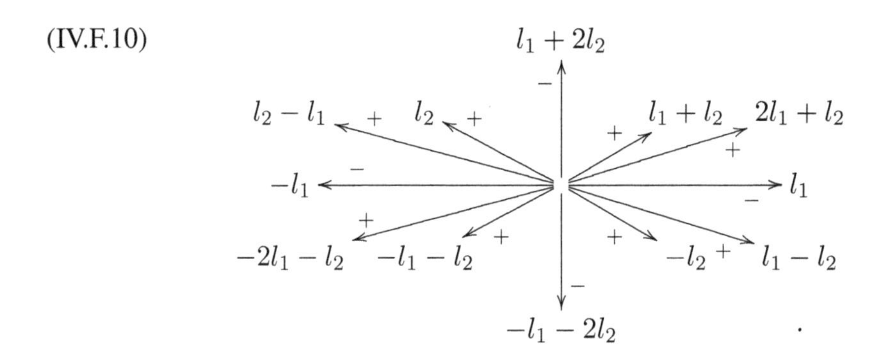

We shall also use the notation that

$$\text{span}\{\alpha_2, -3\alpha_1 - \alpha_2\} = \text{span}\{e_{\alpha_2}, e_{-3\alpha_1-\alpha_2}\}, \text{ etc.}$$

and

$$\begin{aligned} &\text{span}\{H_1, H_2, 2\alpha_2 + 3\alpha_1, -2\alpha_2 - 3\alpha_1\} \\ &\qquad = \text{span}\{H_1, H_2, e_{2\alpha_2+3\alpha_1}, e_{-2\alpha_2-3\alpha_1}\}, \text{ etc.} \end{aligned}$$

In this way, given l_1 and l_2 we can visually see the spaces $\mathfrak{g}_2^{p,q}$.[21]

(IV.F.11a) **Example:** We take

$$l_1 = 4, \quad l_2 = -2.$$

[21]The point of these notations is this: First, the l_j's determine the Hodge numbers in the manner indicated. Secondly, the bracket structure is determined by the root structure, the bracket $[e_\alpha, e_\beta]$ being non-zero exactly when $\alpha + \beta$ is a root (including zero).

For our Hodge representation we take the adjoint representation. This gives a polarized Hodge structure of minimal positive weight six with

$$\begin{aligned}
\mathfrak{g}_2^{0,6} &= \operatorname{span}\{l_2 - l_1, 2l_1 - l_2\} = \operatorname{span}\{\alpha_2, -3\alpha_1 - \alpha_2\} \\
\mathfrak{g}_2^{1,5} &= \operatorname{span}\{-l_1\} = \operatorname{span}\{-\alpha_1\} \\
\mathfrak{g}_2^{2,4} &= \operatorname{span}\{l_2, -l_1 - l_2\} = \operatorname{span}\{\alpha_1 + \alpha_2, -2\alpha_1 - \alpha_2\} \\
\mathfrak{g}_2^{3,3} &= \operatorname{span}\{H_1, H_2, l_1 + 2l_2, -l_1 - 2l_2\} \\
&= \operatorname{span}\{H_2, H_2, 2\alpha_2 + 3\alpha_1, -2\alpha_2 - 3\alpha_1\}.
\end{aligned}$$

The Hodge numbers are

$$h^{6,0} = 2, \quad h^{5,1} = 1, \quad h^{4,2} = 2, \quad h^{3,3} = 4.$$

(IV.F.11b) **Example:** We again take $l_1 = 4$, $l_2 = -2$ but this time for the Hodge representation we take the standard representation of G_2 on $V \cong \mathbb{Q}^7$. As in the proof of Theorem IV.D.1 we write $V = V^- \oplus V^+$ where

$$\begin{cases} H_{\mathfrak{l}_\varphi} \text{ has eigenvalues } \pm l_1 i, 0 \pm 4i \text{ on } V^- \\ H_{\mathfrak{l}_\varphi} \text{ has eigenvalues } \pm(l_1 + l_2)i, \pm l_2 i = \pm 2i, \mp 2i \text{ on } V^+. \end{cases}$$

This gives a polarized Hodge structure of weight four and with Hodge numbers

$$h^{4,0} = 1, \quad h^{3,1} = 2, \quad h^{2,2} = 1.$$

For the induced Hodge structure on $\mathfrak{g}_2$ we have

$$\mathfrak{g}_2^{-1,1} = \operatorname{span}\{\alpha_1 + \alpha_2, -2\alpha_1 - \alpha_2\}.$$

Then

$$\left[e_{\alpha_1+\alpha_2}, e_{-2\alpha_1-\alpha_2}\right] = a e_{-\alpha_1} \qquad a \neq 0$$

so the bracket

$$[\ ,\ ] : \Lambda^2 \mathfrak{g}_2^{-1,1} \to \mathfrak{g}_2^{-2,2}$$

is non-trivial. From

$$\begin{cases} \left[e_{-\alpha_i}, e_{\alpha_1+\alpha_2}\right] = b e_{\alpha_2} & b \neq 0 \\ \left[e_{-\alpha_1}, e_{-2\alpha_1-\alpha_2}\right] = c e_{-3\alpha_1-\alpha_2} & c \neq 0 \end{cases}$$

we see that the triple bracket $[W, [W, W]]$ spans $\mathfrak{g}_2^{-3,3}$. Thus, noting that $\mathfrak{g}_2^{-4,4} = 0$, we conclude that *the infinitesimal period relation is bracket generating.*

(IV.F.12a) **Example:** We consider the adjoint representation of G_2 with $l_1 = 0, l_2 = 2$. This gives a polarized Hodge structure of minimal weight four with

$$\begin{cases} \mathfrak{g}_2^{0,4} = \operatorname{span}\{-l_1 - 2l_2\} = \operatorname{span}\{-3\alpha_1 - 2\alpha_2\} \\ \mathfrak{g}_2^{1,3} = \operatorname{span}\{-2l_1 - l_2, -l_1 - l_2, -l_2, l_1 - l_2\} \\ \qquad = \operatorname{span}\{-3\alpha_1 - \alpha_2, -2\alpha_1 - \alpha_2, -\alpha_1 - \alpha_2, -\alpha_2\} \\ \mathfrak{g}_2^{2,2} = \operatorname{span}\{H_1, H_2, l_1, -l_1\} = \operatorname{span}\{H_1, H_2, -\alpha_1, \alpha_1\} \end{cases}$$

and Hodge numbers

$$h^{4,0} = 1, \quad h^{3,1} = 4, \quad h^{2,2} = 4.$$

(IV.F.12b) **Example:** We again take $l_1 = 0, l_2 = 2$ and for our Hodge representation the standard representation. Then

$$\begin{cases} H_{\mathfrak{l}_\varphi} \text{ has eigenvalues } 0,0,0 & \text{on } V^- \\ H_{\mathfrak{l}_\varphi} \text{ has eigenvalues } \pm 2i, \pm 2i & \text{on } V^+. \end{cases}$$

This gives a polarized Hodge structure of weight two and with Hodge numbers $h^{2,0} = 2, h^{1,1} = 3$.

For the induced polarized Hodge structure on $\mathfrak{g}_2 \subset \operatorname{End}_Q(V)$, the 4×4 matrix of the brackets is:

	$-3\alpha_1 - \alpha_2$	$-2\alpha_1 - \alpha_2$	$-\alpha_1 - \alpha_2$	$-\alpha_2$
$-3\alpha_1 - \alpha_2$	0	0	0	$-3\alpha_1 - 2\alpha_2$
$-2\alpha_1 - \alpha_2$	0	0	$-3\alpha_1 - 2\alpha_2$	0
$-\alpha_1 - \alpha_2$	0	$3\alpha_1 - 2\alpha_2$	0	0
$-\alpha_2$	$3\alpha_1 - 2\alpha_2$	0	0	0

where we recall that our notational correspondence is

$$m_1\alpha_1 + m_2\alpha_2 \leftrightarrow e_{m_1\alpha_1 + m_2\alpha_2}$$

whenever $m_1\alpha_1 + m_2\alpha_2$ is a root. The 0's correspond to sums of roots that are not roots. This matrix is non-singular, and so we may apply the above table to conclude that *the corresponding Mumford-Tate domain has a G_2-invariant contact structure.*

(IV.F.13) **Example:** We take $l_1 = 4, l_2 = 2$, and again for our Hodge representation the standard representation. Then

$$\begin{cases} H_{\mathfrak{l}_\varphi} \text{ has eigenvalues } 0, \pm 4i & \text{on } V^- \\ H_{\mathfrak{l}_\varphi} \text{ has eigenvalues } \pm 2i, \pm 6i & \text{on } V^+. \end{cases}$$

This gives a weight six polarized Hodge structure with Hodge numbers all $H^{p,6-p} = 1$. For the induced polarized Hodge structure on $\mathfrak{g}_2 \subset \mathrm{End}_Q(V)$ we have $\mathfrak{g}$

$$\begin{cases} \mathfrak{g}_2^{-1,1} = \mathrm{span}\,\{l_2 - l_1, -l_2\} = \mathrm{span}\,\{\alpha_2, -\alpha_1 - \alpha_2\} \\ \mathfrak{g}_2^{-2,2} = \mathrm{span}\,\{-l_1\} = \mathrm{span}\,\{-\alpha_1\} \\ \mathfrak{g}_2^{-3,3} = \mathrm{span}\,\{-l_1 - l_2\} = \mathrm{span}\,\{-2\alpha_1 - \alpha_2\} \\ \mathfrak{g}_2^{-4,4} = \mathrm{span}\,\{-l_1 - 2l_2\} = \mathrm{span}\,\{-2\alpha_2 - 3\alpha_1\} \\ \mathfrak{g}_2^{-5,5} = \mathrm{span}\,\{-2l_1 - l_2\} = \mathrm{span}\,\{-3\alpha_1 - \alpha_2\}\,. \end{cases}$$

We note that

$$\begin{cases} [\ ,\ ] : \Lambda^2 \mathfrak{g}_2^{-1,1} \xrightarrow{\sim} \mathfrak{g}_2^{-2,2} \\ [\ ,\ ] : \mathfrak{g}_2^{-1,1} \otimes \mathfrak{g}_2^{-2,2} \twoheadrightarrow \mathfrak{g}^{-3,3} \\ [\ ,\ ] : \mathfrak{g}_2^{-1,1} \otimes \mathfrak{g}_2^{-3,3} \twoheadrightarrow \mathfrak{g}^{-4,4} \\ [\ ,\ ] : \mathfrak{g}_2^{-2,2} \otimes \mathfrak{g}_2^{-3,3} \twoheadrightarrow \mathfrak{g}^{-5,5}. \end{cases}$$

Thus, $\dim W = 2$ and the infinitesimal period relation is bracket generating.

The Cartan-Bryant incidence correspondence. In a remarkable paper [Ca], about which to this day one may say there is probably no higher peak in the theory of exterior differential systems, Cartan studied the geometry associated to a bracket-generating 2-plane field in a 5-dimensional manifold. It was in this paper that the group G_2 was first realized geometrically; heretofore, only its Lie algebra was known. Cartan proved that there are exactly two realizations of G_2 acting transitively on a 5-dimensional manifold X and preserving a non-trivial exterior differential system. One is the "flat" model of the geometry (a Cartan connection) associated to the 2-plane field on X; denote this by X_P (where P stands for 2-plane field). The other is related to a contact structure, which we denote by X_C.[22] We observe that:

> *Cartan's X_P is the compact dual of the Hodge domain in example* (IV.F.10)*, and Cartan's X_C is compact dual of the example* (IV.F.11).

[22]More precisely, it is the geometry associated to a field of twisted cubic curves in a (projectionized) field of contact planes.

In Cartan and Bryant [Br], one finds (in Bryant's notation) an equivariant double fibration of $G_2(\mathbb{C})$ homogeneous spaces

$$\text{(IV.F.13)}_{\text{CB}} \qquad \begin{array}{ccc} & \text{II} & \\ {\scriptstyle\lambda}\swarrow & & \searrow{\scriptstyle\mu} \\ \mathbb{Q}_5 & & \mathbb{N}_5 \end{array}$$

where Bryant's $\mathbb{Q}_5$ and $\mathbb{N}_5$ are exactly the compact duals of X_P and X_C. Moreover, Bryant's II is the compact dual of the Hodge domain $Y =: G_2(\mathbb{R})/T$ corresponding to our example (IV.F.13). Thus, open $G_2(\mathbb{R})$ orbits in (IV.F.13)$_{\text{CB}}$ give a diagram

$$\text{(IV.F.13)}_{\text{H}} \qquad \begin{array}{ccc} & Y & \\ \swarrow & & \searrow \\ X_P & & X_C \end{array}$$

of G_2-Hodge domains. However, *there is no choice of complex structure for Y that makes this a diagram of holomorphic mappings.*

To explain this we shall use Corollary (II.B.11). For G_2, there are 12 Weyl chambers, and thus by Proposition (II.B.8)

> *$G_2(\mathbb{R})/T$ has 12 different homogeneous complex structures (which occur in conjugate pairs).*

The two Mumford-Tate domains correspond to

$$\langle \alpha_1, \mathfrak{l}_\varphi \rangle = 0 \qquad (X_C)$$

and

$$\langle 3\alpha_1 + 2\alpha_2, \mathfrak{l}_\varphi \rangle = 0 \qquad (X_P).$$

Either of these equalities breaks the set of roots $\mathfrak{r}$ into two sets on which $\mathfrak{l}_\varphi$ has the same sign:

$$\begin{aligned} \langle \alpha_1, \mathfrak{l}_\varphi \rangle = 0 \Rightarrow \langle \alpha_2, \mathfrak{l}_\varphi \rangle &= \langle \alpha_1 + \alpha_2, \mathfrak{l}_\varphi \rangle = \langle 2\alpha_1 + \alpha_2, \mathfrak{l}_\varphi \rangle \\ &= \langle 3\alpha_1 + \alpha_2, \mathfrak{l}_\varphi \rangle = \tfrac{1}{2} \langle 3\alpha_1 \pm 2\alpha_2, \mathfrak{l}_\varphi \rangle \\ \langle 3\alpha_1 + 2\alpha_2, \mathfrak{l}_\varphi \rangle = 0 \Rightarrow \langle \alpha_2, \mathfrak{l}_\varphi \rangle &= \tfrac{3}{2} \langle -\alpha_1, \mathfrak{l}_\varphi \rangle = 3 \langle \alpha_1 + \alpha_2, \mathfrak{l}_\varphi \rangle \\ &= 3 \langle -(2\alpha_1 + \alpha_2), \mathfrak{l}_\varphi \rangle = \langle -(3\alpha_1 + \alpha_2), \mathfrak{l}_\varphi \rangle . \end{aligned}$$

Thus from the picture with the value of $\langle \alpha, \mathfrak{l}_\varphi \rangle$ inserted at the roots and where $\mathfrak{p}$ is the Lie algebra of the parabolic subgroup of the compact dual

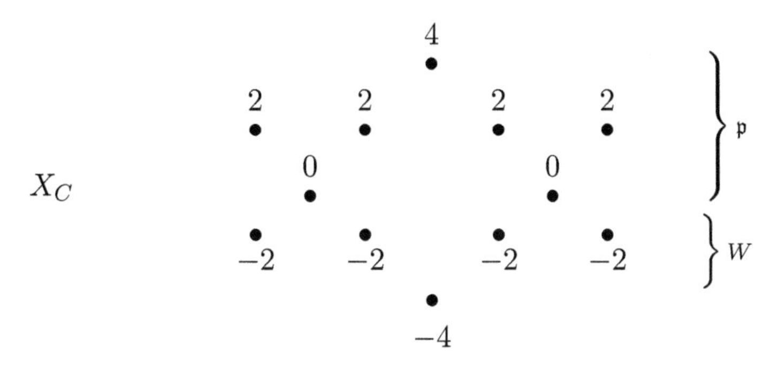

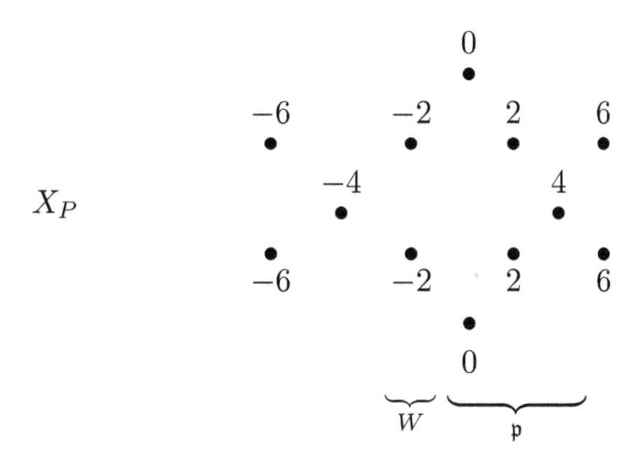

we conclude:

(IV.F.14) *X_P and X_C each have a unique homogeneous complex structure, up to conjugation. There is no complex structure $Y = G_2(\mathbb{R})/T$ that makes both of the maps*

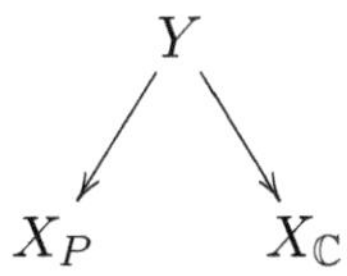

holomorphic.

For X_C,

$$W \leftrightarrow \operatorname{span}(e_{-\alpha_2}, e_{-\alpha_1-\alpha_2}, e_{-2\alpha_1-\alpha_2}, e_{-3\alpha_1-\alpha_2})$$

while for X_P,

$$W \leftrightarrow \operatorname{span}(e_{\alpha_1+\alpha_2}, e_{-2\alpha_1-\alpha_2}).$$

Finally, we remark that the real analog of a variation of Hodge structure (IV.F.11a) is the holonomic mechanical system given by one sphere rolling over another without slipping [BH]. Which, if any, of these arise algebro-geometrically is something one may wonder about.

A somewhat simpler example of the above double fibration analysis is given by the following

Example: The first non-Hermitian symmetric period domain is when $n = 2$ and $h^{2,0} = 2$, $h^{1,1} = 1$. Then

$$D = \mathrm{SO}(4, 1; \mathbb{R})/S(\mathcal{O}(4) \times \mathcal{O}(1)).$$

The root diagram is:

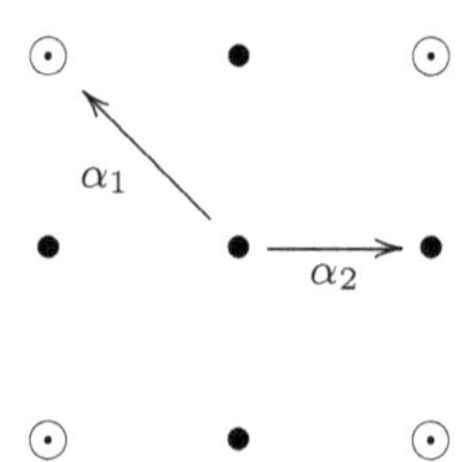

and the Vogan diagram is:

2 1

The compact roots are those that are circled. Circles in the $\mathcal{U}(2)$ that lead to a Hodge representation are given by $\mathfrak{l}_\varphi$ where

$$\begin{cases} \langle \alpha_1, \mathfrak{l}_\varphi \rangle = 4k_1 \\ \langle \alpha_2, \mathfrak{l}_\varphi \rangle = 4k_2 + 2 \end{cases}$$

for integers k_1, k_2. The eigenvalues of the action of the circle S^1 on the root spaces are given by

$$\text{(IV.F.15)} \qquad \begin{cases} \pm 4k_1 & \pm(4k_1 + 8k_2 + 4) & \text{compact roots} \\ \pm(4k_2 + 2) & \pm(4k_1 + 4k_2 + 2) & \text{non-compact roots.} \end{cases}$$

With the notation from the appendix to this chapter

$$\begin{cases} \alpha_1 = e_1 - e_2 \\ \alpha_2 = e_2. \end{cases}$$

The standard action of S^1 for the standard representation of $\mathrm{SO}(4,1)$ has eigenvalues i for e_1 and e_2. In this case

$$\begin{cases} 4k_2 + 2 = 2 \\ 4k_1 + 4k_2 + 2 = 2 \end{cases} \tag{IV.F.16}$$

so that $k_1 = k_2 = 0$ for the standard action. Then (IV.F.15) is

$$\begin{cases} 0, 0, & \pm 4i \\ \pm 2i, & \pm 2i. \end{cases}$$

The Lie algebra of the isotropy group is

$$u(2) = \mathfrak{t} + \mathrm{span}(e_{\alpha_1}, e_{-\alpha_1})_{\mathbb{R}}.$$

The complex tangent space is

$$\mathrm{TD} = \mathrm{span}\Big(\underbrace{e_{-\alpha_2}, e_{-\alpha_1-\alpha_2}}_{W}, e_{-\alpha_1-2\alpha_2}\Big).$$

We note that the bracket on $\Lambda^2 W$ is given by

$$[e_{-\alpha_2}, e_{-\alpha_1-\alpha_2}] = a e_{-\alpha_1-2\alpha_2} \neq 0,$$

confirming the fact that the IPR is a contact system.

A different choice of isotropy group H for a Hodge domain

$$\widetilde{D} =: \mathrm{SO}(4,1;\mathbb{R})/H$$

is obtained by choosing $k_1 = 1$ and $k_2 = -1$. Then

$$\mathfrak{h} = \mathfrak{t} + \mathrm{span}(e_{\alpha_1+2\alpha_2}, e_{-\alpha_1-2\alpha_2})_{\mathbb{R}} \cong u(2),$$

and (IV.F.15) becomes

$$\begin{cases} \pm 4i, & 0, 0 \\ \pm 2i, & \pm 2i \end{cases}$$

giving

$$T\widetilde{D} = \mathrm{span}\Big(e_{-\alpha_1}, \underbrace{e_{\alpha_2}, e_{-\alpha_1-\alpha_2}}_{W}\Big).$$

Again this is a contact structure. As in the G_2-case, we have:

In the diagram

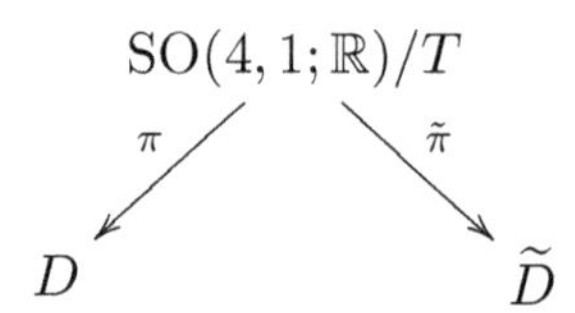

there is no choice of complex structure on $\mathrm{SO}(4,1;\mathbb{R})/T$ *that makes* π *and* $\tilde{\pi}$ *both holomorphic.*

We now give some further examples of low dimensional Hodge domains, referring to Section IV.C for the notations.

It should be mentioned that always $T \subseteq H_\varphi$. If we make a "general" choice of $\mathfrak{l}_\varphi$ we will have $\langle \alpha, \mathfrak{l}_\varphi \rangle \neq 0$ for all $\alpha \in \mathfrak{r}$, and thus:

> *For a general choice of* $\mathfrak{l}_\varphi$ *giving a polarized Hodge structure,* $\mathfrak{h}_\varphi = \mathfrak{t}$.

Likewise, if we arrange that $\langle \alpha, \mathfrak{l}_\varphi \rangle \neq -2i$ for all $\alpha \in \mathfrak{r}$, then in this case:

> *For a general choice of* $\mathfrak{l}_\varphi$ *giving a polarized Hodge structure, there are no tangent vectors to* $D_\mathfrak{m}$ *satisfying the infinitesimal period relations; i.e., variations of Hodge structure on* $D_\mathfrak{m}$ *are necessarily rigid.*

We also know $\mathfrak{h}_\varphi \subset \mathfrak{k}$ with $\mathfrak{k}$ maximal compact and with equality exactly when $D_\mathfrak{m}$ is an Hermitian symmetric domain.

SO(2,2) (cf. (IV.C.4)). In this case, $\dim D_{M_\varphi} = 2$ and the three cases there correspond respectively to

$$\begin{cases} \operatorname{rank} W = 2 & \text{(symmetric space case)} \\ \operatorname{rank} W = 1 & \text{(locally a contact structure in } \mathbb{C}^2\text{)} \\ \operatorname{rank} W = 0 & \text{(rigid case).} \end{cases}$$

Here, we recall that a Hodge domain is *rigid* if there are locally no non-constant variations of Hodge structure. Equivalently, $W = 0$.

SO(2,3) (cf. (IV.C.5)). The three cases have respectively

$$\begin{cases} \dim D_{M_\varphi} = 3, \operatorname{rank} W = 3 & \text{(symmetric space case)} \\ \dim D_{M_\varphi} = 4, \operatorname{rank} W = 1 & \text{(vector field case)} \\ \dim D_{M_\varphi} = 3, \operatorname{rank} W = 0 & \text{(rigid case).} \end{cases}$$

$\mathbf{Sp}(4,\mathbb{R})$ (cf. (IV.A.6)) and (IV.C.6)). The root diagrams for the two methods of displaying them are:

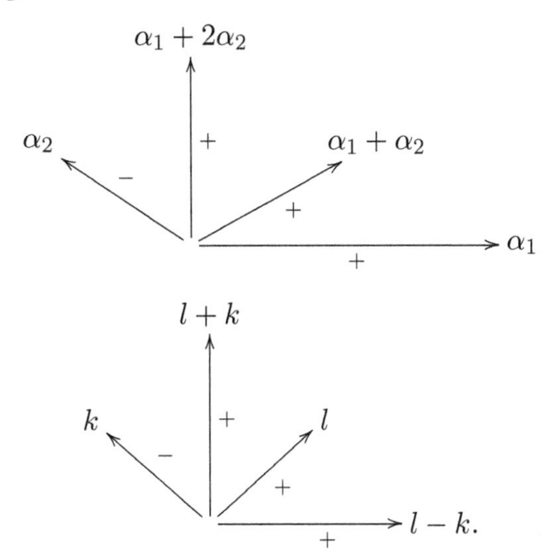

In the first case $k = 0, l = 1$ we obtain a polarized Hodge structure of weight $n = 2$ with Hodge numbers $h^{2,0} = 3$, $h^{1,1} = 4$. Then $\dim D_{M_\varphi} = 3$ and $W = TD_M$, so that D_{M_φ} is a bounded symmetric domain. This is just $\mathrm{Sym}^2 V$ for the standard representation of $\mathrm{Sp}(4)$ and $D_{M_\varphi} \cong \mathcal{H}_2$.

The second case $k = 2, l = 1$ is more interesting. This leads to a polarized Hodge structure of weight $n = 6$ with Hodge numbers $h^{6,0} = 1$, $h^{5,1} = 1$, $h^{4,2} = 2$, $h^{3,3} = 2$. The Hodge domain $D_{M_\varphi} = \mathrm{Sp}(4)/T$ has dimension 4 and is the same as the period domain for weight $n = 3$ polarized Hodge structures of mirror quintic type; i.e., those with $h^{3,0} = h^{2,1} = 1$. In this case (see [CGG]) the exterior differential system is locally equivalent to an *Engel system*

$$\begin{cases} dw - w'dz = 0 \\ dw' - w''dz = 0. \end{cases}$$

We have

$$\begin{cases} \mathfrak{m}^{-1,1} = \mathrm{span}\{-l, -(k-l)\} = \mathrm{span}\{-\alpha_1 - \alpha_2, \alpha_1\} \\ \mathfrak{m}^{-2,2} = \mathrm{span}\{-k\} = \mathrm{span}\{-\alpha_2\} \\ \mathfrak{m}^{-3,3} = \mathrm{span}\{-k-l\} = \mathrm{span}\{-\alpha_1 - 2\alpha_2\}. \end{cases}$$

Then the bracket structure is

$$\begin{cases} [\ ,\ ]\Lambda^2\mathfrak{m}^{-1,1} \xrightarrow{\sim} \mathfrak{m}^{-2,2} \\ [\ ,\ ]\mathfrak{m}^{-1,1} \otimes \mathfrak{m}^{-2,2} \twoheadrightarrow \mathfrak{m}^{-3,3}, \end{cases}$$

exactly as predicted for the Engel system.

Remark: In the introduction we discussed two cases when the Hodge domain is Hermitian symmetric

$$D_{M_\varphi} = M_a(\mathbb{R})/K$$

where K is a maximal compact subgroup of $M_a(\mathbb{R})$ whose center $Z(K)$ is a circle S_0^1. The classical case is when $\operatorname{Ad} S^1$ has only the characters $z, 1, z^{-1}$. This action does *not* give a polarized Hodge structure on $\mathfrak{m}$. If

$$\varphi : S^1 \to S_0^1$$

is the standard two-to-one covering, then $(\mathfrak{m}, B, \operatorname{Ad}\varphi)$ does give a weight two polarized Hodge structure. This mechanism will be illustrated in the case $M = \mathrm{SL}_2$.

In $\mathrm{SL}_2(\mathbb{R})$, for its maximal compact subgroup we have the standard circle

$$S^1 = \left\{ \begin{pmatrix} \cos\theta & -\sin\theta \\ \sin\theta & \cos\theta \end{pmatrix} : 0 \leqq \theta \leqq 2\pi \right\}.$$

The "semi-circle" given by $0 \leqq \theta \leqq \pi$ does not close up in $\mathrm{SL}_2(\mathbb{R})$, but its image does close up to give a circle S_0^1 in the adjoint group

$$\mathrm{SL}_2(\mathbb{R})_a = \mathrm{SL}_2(\mathbb{R})/\{\pm I\}.$$

The adjoint action of S_0^1 on $\mathrm{sl}_2(\mathbb{C})$ has characters $z, 1, z^{-1}$, each occurring with multiplicity one.

If we take $V = \mathbb{Q}^2$ with the standard alternating form Q, then for the natural representation

$$\rho : \mathrm{SL}_2 \to (V, Q)$$

the action of $\rho(S^1)$ gives the polarized weight one Hodge structure on $V \cong H^1(E)$ where $E = \mathbb{C}/\mathbb{Z} + \mathbb{Z}i$. On $(\operatorname{Sym}^2 V, \operatorname{Sym}^2 Q)$ there is an induced weight two polarized Hodge structure. Using the natural identifications $V \cong \check{V}$ and

$$\mathfrak{m} = \operatorname{End}_Q(V) \cong \operatorname{Sym}^2 \check{V} \cong \operatorname{Sym}^2 V,$$

$\operatorname{Sym}^2 Q$ becomes the Cartan-Killing form and we obtain the polarized Hodge structure on $(\mathfrak{m}, B, \operatorname{Ad}\varphi)$ referred to above. Only in the cases of the Hermitian symmetric domains where $M = \mathrm{Sp}(2r)$, $\mathrm{SU}(p,q)$, or $\mathrm{SO}(2,r)$ does the standard polarized Hodge structure on $(\mathfrak{m}, B)$ have a "square root" in the sense illustrated above. These are what we refer to as the classical cases, which arise in the study of Shimura varieties.

Example: SU(2,1) (cf. [C1], [C2], [C3]). The Vogan diagram is

$$\circ\!\!-\!\!\!-\!\!\!-\!\!\bullet$$

thus, $\Psi(\alpha_1) = 0$, $\Psi(\alpha_2) = 2$. Now $\mathfrak{l}_\varphi \in \mathrm{Hom}(R, \mathbb{Z})$ has

$$\langle \alpha_1, \mathfrak{l}_\varphi \rangle = 4k_1, \ \ \langle \alpha_2, \mathfrak{l}_\varphi \rangle = 4k_2 + 2.$$

The roots are $\pm\alpha_1, \pm\alpha_2, \pm(\alpha_1 + \alpha_2)$ and

$$\langle \alpha_1 + \alpha_2, \mathfrak{l}_\varphi \rangle = 4k_1 + 4k_2 + 2.$$

It follows that

$$\mathfrak{h}_\varphi = \begin{cases} \mathfrak{t} & k_1 \neq 0 \\ \mathfrak{k} & k_1 = 0 \end{cases}$$

where $\mathfrak{k}$ is the maximal compact sub-algebra $\mathfrak{k} = \mathfrak{t} + \mathrm{span}\{e_{\alpha_1}, e_{-\alpha_1}\}$.

If $k_1 = 0$, then $\langle \alpha_2, \mathfrak{l}_\varphi \rangle = 4k_2 + 2 = \langle \alpha_1 + \alpha_2, \mathfrak{l}_\varphi \rangle$. Thus

$$\mathfrak{p}_\varphi := \mathfrak{t} + \mathrm{span}\{e_\alpha | \ \langle e_\alpha, \mathfrak{l}_\varphi \rangle \geq 0\}$$

is

$$\mathfrak{p}_\varphi = \begin{cases} \mathfrak{t} + \mathrm{span}\{e_{\alpha_1}, e_{-\alpha_1}, e_{\alpha_2}, e_{\alpha_1+\alpha_2}\} & \text{if } k_2 \geq 0 \\ \mathfrak{t} + \mathrm{span}\{e_{\alpha_1}, e_{-\alpha_1}, e_{-\alpha_2}, e_{-\alpha_1-\alpha_2}\} & \text{if } k_2 < 0. \end{cases}$$

Here

$$M_{\mathfrak{m}} = \mathrm{SU}(2,1)/K \hookrightarrow \mathrm{SL}(3, \mathbb{C})/P_\varphi$$

gives an embedding of the Hermitian symmetric space into its compact dual. The choice of $\mathfrak{p}_\varphi$ determines the complex structure; the two possible complex structures are conjugate.

When $k_1 \neq 0$, the situation is more interesting. There are now six possible outcomes for the inequalities. The root diagram is:

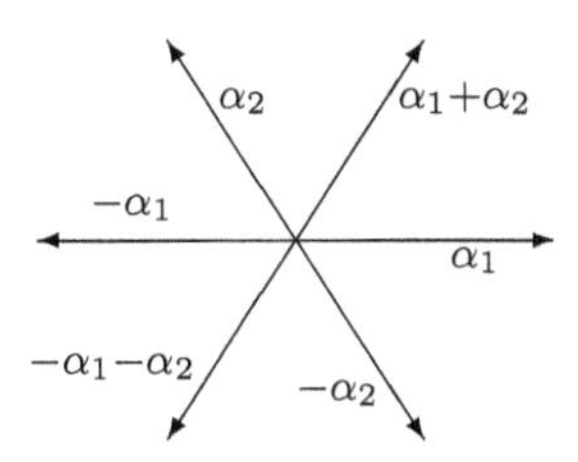

The Weyl chamber diagram is:

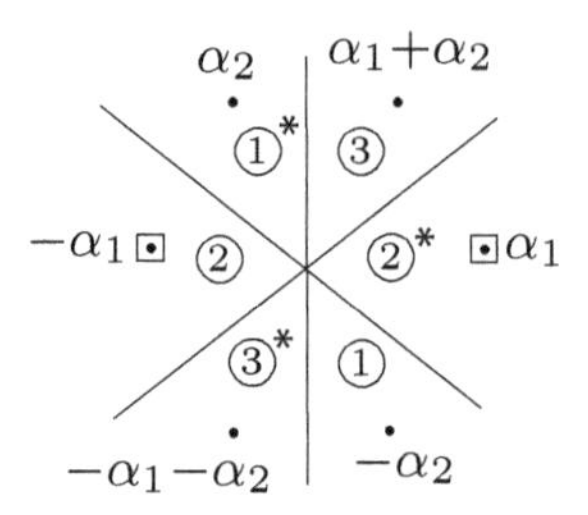

The compact roots have a box around them.

This divides the plane into six Weyl chambers, corresponding to the inequalities

	Inequalities	Basis for $\mathrm{sl}(\mathbb{C})/\mathfrak{p}_\varphi$
(i)	$k_1 > 0,\ k_1 + k_2 > 0$	$e_{\alpha_1}, e_{-\alpha_2}, e_{-\alpha_1-\alpha_2}$
(ii)	$k_2 \geq 0,\ k_1 + k_2 < 0$	$e_{-\alpha_1}, e_{\alpha_2}, e_{-\alpha_1-\alpha_2}$
(iii)	$k_1 > 0,\ k_2 \geqq 0$	$e_{\alpha_1}, e_{\alpha_2}, e_{\alpha_1+\alpha_2}$
(i)*	$k_1 < 0,\ k_1 + k_2 > 0$	$e_{-\alpha_1}, e_{\alpha_2}, e_{\alpha_1+\alpha_2}$
(ii)*	$k_2 \leqq 0,\ k_1 + k_2 > 0$	$e_{\alpha_1}, e_{-\alpha_2}, e_{\alpha_1+\alpha_2}$
(iii)*	$k_1 < 0,\ k_2 \leqq 0$	$e_{-\alpha_1}, e_{-\alpha_2}, e_{-\alpha_1-\alpha_2}.$

Here (i) corresponds to ①, etc., and (i)* is the conjugate complex structure to (i), etc. Now note that φ induces a complex structure on $\mathrm{SU}(2,1)/T$, which is different in each of the six cases.

For (i), (iii), we may choose a complex structure on $\mathrm{SU}(2,1)/K$ so that

$$\mathrm{SU}(2,1)/T \to \mathrm{SU}(2,1)/K$$

is holomorphic, but for (ii) we cannot. The point is that the holomorphic tangent space

$$T\,\mathrm{SU}(2,1)/K \textit{ has basis } e_{\alpha_2}, e_{\alpha_1+\alpha_2} \textit{ or has basis } e_{-\alpha_2}, e_{-\alpha_1-\alpha_2}$$

and for (ii), neither of these is contained in the basis for $\mathrm{sl}_2(\mathbb{C})/\mathfrak{p}_\varphi$.

A discussion of the arithmetic aspects of this example is in [C1], [C2], [C3], whose work inspired this discussion. The complex structure used there is (ii)*.

IV.G MUMFORD-TATE DOMAINS AS PARTICULAR HOMOGENEOUS COMPLEX MANIFOLDS

This section is somewhat in contra-position to the preceding one where aspects of the "universality" of Hodge domains was discussed. Namely, we shall discuss and illustrate the following important

(IV.G.1) **Observation:** *The same complex manifold may appear in multiple ways as a homogeneous complex manifold. These different representations should be regarded as giving distinct Mumford-Tate domains.*

Rather than give a general discussion, we shall illustrate the above observation by a particular example, one that seems to capture the essential features of the general situation.

Example: *Another take on* $\mathcal{SU}(2,1)$.[23] Because of its importance in representation theory [EGW] and in arithmetic automorphic forms [C1], [C2], [C3], and because it is a particularly interesting example of a complex manifold that can be represented in multiple ways as a homogeneous space, we shall give a somewhat detailed discussion of the geometry and Hodge theory associated to it.[24] For this we shall use the following notations.

- e_1, e_2, e_3 will be the standard basis of $\mathbb{C}^3$, represented as column vectors.
- H_0 will be the Hermitian form $\mathrm{diag}(1,1,-1)$.
- $T_{\mathcal{U}} = \left\{ g = \begin{pmatrix} e^{2\pi i\theta_1} & & \\ & e^{2\pi i\theta_2} & \\ & & e^{2\pi i\theta_3} \end{pmatrix} \right\}$ will be the standard maximal torus in $\mathcal{U}(2,1)_{\mathbb{R}}$.
- $T_{\mathcal{SU}} = \{g \text{ as above: } \theta_1+\theta_2+\theta_3 \in \mathbb{Z}\}$ is the standard maximal torus in $\mathcal{SU}(2,1)_{\mathbb{R}}$.
- $T_{\mathrm{ad}} = T_{\mathcal{SU}}/\{g \text{ as above: } \theta_j = \omega \text{ where } \omega^3 = 1\}$ is the standard maximal torus in the adjoint group $\mathcal{SU}(2)_{a,\mathbb{R}}$.
- $\mathfrak{t}_{\mathcal{U}}$, the Lie algebra of $\mathcal{U}(2,1)_{\mathbb{R}}$, will be identified with $\mathbb{R}^3$ with coordinates $\boldsymbol{\theta} = \begin{pmatrix} \theta_1 \\ \theta_2 \\ \theta_3 \end{pmatrix}$.
- $\mathfrak{t}$, the Lie algebra of $\mathcal{SU}(2,1)_{\mathbb{R}}$ and of $\mathcal{SU}(2,1)_{a,\mathbb{R}}$, will be identified with the subspace of $\mathfrak{t}_{\mathcal{U}}$ given by $\mathrm{span}_{\mathbb{R}}\{e_1 - e_2, e_2 - e_3\}$.

[23] *Notational remark.* In this section we shall denote by $\mathcal{SU}(p,q)_{\mathbb{R}}$ and $\mathcal{U}(p,q)_{\mathbb{R}}$ the real Lie groups, as to be distinguished from $\mathcal{SU}(p,q)$ and $\mathcal{U}(p,q)$, which will be the $\mathbb{Q}$-algebraic groups described below and whose associated real Lie groups are $\mathcal{SU}(p,q)_{\mathbb{R}}$ and $\mathcal{U}(p,q)_{\mathbb{R}}$.

[24] The simplest example is the multiple representations of the unit disc as a complex manifold

$$\Delta = \mathrm{SL}_2(\mathbb{R})/\,\mathrm{SO}(2) = \mathcal{SU}(1,1)_{\mathbb{R}}/T_{\mathcal{SU}} = \mathcal{U}(1,1)_{\mathbb{R}}/T_{\mathcal{U}}\,.$$

Here, $T_{\mathcal{SU}}$ and $T_{\mathcal{U}}$ are the maximal tori in $\mathcal{SU}(1,1)_{\mathbb{R}}$ and $\mathcal{U}(1,1)_{\mathbb{R}}$ respectively. We shall revisit this example in a remark at the end of this section.

- $\Lambda_{\mathcal{U}}, \Lambda, \Lambda_a$ will be the lattices so that

$$T_{\mathcal{U}} = \mathfrak{t}_{\mathcal{U}}/\Lambda_{\mathcal{U}}, \quad T_{S\mathcal{U}} = \mathfrak{t}/\Lambda, \quad T_a = \mathfrak{t}/\Lambda_a .$$

- In terms of the above bases

$$\begin{cases} \Lambda_{\mathcal{U}} \cong \mathrm{span}_{\mathbb{Z}}\{e_1, e_2, e_3\} \\ \Lambda \cong \mathrm{span}_{\mathbb{Z}}\{e_1 - e_2, e_2 - e_3\} \\ \Lambda_a \cong \mathrm{span}_{\mathbb{Z}}\left\{\left(\frac{2}{3}\right) e_1 - \left(\frac{1}{3}\right) e_2 - \left(\frac{1}{3}\right) e_3, \left(\frac{1}{3}\right) e_1 + \left(\frac{1}{3}\right) e_2 - \left(\frac{2}{3}\right) e_3\right\}. \end{cases}$$

This gives

$$\begin{cases} 0 \to \Lambda \to \Lambda_{\mathcal{U}} \to \mathbb{Z} \to 0 \\ 0 \to \Lambda \to \Lambda_a \to \mathbb{Z}/3\mathbb{Z} \to 0 . \end{cases}$$

- We denote by $e_1^*, e_2^*, e_3^* \in \check{\mathfrak{t}}_{\mathcal{U}}$ the dual basis to e_1, e_2, e_3. Then $\check{\mathfrak{t}}$ is spanned by e_1^*, e_2^*, e_3^* with the generating relation $e_1^* + e_2^* + e_3^* = 0$.

- The relations among the $e_i^* \in \check{\mathfrak{t}}$ and the roots are[25]

$$\begin{cases} \alpha_1 = e_2^* - e_1^* \\ \alpha_2 = e_3^* - e_2^* \end{cases}$$

and

$$\begin{cases} e_1^* = -\left(\frac{2}{3}\right) \alpha_1 - \left(\frac{1}{3}\right) \alpha_2 \\ e_2^* = \left(\frac{1}{3}\right) (\alpha_1 - \alpha_2) \end{cases}$$

(recalling here that $e_1^* + e_2^* + e_3^* = 0$ in $\check{\mathfrak{t}}$).

- The weight lattice P is generated over $\mathbb{Z}$ by the e_i^*, and the root lattice R has α_1, α_2 as a $\mathbb{Z}$-basis. From

$$e_2^* - e_1^* = \alpha_1$$

[25]In the literature, using the Cartan-Killing form there is frequently an identification $\check{\mathfrak{t}} \cong \mathfrak{t}$, so that one finds notations such as

$$\begin{cases} \alpha_1 = e_2 - e_1 \\ \alpha_2 = e_3 - e_2 . \end{cases}$$

Because $\mathcal{U}(2,1)$ is only reductive so that the Cartan-Killing form is not definite, and also partly for notational clarity, we have chosen not to make this identification.

We remark that the most appropriate setting for the general theory of Mumford-Tate groups and domains is when G is reductive. When G is reductive then $G_{\mathbb{R}}$ is isogeneous to $G_{a,\mathbb{R}} \times A$ where A is compact, as is the case for $\mathcal{U}(2,1)$. There are then fairly evident extensions of the general results, such as Theorem (IV.E.2), that we shall use.

we see that

$$P = \left(\left(\tfrac{2}{3}\right)\alpha_1 + \left(\tfrac{1}{3}\right)\alpha_2\right)\mathbb{Z} + R$$

confirming that

$$P/R \cong \mathbb{Z}/3\mathbb{Z}\,.$$

- The root diagram is:

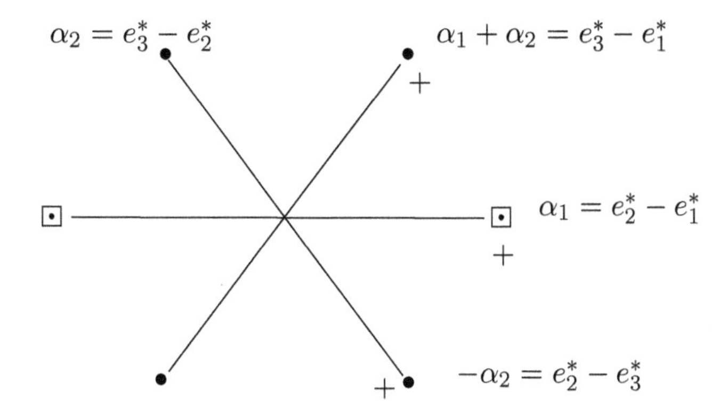

and the compact roots have a box around them. For the complex structure (ii)* at the end of Section IV.F the positive roots have a + sign.

- This complex structure is one that does not fibre holomorphically or anti-holomorphically over an Hermitian symmetric domain. It may be described as follows: Let

$$\check{D} = \mathrm{SL}(3, \mathbb{C})/B$$

be the flag manifold for $\mathbb{C}^3$. As reference flag we choose $[e_1] \subset [e_1, e_3] \subset [e_1, e_3, e_2]$ where $[\ ]$ denotes the span of the indicated vectors. The Borel subgroup fixing the reference flag is

$$B = \left\{ \begin{pmatrix} * & * & * \\ 0 & * & 0 \\ 0 & * & * \end{pmatrix} \right\}.$$

There are the three representations as a homogeneous space

$$\begin{cases} D_{\mathcal{U}} = \mathcal{U}(2,1)_{\mathbb{R}}/T_{\mathcal{U}} \\ D = \mathcal{SU}(2,1)_{\mathbb{R}}/T_{\mathcal{SU}} \\ D_a = \mathcal{SU}(2,1)_{\mathrm{ad},\mathbb{R}}/T_{\mathrm{ad}} \end{cases}$$

of the *same* complex manifold given by the $D = \mathcal{SU}(2,1)_{\mathbb{R}}$-orbit of the reference flag.

- For each of these there is a different set of differentials l_φ of co-characters that satisfy the conditions to give a polarized Hodge structure.

For this we write $l_\varphi \in \Lambda_{\mathcal{U}}$ as

$$l_\varphi = l_1 e_2 + l_2 e_2 + l_3 e_3, \qquad l_i \in \mathbb{Z}.$$

Then

$$l_\varphi \in \Lambda \Leftrightarrow l_1 + l_2 + l_3 = 0. \tag{IV.G.2}$$

We have from the discussion of $S\mathcal{U}(2,1)$ at the end of Section IV.F

$$\begin{cases} \langle \alpha_1, l_\varphi \rangle = l_2 - l_1 \equiv 0 (\operatorname{mod} 4) \Rightarrow l_2 - l_1 = 4k_1 \\ \langle \alpha_2, l_\varphi \rangle = l_3 - l_2 \equiv 2 (\operatorname{mod} 4) \Rightarrow l_3 - l_2 = 4k_2 + 2. \end{cases} \tag{IV.G.3}$$

For the highest weight

$$\lambda = \lambda_1 e_1^* + \lambda_i e_2^* + \lambda_i e_3^*, \qquad \lambda_i \in \mathbb{Z}$$

we have

$$\langle \lambda, l_\varphi \rangle = l_1 \lambda_1 + l_2 \lambda_2 + l_3 \lambda_3.$$

The conditions (IV.G.3) give

$$\begin{cases} l_2 = 4k_1 + l_1 \\ l_3 = 4(k_1 + k_2) + 2 + l_1 \end{cases}$$

so that

$$\langle \lambda, l_\varphi \rangle = \lambda_1 l_1 + 4\lambda_2 k_1 + 2\lambda_3 (2(k_1 + k_2) + 1).$$

<u>$\mathcal{U}(2,1)_{\mathbb{R}}$</u>: We have a polarized Hodge structure of weight n where

$$n \equiv \lambda_1 l_1 \pmod 2.$$

Here, n may be even or odd.

<u>$S\mathcal{U}(2,1)_{\mathbb{R}}$</u>: The constraint (IV.G.3) gives

$$\begin{cases} 2l_2 + l_3 = 4k_1 \\ l_2 + 2l_3 = 4(k_1 + k_2) + 2. \end{cases}$$

This implies that $l_2 \equiv 0 \pmod 2$ and $l_3 \equiv 0 \pmod 2$; hence $\langle \lambda, l_\varphi \rangle \equiv 0 \pmod 2$ and only even weight polarized Hodge structures may occur.

$\underline{\mathcal{SU}(2,1)_{a,\mathbb{R}}}$: There is now a further condition $l_\varphi \in \Lambda_a$. This works out to

$$l_\varphi = m_1 \left(\left(\frac{2}{3} \right) e_1 - \frac{e_2}{3} - \frac{e_3}{3} \right) + m_2 \left(\frac{e_1}{3} + \frac{e_2}{3} - \left(\frac{2}{3} \right) e_3 \right).$$

The conditions analogous to (IV.G.3) are

$$\begin{cases} \langle e_2^* - e_3^*, l_\varphi \rangle = m_2 \equiv 0 \pmod 4 \\ \langle e_1^* - e_2^*, l_\varphi \rangle = m_1 \equiv 2 \pmod 4 . \end{cases}$$

Computation gives that $\langle \lambda, l_\varphi \rangle \equiv 0 \pmod 2$, so that again only even weight polarized Hodge structures occur.

To obtain Hodge representations, we need $\mathbb{Q}$-algebraic groups $\mathcal{SU}(2,1)$ and $\mathcal{U}(2,1)$ whose associated real Lie groups are $\mathcal{SU}(2,1)_\mathbb{R}$ and $\mathcal{U}(2,1)_\mathbb{R}$, together with a representation of $\mathcal{SU}(2,1)$ or $\mathcal{U}(2,1)$ satisfying the conditions of Theorem (4.E.2). For the computation of examples the key computation at the end of step seven in the part **Steps in the analysis of Hodge representations** in Part II of Section IV.A is the crucial ingredient.

In general, for a given Mumford-Tate domain $M_\mathbb{R}/H$, there will be several Hodge representations of M giving $M_\mathbb{R}/H$ as the corresponding Mumford-Tate domain. A natural question is:

(IV.G.4)

For a given $M_\mathbb{R}/H$ is there a "natural" Hodge representation of M?

We do not know the answer to, or even a precise formulation of, this question. One advantage of a precise formulation and positive answer to this question would be that, given the complex homogeneous manifold $M_\mathbb{R}/H$, there will be a natural, or preferred, way to bring Hodge theory into the study of it. For the group $\mathcal{SU}(2,1)$ and $\mathcal{U}(2,1)$ we shall illustrate and provide one answer to the above question.

For the complex, simple Lie group $\mathrm{SL}(3,\mathbb{C})$ given in the standard representation, for the real form $\mathrm{SL}(3,\mathbb{R})$ there is the obvious $\mathbb{Q}$-algebraic group $\mathrm{SL}(3,\mathbb{Q})$ whose associated real Lie group is $\mathrm{SL}(3,\mathbb{R})$. However, for $\mathcal{SU}(2,1)_\mathbb{R}$ given by

$$\{A \in \mathrm{SL}(3,\mathbb{C}) : AH_0\, {}^t\bar{A} = H_0\}$$

where $H_0 = \mathrm{diag}(1,1,-1)$, the $\mathbb{Q}$-form $\mathcal{SU}(2,1)$ whose real Lie group is $\mathcal{SU}(2,1)_\mathbb{R}$ is not evident from this perspective.

A better way to proceed is first to observe that in a Mumford-Tate domain $M_\mathbb{R}/H$ there are always CM points (cf. (VI.C.1)). Next, when $M = \mathcal{SU}(p,q)$ or $\mathcal{U}(p,q)$ there are special conditions on the CM field to have a $\mathbb{Q}$-algebraic torus in M. This result (cf. the appendix to Chapter VI) leads to the following construction, done here only when $p = 2$ and $q = 1$. We shall proceed in three steps:

(i) determine Hodge structures of a certain type;

(ii) put a real polarization on them;

(iii) ensure that the polarization is rational.

Let $\mathbb{F} = \mathbb{Q}(\sqrt{-d})$ where $d > 0$ is a squarefree positive rational number ($d = 1$ will do), and let V be a 6-dimensional $\mathbb{Q}$-vector space with an $\mathbb{F}$-action; i.e., an embedding

$$\mathbb{F} \hookrightarrow \mathrm{End}_{\mathbb{Q}}(V)\,.$$

Setting $V_{\mathbb{F}} = V \otimes_{\mathbb{Q}} \mathbb{F}$, we have over $\mathbb{F}$ the eigenspace decomposition

$$V_{\mathbb{F}} = V_{+} \oplus V_{-}$$

where $\overline{V}_{+} = V_{-}$. We will show how to construct polarized Hodge structures of weights $n = 4$, $n = 3$, and $n = 2$ with respective Mumford-Tate groups $\mathcal{U}(2,1)$, $\mathcal{U}(2,1)$, and $\mathcal{SU}(2,1)$. For this we write $V_{\mathbb{C}} = V_{+,\mathbb{C}} \oplus V_{-,\mathbb{C}}$. We shall do the $n = 4$ case first, and for this we consider the following picture:

$(4,0)$	$(3,1)$	$(2,2)$	$(1,3)$	$(0,4)$	
$*$	$*$	$*$			$V_{+,\mathbb{C}}$
		$*$	$*$	$*$	$V_{-,\mathbb{C}}$

Figure IV.1

The notation means this: Choose a decomposition $V_{+,\mathbb{C}} = V_{+}^{4,0} \oplus V_{+}^{3,1} \oplus V_{+}^{2,2}$ into 1-dimensional subspaces. Then define $V_{-,\mathbb{C}} = V_{-}^{2,2} \oplus V_{-}^{1,3} \oplus V_{-}^{0,4}$ where $V_{-}^{p,q} = \overline{V_{+}^{q,p}}$. Setting $V^{p,q} = V_{+}^{p,q} \oplus V_{-}^{p,q}$ gives a Hodge structure.[26]

Next we define a *real* polarization by requiring $Q(V_{+},V_{+})=0=Q(V_{-},V_{-})$, then choosing a non-zero vector $\omega_{+}^{p,q} \in V_{+}^{p,q}$ and setting

$$\begin{cases} Q(\omega_{+}^{4,0},\overline{\omega}_{+}^{4,0}) = 1, & \overline{\omega}_{+}^{4,0} \in V^{(0,4)} \\ Q(\omega_{+}^{3,1},\overline{\omega}_{+}^{3,1}) = -1, & \overline{\omega}_{+}^{3,1} \in V_{-}^{(1,3)} \\ Q(\omega_{+}^{2,2},\overline{\omega}_{+}^{2,2}) = 1, & \overline{\omega}_{+}^{2,2} \in V_{-}^{(2,2)}\,. \end{cases}$$

All other $Q(*,*) = 0$.

[26]In general, the number of $*$'s in a box will denote the dimension of the complex vector space.

Finally, we may choose the $V_+^{p,q}$ to be defined over $\mathbb{F}$ and $\omega_+^{p,q} \in V_{+,\mathbb{F}}$. Then

$$\begin{cases} \frac{1}{2}(\omega_+^{p,q} + \overline{\omega}_+^{p,q}) = e_{e_5-p} & p = 4,3,2 \\ \frac{1}{2\sqrt{-d}}(\omega_+^{p,q} - \overline{\omega}_+^{p,q}) = e_{7-p} & p = 3,2,1 \end{cases}$$

gives a basis $e_1, \ldots, e_6$ for $V_{\mathbb{R}} \cap V_{\mathbb{F}} = V$. In terms of this basis, the matrix entries of Q are in $\mathbb{R} \cap \mathbb{F} = \mathbb{Q}$.

We observe that, by construction, the action of $\mathbb{F}$ on V preserves the form Q. We set

$$\mathcal{U} = \operatorname{Aut}_{\mathbb{F}}(V, Q).$$

This is an $\mathbb{F}$-algebraic group, and we then set

$$\mathcal{U}(2,1) = \operatorname{Res}_{\mathbb{F}/\mathbb{Q}} \mathcal{U}.$$

(IV.G.5) PROPOSITION: (i) $\mathcal{U}(2,1)$ *is a* $\mathbb{Q}$*-algebraic group whose associated real Lie group is* $\mathcal{U}(2,1)_{\mathbb{R}}$. (ii) *If we operate on the reference polarized Hodge structure conjugated by a generic* $g \in \operatorname{Aut}_{\mathbb{F}}(V_{\mathbb{R}}, Q) \cong \mathcal{U}(\mathbb{R})$, *the resulting polarized Hodge structure has Mumford-Tate group* $\mathcal{U}(2,1)$.

PROOF. Setting $J = \begin{pmatrix} & & 1 \\ & -1 & \\ 1 & & \end{pmatrix}$, the matrix of Q in the $\mathbb{Q}$-basis $e_1, \ldots, e_6$ for V is

$$Q = \begin{pmatrix} J & 0 \\ 0 & \left(\frac{1}{d}\right) J \end{pmatrix}.$$

In terms of this basis, $V_{+,\mathbb{F}}$ is spanned by the columns in the matrix

$$\begin{pmatrix} I \\ \sqrt{-d} I \end{pmatrix}.$$

If $g \in \operatorname{Aut}_{\mathbb{F}}(V)$, then the extension of g to $V_{\mathbb{F}}$ commutes with the projections onto $V_{+,\mathbb{F}}$ and $V_{-,\mathbb{F}}$. A calculation shows that these are equations defined over $\mathbb{Q}$. The conditions that g preserve Q are further equations defined over $\mathbb{Q}$. Thus, $\mathcal{U}$ is a $\mathbb{Q}$-algebraic group. Moreover, g is uniquely determined by its restriction to the induced mapping

$$g_+ : V_{+,\mathbb{F}} \to V_{+,\mathbb{F}}.$$

In terms of the basis $\omega_+^{4,0}, \omega_+^{3,1}, \omega_+^{2,2}$ of $V_{+,\mathbb{C}} \cong \mathbb{C}^3$, g_+ preserves the Hermitian form J; i.e.,

$${}^t\overline{g}_+ J g_+ = J.$$

This shows that the real points $\mathcal{U}(\mathbb{R})$ have an associated Lie group isomorphic to $\mathcal{U}(2,1)_{\mathbb{R}}$, and therefore proves (i). The proof of (ii) is similar to that of (IV.A.9). The reason that the Mumford-Tate is $\mathcal{U}(2,1)$ and not $S\mathcal{U}(2,1)$ is that the circle $\{z \in \mathbb{C} : |z| = 1\}$ acts on $\omega_+^{p,q}$ by z^{p-q} and $z^4 \cdot z^2 \cdot z^0 = z^6 \neq 1$. $\square$

To obtain a polarized Hodge structure of weight $n = 2$ with Mumford-Tate group $\mathcal{SU}(2,1)$ we do the construction as shown in Figure IV.2.

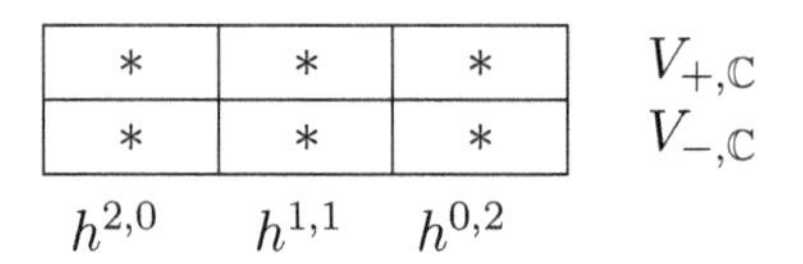

Figure IV.2

We are in $\mathcal{SU}(2,1)$ because $z^2 \cdot z^0 \cdot z^{-2} = 1$.

To obtain a polarized Hodge structure of weight $n = 3$ with Mumford-Tate group $\mathcal{U}(2,1)$ we do a similar construction (Fig. IV.3).

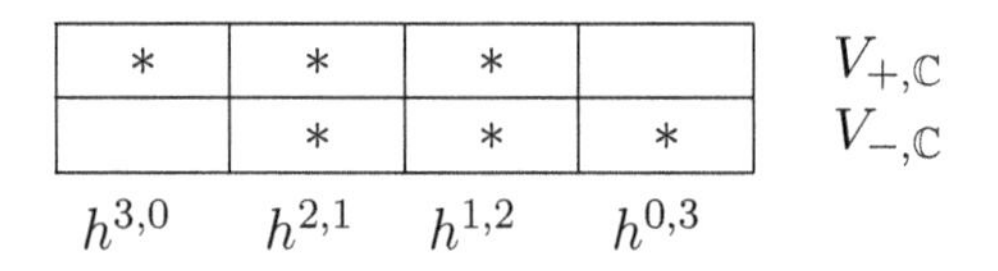

Figure IV.3

A difference is that, in order to have Q alternating, we set

$$iQ(\omega_+^{3,0}, \overline{\omega}_+^{3,0}) = 1\,.$$

All of the above give Mumford-Tate domains that are of the form $M_{\mathbb{R}}/T$ where T is a compact maximal torus. The picture when $n = 1$ (Fig. IV.4)

$h^{1,0}$	$h^{0,1}$	
**	*	$V_{+,\mathbb{C}}$
*	**	$V_{-,\mathbb{C}}$

Figure IV.4

gives a Mumford-Tate domain $\mathcal{U}(2,1)/\mathcal{U}(2) \times \mathcal{U}(1)$, which as a complex manifold is $\mathcal{SU}(2,1)/S(\mathcal{U}(2) \times \mathcal{U}(1))$. It is an Hermitian symmetric domain parametrizing polarized abelian varieties of dimension 3 and with an $\mathbb{F}$-action.

We now realize $\mathcal{U}(2,1)_{\mathbb{R}}$ as unitary matrices acting on $V_{+,\mathbb{C}} \cong \mathbb{C}^3$ with the unitary basis $\omega_1^{3,0} = v_1, \omega_{+1}^{2,1} = v_2$, $\omega_+^{1,2} = v_3$ described in Figure IV.3. The Hermitian matrix H is now $\operatorname{diag}(1,-1,1)$, and the $l_\varphi = l_1 v_1 + l_2 v_2 + l_3 v_3$ that gives the differential of the co-character giving the circle in Figure IV.1 has

$$\begin{cases} l_1 = 3 \\ l_2 = -1 \\ l_3 = 1. \end{cases}$$

Using our previous notation this gives

$$\begin{cases} k_1 = -1 \\ k_2 = 0. \end{cases}$$

This is the "minimal" pair k_1, k_2 satisfying the inequalities in (ii)* in the table at the end of Section IV.F.

We now discuss the question (IV.G.3). To begin, as illustrated above we first must identify the Mumford-Tate domain as a homogeneous space $D = M_{\mathbb{R}}/H$. It is understood that we want a generic point of D to have Mumford-Tate group M, and not a subgroup or quotient group of M. Next, we have to specify the complex structure. The following illustrates the case of $\mathcal{U}(2,1)$.

Example: We take $M = \mathcal{U}(2,1)$ and use the complex structure (ii)*. This choice of complex structure rules out Figures IV.1 and IV.4. That M is $\mathcal{U}(2,1)$ and not $S\mathcal{U}(2,1)$ rules out Figure IV.2. This leaves Figure IV.3 as the "natural" choice for $\mathcal{U}(2,1)$.

We note that this $D_{\mathcal{U}}$ with the complex structure (ii)* is the main object of study in [C1], [C2], [C3]. Although Carayol did not make explicit use of $D_{\mathcal{U}}$ as a Mumford-Tate domain, it is implicit in [C3] where he used the Kato-Usui method of extending, or partially compactifying, $D_{\mathcal{U}}$ by adding spaces of nilpotent orbits, which are Hodge-theoretic objects. In [C3] the Hodge theory that entered is exactly that in Figure IV.3.

Remark: For $\mathcal{U}(1,1)$ and $S\mathcal{U}(1,1)$ we take V of dimension 4 with an $\mathbb{F}$-action and do the above construction in the cases:

(i)

	$h^{1,0}$	$h^{0,1}$
$V_{+,\mathbb{C}}$	*	*
$V_{-,\mathbb{C}}$	*	*

(ii)

	$h^{2,0}$	$h^{1,1}$	$h^{0,2}$
$V_{+,\mathbb{C}}$	*	*	
$V_{-,\mathbb{C}}$		*	*

(iii)

	$h^{3,0}$	$h^{2,1}$	$h^{2,2}$	$h^{0,3}$
$V_{+,\mathbb{C}}$	*	*		
$V_{-,\mathbb{C}}$			*	*

In case (i) there is no distinction between $\mathcal{U}(1,1)$ and $S\mathcal{U}(1,1)$; $S\mathcal{U}(1,1)$ is a Mumford-Tate group. In case (ii) the Mumford-Tate group is $\mathcal{U}(1,1)$; the corresponding Mumford-Tate domain parametrizes $K3$ surfaces with Picard number $\rho \geqq 18$. In case (iii) one has polarized Hodge structures of the type that are associated to Calabi-Yau varieties of mirror quintic type. We suspect, but do not know, if any of these are motivic.

APPENDIX TO CHAPTER IV: NOTATION FROM THE STRUCTURE THEORY OF SEMI-SIMPLE LIE ALGEBRAS

• T will denote a compact, real torus of dimension r and with Lie algebra $\mathfrak{t}$. Via the exponential map we have an identification

$$T = \mathfrak{t}/\Lambda$$

where Λ is a lattice in $\mathfrak{t}$. Choosing an isomorphism $\Lambda \cong \mathbb{Z}^r$ gives $\mathfrak{t} \cong \mathbb{R}^r$ and an identification

$$T = \underbrace{S^1 \times \cdots \times S^1}_{r} \tag{A.1}$$

where $S^1 \subset \mathbb{C}$ is the unit circle. Using this identification we write elements of T as

$$t = (t_1, \ldots, t_r) = (e^{2\pi i\theta_1}, \ldots, e^{2\pi i\theta_r})$$

where $|t_i| = 1$ and where $\boldsymbol{\theta} = (\theta_1, \ldots, \theta_r) \in \mathbb{R}^r \cong \mathfrak{t}$.

• A *character* of T is a homomorphism

$$\chi : T \to S^1.$$

Using the identification (A.1), we will have

$$\chi(t) = t_1^{w_1} \ldots t_r^{w_r} = e^{2\pi i\langle\omega,\theta\rangle}$$

where $\omega = (w_1, \ldots, w_r)$ with $w_i \in \mathbb{Z}$. The correspondence $\chi \to \omega$ gives a natural identification

$$X(T) \cong \operatorname{Hom}(\Lambda, \mathbb{Z}).$$

• A *co-character*, or equivalently a 1-parameter subgroup, is a homomorphism

$$\varphi : S^1 \to T.$$

The co-characters form a group $\check{X}(T)$. Writing $z = e^{2\pi i\xi} \in S^1$, we have

$$\varphi(z) = \left(z^{l_1}, \ldots, z^{l_r}\right), \quad l_i \in \mathbb{Z}$$
$$\boldsymbol{\theta} = (l_1\xi, \ldots, l_r\xi).$$

Setting $\mathfrak{l}_\varphi = (l_1, \ldots, l_r)$, the map $1 \to \mathfrak{l}_\varphi$ gives a natural identification

$$\check{X}(T) \cong \operatorname{Hom}(\mathbb{Z}, \Lambda).$$

The composition of a character χ and co-character φ is given by

$$(\chi \circ \varphi)(z) = e^{2\pi i \langle \omega, \mathfrak{l}_\varphi \rangle \xi}.$$

• Since the real Lie group $M(\mathbb{R})$ contains a compact maximal torus T (Proposition (IV.A.2)), we will have

$$\mathfrak{t} \subset \mathfrak{m}_{\mathbb{R}}.$$

The complexification $\mathfrak{m}_{\mathbb{C}}$ is a complex semi-simple Lie algebra with Cartan sub-algebra

$$\mathfrak{t}_{\mathbb{C}} = \mathfrak{t} \otimes \mathbb{C}.$$

We will denote conjugation of $\mathfrak{m}_{\mathbb{C}}$ with respect to the real form $\mathfrak{m}_{\mathbb{R}}$ by

$$x \to \overline{x}.$$

With the identification (A.1), we let $H_1, \ldots, H_r$ be the basis for Λ and set

$$H_j = \sqrt{-1} h_j$$

where $h_1, \ldots, h_r$ give a basis for $\mathfrak{h}$.

• Under the adjoint action of $\mathfrak{h}$ on $\mathfrak{m}_{\mathbb{C}}$, we have the usual direct sum decomposition

$$\mathfrak{m}_{\mathbb{C}} = \mathfrak{h} \oplus \left(\underset{\alpha \in \mathfrak{r}}{\oplus} \mathfrak{m}_\alpha \right)$$

where $\mathfrak{r} \subset i\check{\mathfrak{t}} \subset \check{\mathfrak{h}}$ are the set of *roots*. The root spaces $\mathfrak{m}_\alpha$ are 1-dimensional, and for $h \in \mathfrak{h}$ and $x \in \mathfrak{m}_\alpha$ we have

$$[h, x] = \langle \alpha, h \rangle \, x.$$

• Since the roots are purely imaginary on $\mathfrak{t}$; thus, taking $h \in \mathfrak{t}$ and conjugating this relation using $\overline{h} = h$ gives that $-\alpha$ is a root and

$$\mathfrak{m}_{-\alpha} = \overline{\mathfrak{m}}_\alpha.$$

• We have $[\mathfrak{m}_\alpha, \mathfrak{m}_{-\alpha}] \subset \mathfrak{h}$, and it is possible to choose $e_\alpha \in \mathfrak{m}_\alpha$, $e_{-\alpha} \in \mathfrak{m}_{-\alpha}$ and *co-roots* $h_\alpha \in \mathfrak{h}$ such that:

$$\begin{cases} [h_\alpha, e_\alpha] &= 2e_\alpha \\ [h_\alpha, e_{-\alpha}] &= -2e_{-\alpha} \\ [e_\alpha, e_{-\alpha}] &= h_\alpha. \end{cases}$$

Thus $\{h_\alpha, e_\alpha, e_{-\alpha}\}$ span an sl_2 in $\mathfrak{m}_{\mathbb{C}}$.

• If $\alpha+\beta \in \mathfrak{r}$, then

$$[e_\alpha, e_\beta] = N_{\alpha\beta} e_{\alpha+\beta}$$

where the $N_{\alpha\beta}$ are non-zero real constants such that

$$N_{-\alpha,-\beta} = -N_{\alpha,\beta} = N_{-\beta,\alpha+\beta} = N_{\alpha+\beta,-\alpha}.$$

If $\alpha+\beta$ is not a root, then $[e_\alpha, e_\beta] = 0$ (where we consider 0 as a root).

• The *Cartan-Killing form*

$$B(x,y) = \text{Trace}(\operatorname{ad} x \operatorname{ad} y) \qquad x, y \in \mathfrak{m}_\mathbb{C}$$

is symmetric, non-singular on $\mathfrak{m}_\mathbb{C}$, positive definite on it and therefore determines an inner product $(\ ,\ )$ on $i\check{\mathfrak{t}}$. The hyperplanes $P_\alpha = \{\lambda \in i\check{\mathfrak{t}} : (\lambda, \alpha) = 0\}$, $\alpha \in \mathfrak{r}$, divide $i\check{\mathfrak{t}}$ into a finite number of closed convex cones, the *Weyl chambers*. The reflections in the P_α generate the *Weyl group* W, which leaves $\mathfrak{r}$ invariant and permutes the Weyl chambers simply and transitively. A system of *positive roots* $\mathfrak{r}^+$ is a subset of $\mathfrak{r}$ such that:

(i) for $\alpha \in \mathfrak{r}$, either α or $-\alpha \in \mathfrak{r}^+$;

(ii) if $\alpha, \beta \in \mathfrak{r}^+$ and $\alpha+\beta \in \mathfrak{r}$, then $\alpha+\beta \in \mathfrak{r}^+$.

Associated to $\mathfrak{r}^+$ is the *dominant Weyl chamber*

$$C^* = \{\lambda \in i\check{\mathfrak{t}} : (\alpha, \lambda) \geqq 0 \ \text{ for all } \ \alpha \in \mathfrak{r}^+\}.$$

The Weyl group acts simply on the sets of positive roots and on the set of Weyl chambers. In particular, there is a unique $w_0 \in W$ with $w_0(\mathfrak{r}^+) = \mathfrak{r}^-$.

• The Cartan-Killing form has the properties

$$\text{(A.2)} \qquad \begin{cases} B(e_\alpha, e_\beta) = \delta_{\alpha,-\beta} \\ B(h_\alpha, h) \ = \langle \alpha, h\rangle \qquad \text{for } h \in \mathfrak{h}. \end{cases}$$

• For Mumford-Tate groups, there will be singled out a unique maximal compact subgroup $K \subset M(\mathbb{R})$ with $T \subset K$. There is an involution

$$\theta : \mathfrak{m}_\mathbb{R} \to \mathfrak{m}_\mathbb{R}, \qquad \theta^2 = 1$$

and a direct sum decomposition

$$\mathfrak{m}_\mathbb{R} = \mathfrak{k} \oplus \mathfrak{p}$$

where $\theta = 1$ on $\mathfrak{k}$ and $\theta = -1$ on $\mathfrak{p}$.

The roots decompose

$$\mathfrak{r} = \mathfrak{r}_c \oplus \mathfrak{r}_{nc}$$

into the *compact roots*, i.e., those α with $e_\alpha \in \mathfrak{k}_{\mathbb{C}}$, and the *non-compact roots*; i.e., those α with $e_\alpha \in \mathfrak{p}_{\mathbb{C}}$. Then

$$\text{(A.3)} \qquad \overline{e_\alpha} = \epsilon_\alpha e_{-\alpha} \;\text{ where }\; \begin{cases} \epsilon_\alpha = -1 & \text{if } \alpha \in \mathfrak{r}_c \\ \epsilon_\alpha = 1 & \text{if } \alpha \in \mathfrak{r}_{nc}. \end{cases}$$

The relation $[\mathfrak{k}, \mathfrak{p}] \subseteq \mathfrak{p}$ gives

$$\epsilon_{\alpha+\beta} = -\epsilon_\alpha \epsilon_\beta \;\text{ when }\; \alpha, \beta \;\text{ and }\; \alpha + \beta \in \mathfrak{r}.$$

It follows that from (A.2) and (A.3)

$$\text{(A.4)} \qquad B(e_\alpha, \overline{e_\alpha}) = \epsilon_\alpha = \begin{cases} 1 & \text{if } \alpha \text{ is non-compact} \\ -1 & \text{if } \alpha \text{ is compact.} \end{cases}$$

In light of (A.4), we shall write

$$\left\{ \begin{array}{l} \mathfrak{m}_{\mathbb{R}}^{+} = \operatorname{span}\{(\mathfrak{m}_\alpha \oplus \mathfrak{m}_{-\alpha})_{\mathbb{R}} : \alpha \text{ non-compact} \\ \mathfrak{m}_{\mathbb{R}}^{-} = \operatorname{span}\{(\mathfrak{m}_\alpha \oplus \mathfrak{m}_{-\alpha})_{\mathbb{R}} : \alpha \text{ compact} \end{array} \right\}.$$

• A root is *simple* if it is not a non-trivial sum of roots. Given a choice of positive roots, there is then determined a set

$$\alpha_1, \ldots, \alpha_r$$

of simple, positive roots.

• We shall denote by $R \subset i\check{\mathfrak{t}}$ the *root lattice*, defined as span over $\mathbb{Z}$ of a set of simple roots.

• An element $\lambda \in i\check{\mathfrak{t}}$, which is the differential of a character of T is a *weight*. The weights form a lattice

$$P = \left\{ \lambda \in i\check{\mathfrak{t}} : \frac{2(\lambda, \alpha)}{(\alpha, \alpha)} \in \mathbb{Z} \;\text{ for all }\; \alpha \in R \right\}$$

and

$$R \subset P.$$

• The restriction to $\mathfrak{h}$ of the extension to $\mathfrak{m}_{\mathbb{C}}$ of an irreducible real representation

$$\rho : \mathfrak{m}_{\mathbb{R}} \to \operatorname{End}(V_{\mathbb{R}})$$

decomposes into *weight spaces*

$$V_{\mathbb{C}} = \bigoplus_{\omega \in W(\rho)} V_\omega$$

where $W(\rho) \subset P$ is the set of weights associated to ρ and

$$V_\omega = \{v \in V_{\mathbb{C}} : \rho(h)(v) = \langle \omega, h \rangle \, v \ \text{ for } \ h \in \mathfrak{h}\}\,.$$

There is a unique *highest weight* $\lambda \in C$ characterized by

$$\rho(e_\alpha) V_\lambda = 0 \qquad \alpha \in \mathfrak{r}^+.$$

• Alternatively the *weight lattice* P is defined as the set of $\lambda \in i\check{\mathfrak{t}}$ such that $\langle \lambda, h_\alpha \rangle \in \mathbb{Z}$ for all $\alpha \in R$. The root and weight lattices satisfy

$$R \subseteq \mathrm{Hom}(\Lambda, \mathbb{Z}) \subseteq P.$$

• The *Cartan matrix* C is defined by

$$C_{ij} = \langle \alpha_j, h_{\alpha_i} \rangle\,.$$

If $\omega = \sum_j q_j \alpha_j$ is a weight, then

$$\langle \omega, h_{\alpha_i} \rangle = \sum_j C_{ij} q_j.$$

Thus, if $D = \det \|C_{ij}\|$ then $\omega \in \left(\frac{1}{D}\right) R$; i.e.,

$$P = \left(\tfrac{1}{D}\right) R.$$

• The finite coverings groups of a simple algebraic $\mathbb{Q}$-algebraic group M and those of the associated real Lie group $M(\mathbb{R})$ are in one-to-one correspondence.

• Denoting by M_a and M_s the adjoint group and universal covering group respectively, for $T_a = \mathfrak{t}/\Lambda_a$ and $T_s = \mathfrak{t}/\Lambda_s$ a compact maximal torus we have $\Lambda_s \subset \Lambda_a$ and

$$\begin{cases} R = \mathrm{Hom}(\Lambda_a, \mathbb{Z}) \\ P = \mathrm{Hom}(\Lambda_s, \mathbb{Z}). \end{cases}$$

There are $\mathbb{Q}$-algebraic groups with any lattice Λ between Λ_s and Λ_a.

• Given a choice of positive roots we set

$$h_0 = \left(\sum_{\alpha \in R+} \alpha \right).$$

This is a non-singular weight and lies in the Weyl chamber C^*.

Notation from [GW].

- e_{ij} is the matrix with 1 in i, j (i^{th} row, j^{th} column) position, zeroes elsewhere;
- $\mathfrak{h}$ is a Cartan sub-algebra of a complex simple Lie algebra $\mathfrak{g}$.

$\mathrm{sl}(l+1)$:

- $\mathfrak{h} =$ traceless diagonal matrices;
- $\epsilon_i(A) = a_i$ where $A = \mathrm{diag}(a_1, \ldots, a_{l+1})$;
- $\epsilon_i - \frac{\epsilon_1 + \cdots + \epsilon_{l+1}}{l+1}$, $i = 1, \ldots, l$, is a basis for $\check{\mathfrak{h}}$.

$\mathrm{sp}(2l)$ *and* $\mathrm{so}(2l)$:

- $\mathfrak{h} =$ matrices of the form $A = \mathrm{diag}(a_1, \ldots, a_l, -a_l, \ldots, -a_1)$;
- $\epsilon_i(A) = a_i$ where A is as above;
- $\epsilon_1, \ldots, \epsilon_l$ is a basis for $\check{\mathfrak{h}}$.

$\mathrm{so}(2l+1)$:

$\mathfrak{h} =$ matrices of the form $A = \mathrm{diag}(a_1, \ldots, a_l, 0, -a_l, \ldots, -a_1$;

$\epsilon_i(A) = a_i$;

$\epsilon_1, \ldots, \epsilon_l$ is a basis for $\check{\mathfrak{h}}$.

Roots:

$$\begin{aligned}
\mathrm{sl}(l+1)\colon &\quad \pm(\epsilon_i - \epsilon_j),\ 1 \leqq i < j \leqq l+1;\\
\mathrm{sp}(2l)\colon &\quad \pm(\epsilon_i - \epsilon_j), \pm(\epsilon_i + \epsilon_j),\ 1 \leqq i < j \leqq l \text{ and } 2\epsilon_i,\ i = 1, \ldots, l;\\
\mathrm{so}(2l)\colon &\quad \pm(\epsilon_i - \epsilon_j), \pm(\epsilon_i + \epsilon_j),\ 1 \leqq i < j \leqq l;\\
\mathrm{so}(2l+1)\colon &\quad \pm(\epsilon_i - \epsilon_j), \pm(\epsilon_i + \epsilon_j),\ 1 \leqq i < j \leqq l \text{ and } \pm\epsilon_i,\ 1 \leqq i \leqq l.
\end{aligned}$$

$\mathfrak{r} =$ set of roots, $\mathfrak{r}^+ =$ one choice of positive roots, and $\Delta =$ set of primitive roots.[27]

[27] [GW] uses Φ, Φ^+ for $\mathfrak{r}, \mathfrak{r}+$; our choice is because of the use of Φ to denote a variation of Hodge structure elsewhere in the text.

$\mathrm{sl}(l+1)$: $\Sigma^+ = \{\epsilon_i - \epsilon_j, i < j\}$

$\alpha_i = \epsilon_i - \epsilon_{i+1}, \Delta = \{\alpha_1, \dots, \alpha_l\}$;

$\mathrm{so}(2l)$: $\Sigma^+ - \{\epsilon_i - \epsilon_j, \epsilon_i + \epsilon_j \text{ for } i < j\}$

$\alpha_i = \epsilon_i - \epsilon_{i+1}, 1 \leqq i \leqq l-1$ and $\alpha_l = \epsilon_{l-1} + \epsilon_l$;

$\Delta = \{\alpha_1, \dots, \alpha_l\}$;

$\mathrm{sp}(2l)$: $\Sigma^+ = \{\epsilon_i - \epsilon_j, \epsilon_i + \epsilon_j \text{ for } i < j \text{ and } 2\epsilon_i\}$;

$\alpha_i = \epsilon_i - \epsilon_{l+1}, 1 \leqq i \leqq l-1$ and $\alpha_l = 2\epsilon_l$;

$\Delta = \{\alpha_1, \dots, \alpha_l\}$;

$\mathrm{sp}(2l+1)$: $\Sigma^+ = \{\epsilon_i - \epsilon_j, \epsilon_i + \epsilon_j \text{ where } i < j, \text{all } \epsilon_i\}$;

$\alpha_i = \epsilon_i - \epsilon_{i+1}, 1 \leqq i \leqq l-1$ and $\alpha_l = \epsilon_l$;

$\Delta = \{\alpha_1, \dots, \alpha_l\}$.

For each root α there is a root space $\mathfrak{g}_\alpha$, the α-eigenspace for the action of $\mathfrak{h}$, and for $\alpha \in \mathfrak{r}+$ a unique $e_\alpha \in \mathfrak{g}_\alpha$, $f_\alpha \in \mathfrak{g}_{-\alpha}$, and co-root associated to α $h_\alpha \in \mathfrak{h}$ such that

$$\begin{cases} [h_\alpha, e_\alpha] = 2e_\alpha \\ [h_\alpha, f_\alpha] = -2f_\alpha \\ [e_\alpha, f_\alpha] = h_\alpha \text{ and } \langle \alpha, h_\alpha \rangle = 2. \end{cases}$$

The simple co-roots are $H_1, \dots, H_l$ associated to $\alpha_1, \dots \alpha_l$.

- W is the *Weyl group*; there is a unique $w_0 \in W$ with $w_0(\mathfrak{r}^+) = \mathfrak{r}^-$.

$\mathrm{sl}(l+1)$: $w_0(\alpha_i) = \alpha_{l+2-i}$;

$\mathrm{so}(2l)$: $w_0(\alpha_i) = -\alpha_i$ for $1 \leqq i \leqq l-2$;

$w_0(\alpha_{l-1}) = -\alpha_l$ and $w_0(\alpha_l) = -\alpha_{l-1}$;

$\mathrm{sp}(2l)$: $w_0 = -I$;

$\mathrm{so}(2l+1)$: $w_0 = -I$.

- The *fundamental weights* $w_1, \dots, w_l \in \check{\mathfrak{h}}$ are the basis for $\check{\mathfrak{h}}$ dual to the basis $H_1, \dots, H_l$ for $\mathfrak{h}$.

$$\mathrm{sl}(l+1): \quad \omega_i = \epsilon_1 + \cdots + \epsilon_i - \tfrac{i}{l+1}(\epsilon_1 + \cdots + \epsilon_{l+1}),\ 1 \leqq i \leqq l;$$

$$\mathrm{so}(2l): \quad \omega_i = \epsilon_i + \cdots + \epsilon_i,\ 1 \leqq i \leqq l-2;$$

$$\omega_{l-1} = \tfrac{1}{2}(\epsilon_1 + \cdots + \epsilon_{l-1} - \epsilon_l);$$

$$\omega_l = \tfrac{1}{2}(\epsilon_1 + \cdots + \epsilon_{l-1} + \epsilon_l);$$

$$\mathrm{sp}(2l): \quad \omega_i = \epsilon_1 + \epsilon_i,\ 1 \leqq i \leqq l; \text{ or } \epsilon_1; \epsilon_1 + \epsilon_i,\ 2 \leqq i \leqq l;$$

$$\mathrm{so}(2l+1): \quad \omega_i = \epsilon_i + \cdots + \epsilon_i,\ 1 \leqq i \leqq l-1 \text{ and } \omega_l = \tfrac{1}{2}(\epsilon_1 + \cdots + \epsilon_l).$$

Chapter V

Hodge Structures with Complex Multiplication

This largely expository chapter deals with Hodge structures having "lots" of endomorphisms — so many Hodge 2-tensors that the Mumford-Tate group is forced inside the isotropy group, or equivalently:

(V.1) A Hodge structure (V, φ) has *complex multiplication*, or *is a CM-Hodge structure*, if and only if M_φ is a torus.

CM-Hodge structures will play a pivotal role in Chapters VI–VII as convenient "base points" in period domains, or as one-point period domains, and in Chapter VIII as a focal point for conjectures on the arithmetic properties of variation of Hodge structure. They generalize isogeny classes of CM abelian varieties and, because they are motivic when polarizable [Ab], perhaps point toward an interesting generalization of class field theory.

The aim of this chapter is to reorganize and lay out clearly what is known about CM-Hodge structures in a fair amount of generality, and so much of it (with the notable exception of Section V.F) is done in the form of a "crash course" without proofs, as we do not wish to bog the reader down in excessive Galois theory. But we are going to start with a careful proof that, while not of the most general result, captures exactly how the arithmetic begins to emerge from Definition (V.1). Recall that a *CM-field* is a totally imaginary extension L of $\mathbb{Q}$ having a totally real subfield K with $[L : K] = 2$.

(V.2) PROPOSITION: *Suppose (V, φ) is an irreducible Hodge structure polarized by Q, and $M = M_\varphi$ is abelian but nontrivial. Then $E := \mathrm{End}(V, \varphi)$ is isomorphic to a CM field.*

PROOF. Write $G = \mathrm{Aut}(V, Q)$, $\widetilde{G} = \mathrm{GL}(V)$. Since (V, φ) is simple, $E^* := E\backslash\{0\} \subset \widetilde{G}(\mathbb{Q})$; in fact, E^* is just the $\mathbb{Q}$-points of the centralizer $Z := Z_{\widetilde{G}}(M)$. This is a $\mathbb{Q}$-algebraic group, hence among its maximal tori contains one defined $/\mathbb{Q}$ [Bor] — call this T. As this is maximal in $\widetilde{G}$ and centralizes M, we have $T \supset M$.

Now, any non-zero sum of elements of $T(\mathbb{Q}) \subset E^*$ belongs to $E^* \subset \widetilde{G}(\mathbb{Q})$ and commutes with T, hence by maximality belongs to $T(\mathbb{Q})$. This makes $T(\mathbb{Q})$ the non-zero points of a field; more precisely, we have $L \overset{\eta}{\hookrightarrow} E$ with

$\eta(L^*) = T(\mathbb{Q})$. Furthermore, for any $w \in V$, the subspace $\eta(L).w =: W \subset V$ is stabilized by $M(\mathbb{Q})$. By (I.B.5) and Zariski-density of $M(\mathbb{Q})$ in M, W underlies a sub-Hodge structure of (V, φ), hence by simplicity of the latter equals V. We conclude that, via η, V is a 1-dimensional vector space over L; in particular, $[L : \mathbb{Q}] = \dim(V) = \sum_{p+q=n} h^{p,q} = r$.

If $E^* \supsetneq T(\mathbb{Q})$, then the entire $T(\mathbb{Q})$-orbit of any $\xi \in E^* \backslash T(\mathbb{Q})$ avoids $T(\mathbb{Q})$. If $v \in V\backslash\{0\} = V \backslash \ker \xi$, then there exists $\ell \in L$ with $\eta(\ell)\xi v = v$. Since $\eta(\ell)\xi \notin T(\mathbb{Q})$, in particular $\neq \mathbb{I}$, $\mathbb{I} - \eta(\ell)\xi$ belongs to E^* but annihilates v, contradicting our assumed simplicity of (V, φ). Hence $E^* = T(\mathbb{Q})$ and $E = \eta(L)$; moreover, $Z = T$.

Referring to (VI.B.5), we find that, being its own maximal torus, M diagonalizes with respect to a Q-quasi-unitary self-complex-conjugate basis ω ($\omega_{r-i} = \bar{\omega}_i$, where $r = \dim(V)$). This diagonalizes $\varphi(\mathbb{U}) \subset M$ hence is also a Hodge basis for φ, and satisfies $\sqrt{-1}^{p_i - q_i} Q(\omega_i, \omega_{r-i}) = \delta_{ij}$, where $\omega_i \in V^{p_i, q_i}$. This basis defines and diagonalizes a maximal torus of $\widetilde{G}$ centralizing M and a priori defined over $\mathbb{R}$; but since $Z = T$, this torus is necessarily T. So ω diagonalizes $\eta(L)$.

Let $\gamma \in L$ be a primitive element for the extension $L/\mathbb{Q}$, and $p(\lambda) = \prod_{i=1}^r (\lambda - \eta_i(\gamma))$ be its minimal polynomial over $\mathbb{Q}$, where $\eta_i : L \hookrightarrow \mathbb{C}$ are the embeddings. This is also the minimal polynomial for $\eta(\gamma)$, and so up to reordering we must have $\eta(\gamma)\omega_i = \eta_i(\gamma)\omega_i$, or more generally, $[\eta(\gamma)]_\omega = \operatorname{diag}\{\eta_1(\ell), \ldots, \eta_r(\ell)\}$ for all $\ell \in L$. Moreover, since $\eta(L) \subset \widetilde{G}(\mathbb{R})$, the eigenvalues of $\eta(\gamma)$ on ω_i and $\omega_{r-i} = \bar{\omega}_i$ must be conjugate: $\eta_{r-i}(\gamma) = \bar{\eta}_i(\gamma)$, which implies that $\eta_{r-i} = \bar{\eta}_i$. Except where n is even *and* r is odd, where there is the possibility that $\omega_{\frac{r+1}{2}}$ (and *only* this eigenvector; see (VI.B.5)) is self-conjugate, this already shows that L is totally imaginary.

Next we consider the *Rosati involution* $\dagger : E \to E^{\mathrm{op}}$ defined in general on the endomorphism algebra by

$$Q(\xi^\dagger v, w) = Q(v, \xi w) \qquad \text{for all } v, w \in V.$$

In our setting $E = \eta(L)$, this automatically produces a field automorphism $\rho := \eta^{-1} \circ \dagger \circ \eta \in \operatorname{Gal}(L/\mathbb{Q})$, and

$$\begin{aligned}
\eta_{r-i}(\ell) Q(\omega_i, \omega_{r-i}) &= Q(\omega_i, \eta(\ell)\omega_{r-i}) \\
&= Q(\eta(\ell)^\dagger \omega_i, \omega_{r-i}) = Q(\eta(\rho(\ell))\omega_i, \omega_{r-i}) \\
&= \eta_i(\rho(\ell)) Q(\omega_i, \omega_{r-i})
\end{aligned}$$

implies

$$\eta_i \circ \rho = \bar{\eta}_i \qquad \text{for all } i.$$

If L is totally imaginary, the existence of such an involution ρ is known to be the defining property for its being a CM field.

Finally, if K is the fixed field of ρ, then (regardless of whether $L/\mathbb{Q}$ is Galois) we have $[L : K] = 2$. This must divide $[L : \mathbb{Q}] = r$, so r cannot be odd and the exceptional case above does not exist. □

Conversely one has:

(V.3) PROPOSITION: *Suppose the endomorphism algebra E of a Hodge structure (V, φ) has an embedded field*[1] $L \overset{\eta}{\hookrightarrow} E$ *of degree* $r = \dim V$. *Then* $M = M_\varphi$ *is abelian.*

PROOF. M must centralize $\eta(L)$ since it consists of rational $(0,0)$-tensors in $T^{1,1}V$. Furthermore, $\eta(L^*) \subset \widetilde{G}(\mathbb{Q})$ gives the $\mathbb{Q}$-points of a torus of dimension $[L : \mathbb{Q}]$. Since $[L : \mathbb{Q}] = \dim(V)$, the torus is maximal, hence contains M. □

We will show how to *construct* Hodge structures with such an endomorphism algebra, give algorithms for computing the rank and $\mathbb{Q}$-points of M, and explain why they arise algebro-geometrically in the polarizable case. Some of the omitted proofs may be found in [Ab].

Warning: In the even weight case $n = 2m$, in this chapter we assume that our Hodge structures do not have a nontrivial sub-Hodge structure of pure type $(\frac{n}{2}, \frac{n}{2})$. This simplifies some statements, and we trust that the reader can make the appropriate modifications.

Remark: Following Proposition (II.A.7) we mentioned the result of [Bo]:

(V.4) *If $M_\varphi(\mathbb{R})$ is contained in the isotropy group, then M_φ is a torus and φ is a CM-Hodge structure.*

Here is the proof: First, from the definitions it follows that any Mumford-Tate group M_φ commutes with $\mathrm{End}(V, \varphi)$. Secondly, since $M_\varphi(\mathbb{R}) \subset H_\varphi$ it follows that $M_\varphi(\mathbb{Q}) \subset \mathrm{End}(V, \varphi)$. Thus $M_\varphi(\mathbb{Q})$ is commutative, and by Zariski-density of $\mathbb{Q}$-points in a connected linear algebraic group defined over $\mathbb{Q}$, so is M_φ. So it must be a torus, and then φ is CM by definition (V.1). □

V.A ORIENTED NUMBER FIELDS

For a number field F, let $\mathcal{S}_F(\mathbb{C})$ (resp. $\mathcal{S}_F(\mathbb{R})$) be the set of complex (resp. real) embeddings. When F is totally imaginary, or equivalently $\mathcal{S}_F(\mathbb{R})$ is empty, one defines a "type" $\Theta \subset \mathcal{S}_F(\mathbb{C})$ of F by the criterion

$$\Theta \sqcup \bar{\Theta} = \mathcal{S}_F(\mathbb{C}),$$

[1] *It turns out that this must be totally imaginary, and not necessarily CM.*

where the bar simply conjugates all embeddings. If F is a CM field, Θ is called a *CM type* in the literature. More generally, for $n \in \mathbb{Z}^+$ a n*-orientation of* F is a partition $\Pi = \{\Pi^{p,q}\}_{p+q=n}$ of $\mathcal{S}_F(\mathbb{C})$:

$$\sqcup_{p+q=n}\Pi^{p,q} = \mathcal{S}_F(\mathbb{C}), \quad \overline{\Pi^{p,q}} = \Pi^{q,p}.$$

It is *effective* if $\Pi^{p,q} = \emptyset$ for p or $q < 0$; clearly a type of F is just an effective 1-orientation.

(V.A.1) **Definition:** (i) OIF(n) is the category of n-oriented totally imaginary number fields, with typical element (F, Π).

(ii) (K, Υ) is a subOIF of (F, Π) if, any only if, $K \subseteq F$ is a subfield and $\Upsilon = \Pi|_K$. The context here is that if $\theta_1, \theta_2 \in \mathcal{S}_F(\mathbb{C})$ restrict to the same embedding on K, then they belong to the same $\Pi^{p,q}$. A *primitive* OIF(n) has no proper subOIF.

(iii) OCMF(n) is the subcategory of OIF(n) consisting of (F, Π) with F a CM-field; that is, we have one of the equivalent conditions: (a) There exists $F_0 \subseteq F$ such that $[F : F_0] = 2$ and F_0 is totally real OR (b) there exists $\rho \in \mathrm{Gal}(F/\mathbb{Q})$ (F need not be Galois) such that for all $\theta \in \mathcal{S}_F(\mathbb{C})$ $\theta\rho = \bar{\theta}$. In this case it follows that moreover $\rho \in Z(\mathrm{Gal}(F/\mathbb{Q}))$.

(iv) $\widetilde{\mathrm{OCMF}}(n)$ is that "intermediate" subcategory consisting of $(F, \Pi) \in$ OIF(n) that has a subOIF $(K, \Upsilon) \in$ OCMF(n). We say (F, Π) is *induced* from K. (One does not get a smaller category by asking for (K, Υ) to be primitive.)

(V.A.2) **Remark:** We will still refer to an effective 1-orientation as simply a type, and write Θ, $\bar{\Theta}$ in lieu of $\Pi^{1,0}$, $\Pi^{0,1}$. For an effective n-orientation Π of odd weight, (i.e. n odd), the associated *W-type* is

$$\Theta_\Pi^W := \Pi^{0,n} \sqcup \Pi^{2,n-2} \sqcup \cdots \sqcup \Pi^{n-1,1}$$

and *G-type*

$$\Theta_\Pi^G := \Pi^{n,0} \sqcup \Pi^{n-1,1} \sqcup \cdots \sqcup \Pi^{\frac{n+1}{2},\frac{n-1}{2}}.$$

So we pass in one of these two ways from higher weight to weight 1. More generally, let

$$\Theta(\Pi) = \text{ set of types of } F \text{ refined by } \Pi \text{ (unions of } \Pi^{p,q}\text{'s)}.$$

That is, for $\Theta \in \Theta(\Pi)$ and $p+q = n$, exactly one of $\Pi^{p,q}$ and $\Pi^{q,p}$ is contained in Θ.

Next some rather technical-seeming Galois-theoretic ideas will enter, which are necessary if we want to classify all CM-Hodge structures. The point is that if one just assumes a Hodge structure is polarizable and has commutative Mumford-Tate group, then the field of complex multiplication need not be an abelian or even a Galois extension of $\mathbb{Q}$; in fact it may not precisely be a CM-field. So to prove anything about such Hodge structures in general, and also in even the nicest cases (beyond rank 2) to compute M_φ, this material must be dealt with.

Now let $(F,\Pi) \in \widetilde{\text{OCMF}}(n)$, and consider the Galois closure $(F^c,\tilde{\Pi})$ of $(F,\Pi) \in \text{OIF}(n)$, where $\tilde{\Pi}^{p,q}$ = all embeddings restricting to $\Pi^{p,q}$ on F. Fix a reference $\tilde{\theta}_1 \in \mathcal{S}_{F^c}(\mathbb{C})$. Sending

$$\sigma \mapsto \tilde{\theta}_1\sigma =: \tilde{\theta}_\sigma$$

induces an isomorphism of sets

$$\text{(V.A.3)} \qquad \text{Gal}(F^c/\mathbb{Q}) \longrightarrow \sqcup_{p+q=n}\tilde{\Pi}^{p,q};$$

and we put

$$\tilde{\theta}_\sigma^{\{-1\}} := \tilde{\theta}_{\sigma^{-1}} = \tilde{\theta}_1\tilde{\theta}_\sigma^{-1}\tilde{\theta}_1,$$

taking $\tilde{\Pi}^{-1}$ to be the partition with $(\tilde{\Pi}^{-1})^{p,q} := \{\tilde{\theta}^{\{-1\}} \,|\, \tilde{\theta} \in \tilde{\Pi}^{p,q}\}$. Here we are using the fact that since $F^c/\mathbb{Q}$ is Galois, all embeddings in $\mathbb{C}$ have the same image — so we can consider the inverse of one embedding applied to the image of another. Next, define as usual traces and norms on F:

$$Tr_\Pi^{p,q} := \sum_{\theta\in\Pi^{p,q}} \tilde{\theta}_1^{-1}\theta \,:\, F \to F^c,$$

$$N_\Pi^{p,q} := \prod_{\theta\in\Pi^{p,q}} \tilde{\theta}_1^{-1}\theta \,:\, F \to F^c$$

where $\theta \mapsto \tilde{\theta}_1^{-1}\theta$ should be thought of as the inverse of (V.A.3). The "$\prod_{\theta\in\Pi^{p,q}}$" in the definition of $N_\Pi^{p,q}$ is a *product*, and *not* composition.

(V.A.4) **Definition:** The *reflex field* is generated by such traces taken over all (p,q):

$$F' := \mathbb{Q}\left(\left\{Tr_\Pi^{p,q}(f) \,|\, f \in F\right\}_{p+q=n}\right).$$

This is equal to the fixed field of

$$\left\{\sigma \in \text{Gal}(F^c/\mathbb{Q}) \,|\, (\tilde{\Pi}^{-1})^{p,q}\sigma = (\tilde{\Pi}^{-1})^{p,q} \text{ for all } p+q=n\right\}$$

in F^c. This has accompanying *reflex type* $\Pi' := \tilde{\Pi}^{-1}|_{F'}$, and $(F',\Pi') \in \text{OCMF}(n)$ is primitive.

Warning: The reflex field depends on Π, and reflex fields for weight 1 and higher weight may be quite different, and this is not captured by the notation F'.

A *group* homomorphism we will use later is the *reflex norm*

$$N_{\Pi'} := \prod_{p+q=n} (N_{\Pi'}^{p,q})^p : (F')^* \to F^*,$$

which factors through $(F'')^*$ and has the same image as the similarly defined $N_{(\tilde{\Pi}^{-1})} : (F^c)^* \to F^*$. Note that (F'', Π'') ($\in \mathrm{OCMF}(n)$) is the primitive subOIF from which (F, Π) is induced; if the latter was primitive already, they are equal.

In the very special case where $F/\mathbb{Q}$ is an abelian Galois extension, which implies that F is a CM subfield of a cyclotomic field, $F' = F = F^c$. Fixing $\theta_1 : \zeta_m \mapsto e^{\frac{2\pi i}{m}}$ when $F = \mathbb{Q}(\zeta_m)$ is cyclotomic, we write

$$\theta_j := \theta_1 \sigma_j \quad \text{where} \quad \sigma_j(\zeta_m) := (\zeta_m)^j,$$

and then $\theta_j^{\{-1\}} = \theta_{j^{-1}}$ where j^{-1} is computed multiplicatively mod m. This then makes $(\Pi')^{p,q} = (\Pi^{p,q})^{\{-1\}}$ easy to compute.

(V.A.5) **Example:**

(i) $F = \mathbb{Q}(\zeta_5)$, $n = 1$, $\Theta = \{\theta_1, \theta_2\}$; $\Theta' = \{\theta_1, \theta_3\}$.

(ii) $F = \mathbb{Q}(\zeta_5)$, $n = 3$, $\Pi^{3,0} = \{\theta_1\}$, $\Pi^{2,1} = \{\theta_2\}$, etc.;
$(\Pi')^{3,0} = \{\theta_1\}$, $(\Pi')^{2,1} = \{\theta_3\}$, etc.

(iii) $F = \mathbb{Q}(\zeta_{13})$, $n = 1$, $\Theta = \{\theta_1, \theta_2, \theta_3, \theta_5, \theta_6, \theta_9\}$;
$\Theta' = \{\theta_1, \theta_7, \theta_9, \theta_8, \theta_{11}, \theta_3\}$.

(iv) $F = \mathbb{Q}(\zeta_{13})$, $n = 1$, $\Theta = \{\theta_{12}, \theta_2, \theta_3, \theta_5, \theta_6, \theta_9\}$;
$\Theta' = \{\theta_{12}, \theta_7, \theta_9, \theta_8, \theta_{11}, \theta_3\}$.

(v) $F = \mathbb{Q}(\zeta_{13})$, $n = 3$, $\Pi^{3,0} = \{\theta_1\}$, $\Pi^{2,1} = \{\theta_2, \theta_3, \theta_5, \theta_6, \theta_9\}$, etc.;
$(\Pi')^{3,0} = \{\theta_1\}$, $(\Pi')^{2,1} = \{\theta_7, \theta_9, \theta_8, \theta_{11}, \theta_3\}$.

Now in (iii), σ_3 fixes Θ and Θ', which implies that (F, Θ) and (F, Θ') are not primitive. In contrast, (F, Θ) in (iv) is primitive because no σ_j except for $\sigma_1 = \mathrm{id}$ fixes Θ; the same thing goes for (F, Π) in (v) (and (i), and (ii)). Note that (iii) (respectively (iv)) is the G (resp. W) type associated to Π from (v). □

(V.A.6) **Example:** (i) $F = \mathbb{Q}(\zeta_{32})$, $\Theta = \{\theta_1, \theta_7, \theta_9, \theta_{11}, \theta_{13}, \theta_{15}, \theta_{27}, \theta_{29}\}$. (ii) $F = \mathbb{Q}(\zeta_{32})$, $\Pi^{3,0} = \{\theta_1\}$, $\theta^{2,1} = \{\theta_7, \theta_9, \theta_{11}, \theta_{13}, \theta_{15}, \theta_{27}, \theta_{29}\}$. Clearly (ii) refines (i). □

We conclude this part with one more definition and some basic related results. Let $\mathcal{G} = \mathrm{Gal}(F^c/\mathbb{Q})$, with $\mathcal{G}^{p,q} \subset \mathcal{G}$ the preimage of $\widetilde{\Pi}^{p,q}$ under (V.A.3). Every $g_0 \in \mathcal{G}$ belongs to some $\mathcal{G}^{p_0,q_0}$, and then we write $p(g_0) = p_0$.

(V.A.7) **Definition:** The *generalized Kubota rank* $\mathcal{R}(F,\Pi)$ of the $\widetilde{\text{OCMF}}$ (F,Π) is the rank of the linear transformation

$$\tau : \mathbb{Z}[\mathcal{G}] \to \mathbb{Z}[\mathcal{G}]$$

given by

$$\tau([g_0]) := \sum_{g \in \mathcal{G}} p(g)[gg_0].$$

We are thinking of this as a map of vector spaces. (F,Π) is said to be *nondegenerate* if $\mathcal{R}(F,\Pi) = \frac{1}{2}[F'' : \mathbb{Q}]+1$ and *strongly nondegenerate* if $\mathcal{R}(F,\Pi) = \frac{1}{2}[F : \mathbb{Q}] + 1$.

In general $\mathcal{R}(F,\Pi) = \mathcal{R}(F',\Pi')$, and

$$\text{(V.A.8)} \qquad \log_2[F'' : \mathbb{Q}] \subseteq \mathcal{R}(F,\Pi) - 1 \leq \tfrac{1}{2}[F'' : \mathbb{Q}].$$

The proof of this is practically the same as for $n = 1$ (from Ribet and Kubota), for which see [La, p. 150]. It leads easily to (i)–(iii) of the following

(V.A.9) PROPOSITION: (i) *Strong nondegeneracy implies primitivity.*

(ii) *If* $\mathcal{R}(F,\Pi) - 1$ *does not divide* $\frac{1}{2}[F : \mathbb{Q}]$, *then* (F,Π) *is degenerate.*

(iii) *If* $\frac{1}{2}[F'' : \mathbb{Q}] \leq 3$, *then* (F,Π) *is nondegenerate.*

(iv) (*Ribet's nondegeneracy theorem, cf.* [Do, Th. 1.0].) *For* $n = 1$ *and* $\frac{1}{2}[F : \mathbb{Q}]$ *prime, primitivity implies (strong) nondegeneracy.*

(V.A.10) **Question:** If $\Theta \in \Theta(\Pi)$ as in (V.A.2), is $\mathcal{R}(F,\Pi) \geq \mathcal{R}(F,\Theta)$? This would effectively (for $n > 1$) strengthen the lower bound in (V.A.8), and in particular extend (V.A.9)(iv) to higher weight.

The above examples will be revisited, and their ranks addressed, in Section V.D.

V.B HODGE STRUCTURES WITH SPECIAL ENDOMORPHISMS

Given a Hodge structure (V,φ) we have M_φ and $M_{\tilde{\varphi}}$ as in Chapter I; write $E_\varphi := \mathrm{End}(V_{\mathbb{Q}},\varphi)$ for the endomorphism algebra. We recall that (V,φ) is a CM-Hodge structure if one of the equivalent conditions

(i) M_φ abelian;

(ii) $M_\varphi(\mathbb{Q}) \subseteq E_\varphi$

holds. When (V, φ) is *also* irreducible, we find that E_φ is a field with elements consisting of the rational points of a maximal subtorus of $\mathrm{GL}(V)$.

We want to classify the polarizable CM-Hodge structure of higher weight. To this end make the following

(V.B.1) **Definition:** (i) The category of WCMHS's, where W stands for "weak", has objects (V, φ, F, η) with F a field, $F \overset{\eta}{\hookrightarrow} E_\varphi$ a ring homomorphism, and (V, φ) a Hodge structure. We say "V has CM by F." Morphisms $(\mathcal{V}, \psi, \mathcal{F}, \mu) \to (V, \varphi, F, \eta)$ are given by a morphism of Hodge structures together with a diagram

(V.B.2)
$$\begin{array}{ccc} \mathcal{F} & \hookrightarrow & F \\ \downarrow{\scriptstyle \mu} & & \downarrow{\scriptstyle \eta} \\ E_\psi & \longrightarrow & E_\varphi . \end{array}$$

(ii) The subcategory SCMHS of WCMHS consists of those objects with $[F:\mathbb{Q}] = rk(V)$. Here, S stands for "strong."

(iii) There are obvious notions of sub(S or W)CMHS arising from the above definition of morphism.

(iv) One can also consider the subcategory of *polarizable* objects, (S or W)CMpHS; or one can "refine" this category to (S or W)CMPHS, where P means polarized, by taking objects (V, φ, F, η, Q), where Q is a choice of polarization.

(v) Given a WCMHS $(V, \varphi, F, \eta) =: V$ and a type Θ of F, the $\{-\frac{a}{2}\}_\Theta$-*twist* of V, which is still a WCMHS, has by definition the same underlying V, F, and η — only φ changes. Namely, thinking of $\mathbb{S}(\mathbb{R}) \cong \mathbb{C}^*$, if $\chi_\Theta : \mathbb{C}^* \to \mathrm{End}(V_\mathbb{C})$ is the co-character with eigenspaces

$$E_\alpha(\chi_\Theta(\alpha)) \oplus E_{\bar{\alpha}}(\chi_\Theta(\alpha)) = \left(\bigoplus_{\theta \in \Theta} E_\theta(\eta) \right) \oplus \left(\bigoplus_{\theta \in \bar{\Theta}} E_\theta(\eta) \right) = V_\mathbb{C},$$

then $\tilde{\varphi} \otimes \chi_\Theta^{\frac{a}{2}}$ defines $V\{-\frac{a}{2}\}$. So $weight(V\{-\frac{a}{2}\}) = weight(V) + a$. (These are not to be confused with Tate twists!)

Basic facts:

- A SCMHS is a CM-Hodge structure;[2] and any *irreducible* CM-Hodge structure underlies a SCMHS.

[2] One should think of this as a forgetful functor SCMHS$\to$CMHS: we forget the field, which cannot be recovered if the Hodge structure is not irreducible.

- Any SCMpHS has a decomposition $V = V_0^{\oplus m}$, where $(V_0, \varphi_0, \text{etc.})$ is an irreducible subSCMpHS; more generally, a SCMHS with an irreducible subSCMHS has such a decomposition. In this case E_{φ_0} is obviously a field and $E_\varphi \cong \mathrm{Mat}_m(E_{\varphi_0})$.

- Any CMpHS has a unique decomposition $V = V_1^{\oplus m_1} \oplus \cdots \oplus V_\ell^{\oplus m_\ell}$ into irreducible SCMpHS's. $E_\varphi \cong \oplus_i \mathrm{Mat}_{m_i}(K_i)$, where $K_i \cong E_{\varphi_i}$.

- A WCMHS with abelian M_φ need not be a SCMHS, even in the polarizable case.

- As far as we know, an irreducible sub-Hodge structure of a SCMHS need not be a SCMHS (or CM-Hodge structure), and an irreducible SCMHS sub-Hodge structure (of a SCMHS) need not be a subSCMHS. In the polarizable case, both of these problems disappear.

(V.B.3) **Remark:** "Irreducible" will mean "as Hodge structure" for CMHS but for SCMHS it means not having a proper nontrivial subSCMHS. In the polarizable case there is no difference.

We want to explain the *half-twists* of weak CM Hodge structures mentioned above — a notion introduced by van Geemen [vG] (and further developed in his paper with Izadi [IvG]) — in a little more depth.

Let K be a totally imaginary field, $\Sigma = (\sigma_1, \ldots, \sigma_g)$ a type for K and $\sigma_{i+g} = \bar{\sigma}_i$ for the remaining embeddings. Given a WCMHS (V, φ, F, η), the action of K decomposes each $V_{\mathbb{C}}^{p,q}$ into eigenspaces $V_j^{p,q}$ where $\alpha \in K$ acts via $v \mapsto \sigma_j(\alpha)v$. Write $V_+^{p,q} := \oplus_{j=1}^{g} V_j^{p,q}$, $V_-^{p,q} := \oplus_{j=g+1}^{2g} V_j^{p,q}$. Define a new WCMHS $V\{-\frac{b}{2}\}$, $b \in \mathbb{Z}$, with the same underlying rational vector space and action of K and choice of K-type, by

$$V\left\{-\frac{b}{2}\right\}^{P,Q} := V_+^{P-b,Q} \oplus V_-^{P,Q-b}.$$

Obviously $V\{-\beta_1\}\{-\beta_2\} = V\{-(\beta_1+\beta_2)\}$. Assuming $V = F^0V$, at issue for *negative* b is whether the twist preserves "effectivity", i.e., has $F^0V\{-\frac{b}{2}\} = V\{-\frac{b}{2}\}$; this really does depend on V and the choice of Σ.

This is true for $b = -1$ — that is, the the half-twist $V\{\frac{1}{2}\}$ is an "effective" Hodge structure — precisely if $V_-^{n,0} = 0$. This can always be arranged in the "Calabi-Yau-Hodge structure" case where $h^{n,0}(V) = 1$, simply by choosing the CM-type to include the corresponding embedding of K. For $n = -2$, i.e., starting from a Hodge structure of "K3 type," this gives a summand of the Kuga-Satake construction. One is interested in trying to do something comparable, i.e., to twist down to an effective *weight* 1 *Hodge structure*, for higher weight n.

From the type (K, Σ) we can construct a weight 1 Hodge structure $V^1_{(K,\Sigma)}$ by taking $V_{\mathbb{Q}} = K$ and $V^{1,0}_{\mathbb{C}} = \oplus_{\sigma\in\Sigma} E_\sigma(V_{\mathbb{C}})$. Let $\mathfrak{a} \in \mathcal{J}(K)$ be a fractional ideal, $\Sigma(\mathfrak{a}) \subset \mathbb{C}^g$ be the lattice consisting of all vectors ${}^t(\sigma_1(\alpha), \ldots, \sigma_g(\alpha))$ for $\alpha \in \mathfrak{a}$, and $A^{(K,\Sigma)}_{\mathfrak{a}}$ be the complex torus $\mathbb{C}^g/\Sigma(\mathfrak{a})$. Then in fact $V^1_{(K,\Sigma)} = H^1(A^{(K,\Sigma)}_{\mathfrak{a}})$ as $\mathbb{Q}$-Hodge structures, independently of $\mathfrak{a}$. Integrally the situation is a bit more subtle: if we put $V_{\mathbb{Z}} := \mathcal{O}_K$ (and $\mathfrak{a} = \mathcal{O}_K$), then V and $H^1(A)$ are actually dual (but isogenous), cf. Section V.E.

(V.B.4) **Remark:** Anticipating Section V.D, we point out that if K is CM, then $V^1_{(K,\Sigma)}$ is polarizable. Consequently, the map $\mathfrak{a} \mapsto A^{(K,\Sigma)}_{\mathfrak{a}}$ induces an isomorphism of sets[3]

$$\mathrm{Cl}(K) \xrightarrow{\cong} \mathfrak{Ab}(K, \Sigma)$$

where $\mathrm{Cl}(K) = \frac{\mathcal{J}(K)}{(K^*)}$ is the ideal class group and $\mathfrak{Ab}(K, \Sigma)$ is the set of isomorphism classes of Abelian varieties with CM by $(\mathcal{O}_K, \Sigma)$. That is, the action of $\mathrm{End}(A) \cong \mathcal{O}_K$ on T_0A has eigenvalues $\sigma_1(\cdot), \ldots, \sigma_g(\cdot)$.

Now, take V to be an effective SCMHS of weight n and rank r, with CM by K, where $[K : \mathbb{Q}] = 2g = r$. We can choose a *type* Σ of K so that $V^{n-i,i}_{-} = \{0\}$ for $i \leq \lfloor \frac{n}{2} \rfloor$, and then apply the $\{\frac{1}{2}\}$-twist $\lfloor \frac{n+1}{2} \rfloor$ times to arrive at an effective SCMHS $\tilde{V} := V\left\{\frac{\lfloor \frac{n+1}{2} \rfloor}{2}\right\}$ of weight $\lfloor \frac{n}{2} \rfloor$. It is not hard to see that one has an embedding $V \hookrightarrow \tilde{V} \otimes \left(V^1_{(K,\Sigma)}\right)^{\otimes \lfloor \frac{n+1}{2} \rfloor}$. Keeping track of where the K-eigenspaces have migrated, we choose a new K-type Σ' and repeat the procedure for $\tilde{V}$, then $\tilde{\tilde{V}}$, etc. — eventually reaching a weight 1 SCMHS. The consequence is that we have an inclusion $V \hookrightarrow \otimes_\gamma \left(V^1_{(K,\Sigma_\gamma)}\right)^{\otimes m_\gamma}$ where the $\{\Sigma_\gamma\}$ are various K-types. This seems especially interesting for (SCM)CY-HS; see Example (V.D.3).

V.C A CATEGORICAL EQUIVALENCE

Fix a weight n and rank r, and an object $(F, \Pi) \in \mathrm{OIF}(n)$ with $[F : \mathbb{Q}] = r$. To construct a SCMHS $V^n_{(F,\Pi)} = (V, \varphi, F, \eta)$,[4] put $V_{\mathbb{Q}} := F/\mathbb{Q}$ and describe this as a map by $F \xrightarrow{\beta} V_{\mathbb{Q}}$, so that

$$V_{\mathbb{C}} \cong F \otimes_{\mathbb{Q}} \mathbb{C} \cong \oplus_{\theta\in\mathcal{S}_F(\mathbb{C})} E_\theta(V_{\mathbb{C}}) = \bigoplus_{p+q=n} (\oplus_{\theta\in\Pi^{p,q}} E_\theta(V_{\mathbb{C}})) =: \oplus_{p+q=n} V^{p,q}_{\mathbb{C}},$$

[3]Actually, of $\mathrm{Cl}(K)$-torsors.

[4]Occasionally we shall write $\varphi^n_{(F,\Pi)}$ also.

where $E_\theta(V_\mathbb{C})$ are the 1-dimensional F-eigenspaces. The corresponding basis $\omega = \{\omega_\theta\}_{\theta\in\mathcal{S}_F(\mathbb{C})}$, scaled so that

$$[\beta(f)]_\omega = {}^t(\theta_1(f), \ldots, \theta_r(f)),$$

is an eigenbasis for the regular representation of F on itself by multiplication $\eta : F \hookrightarrow \mathrm{End}(H_\mathbb{Q})$; that is,

$$[\eta(f)]_\omega = \mathrm{diag}\{\theta_1(f), \ldots, \theta_r(f)\}.$$

This construction defines a functor

$$\mathrm{OIF}(n) \xrightarrow{\Xi} \mathrm{SCMHS}(n),$$

which is an equivalence of categories:

(i) it is a bijection on objects, with inverse given by $(V, \varphi, F, \eta) \mapsto (F, \Pi)$ with $\Pi^{p,q} :=\{\text{eigenvalues of } \eta(F) \text{ on } V^{p,q}\}$.

(ii) the subobjects are equivalent: there is a one-to-one correspondence between (a) subSCMHS of $V_{(F,\Pi)}$, (b) subOIF's of (F, Π), and also (c) subgroups $\mathcal{G}' \subseteq \mathrm{Gal}(F^c/\mathbb{Q})$ containing $\mathrm{Gal}(F^c/F)$ and satisfying $\widetilde{\Pi}^{p,q}g = \widetilde{\Pi}^{p,q}$ for all $g \in \mathcal{G}'$ and $p + q = n$.

(V.C.1) **Remark:** In the non-polarizable case, we are not saying that Ξ(primitive OIF) yields an irreducible Hodge structure. It is only necessarily irreducible as an SCMHS.

Under Ξ, the following objects also correspond bijectively:

- irreducible SCMHS and primitive OIF;
- irreducible SCMpHS and primitive OCMF;
- SCMpHS and $\widetilde{\mathrm{OCMF}}$.

Note that the second bullet implies that for an irreducible, polarizable, CMHS (V, φ) of any weight (including even weight), E_φ is a CM-field. But for Hodge structures of "CY-type" undoubtedly the most important equivalence under Ξ is:

(V.C.2) PROPOSITION: *For any weight n, there is a one-to-one correspondence between irreducible CMpHS (V, φ) with relatively prime $\{h^{p,q}(V)\}_{p+q=n}$ and $(F, \Pi) \in$ OCMF(n) with $|\Pi^{p,q}| = h^{p,q}$.*

In particular, a SCMpHS with relatively prime $h^{p,q}$ cannot have a proper nontrivial irreducible sub-Hodge structure, because the corresponding field F_0 must satisfy $[F : F_0] | h^{p,q}$ for all $p+q=n$. So part of the point here is that Ξ applied to an OCMF(n) with $\{|\Pi^{p,q}|\}_{p+q=n}$ relatively prime, always yields something irreducible *as a Hodge structure*. In any case, we see that a polarized CY Hodge structure with abelian M_H is always of the form $V_{(F,\Pi)} \oplus \{$lower level PHS$\}$ with F a CM field and $V_{(F,\Pi)}$ irreducible. For fixed weight and rank this will lead to a complete classification (not done in this monograph).

V.D POLARIZATION AND MUMFORD-TATE GROUPS

One item not explained above, is how one obtains a polarization on V given by Ξ of an OCMF (F,Π). Write the rank $r = 2g$, $f, f' \in F$, $[\beta(f)]_\omega = {}^t(z_1, \ldots, z_{2g})$, $[\beta(f')]_\omega = {}^t(z'_1, \ldots, z'_{2g})$, and assume $\omega_{i+g} = \overline{\omega_i}$, where all subscripts are mod $2g$. Note that here $z_j^{(')} = \theta_i(f^{(')})$.

Now, we have the "complex conjugation" $\rho \in \mathrm{Gal}(F/\mathbb{Q})$ with totally real fixed field F_0. If n is even, take $\xi \in F_0$ such that $\sqrt{-1}^{p-q}\theta_i(\xi) > 0$ for all $\theta_i \in \Pi^{p,q}$; if n is odd, take $\delta \in F_0$ with all $\theta_i(\delta) < 0$ and $\xi = \sqrt{\delta} \in F$, such that $\sqrt{-1}^{p-q}\theta_i(\xi) > 0$ for all $\theta_i \in \Pi^{p,q}$. Using independence of the embeddings, one easily sees that this is always possible. The polarization is then

$$Q(\beta(f), \beta(f')) := Tr_{F/\mathbb{Q}}(\xi f \rho(f')) = \sum_{j=1}^{2g} \theta_j(\xi) z_j z'_{j+g},$$

which is defined over $\mathbb{Q}$ and satisfies the Hodge-Riemann bilinear relations. In particular, note that the complex conjugation of a vector $[\vec{v}]_\omega = {}^t(v_1, \ldots, v_g; v_{g+1}, \ldots, v_{2g})$ is ${}^t(\overline{v_{g+1}}, \ldots, \overline{v_{2g}}; \overline{v_1}, \ldots, \overline{v_g})$. It is symmetric (resp. antisymmetric) for n odd (resp. even) because our characterization of ξ above implies that $\theta_{j+g}(\xi) = (-1)^n \theta_j(\xi)$. In the odd weight case, by varying the choice of ξ, we see that once one has a SCMpHS, and hence a CM field, any "related" Jacobians $J_\Theta(V) := J(V^1_{(F,\Theta)})$ $(\Theta \in \Theta(\Pi))$ will be polarizable.[5]

(V.D.1) PROPOSITION: *Let n be odd. An effective SCMHS $V = V^n_{(F,\Pi)}$ is polarizable if, and only if, $V^1_{(F,\Theta)}$ is polarizable for any F-type $\Theta \in \Theta(\Pi)$. In this case, both $J_{\mathrm{W}}(V)$ and $J_{\mathrm{G}}(V)$ are abelian varieties.*[6]

[5] Unless otherwise mentioned, Jacobians/abelian varieties are up to isogeny, as the Hodge structures under consideration are $\mathbb{Q}$-Hodge structures.

[6] That W is "no more frequently" polarizable than G (in the CM setting) may seem strange, in light of the non-CM situation. However, the polarization on $V^1_{(F,\Theta^W_\Pi)} = H^1(J_{\mathrm{W}}(V))$ uses the same ξ as that on V; so the passage to its polarizability is at least more direct.

The $\Longleftarrow$ direction is proved using repeated half-twists $\{\frac{1}{2}\}_\Theta$ together with the rule

$$V \text{ is a sub-Hodge structure of } V\{\frac{1}{2}\}_\Theta \otimes V^1_{(F,\Theta)}$$

applied inductively (varying Θ) until $H\{\frac{1}{2}\}_\Theta$ has weight 1. As a byproduct one recovers the recent theorem from Abdulali:

(V.D.2) PROPOSITION: [Ab] *An effective weight n (even or odd) CMpHS V is a sub-Hodge structure of $H^n(\times CM$ abelian varieties$)$, hence is "motivic." If V is a SCMpHS (e.g., this is implied if V is irreducible) $V^n_{(F,\Pi)}$, then*

$$V \subset H^n\left(\times_{\Theta\in\Theta(\Pi)} J_\Theta(V)^{\times m_\Theta}\right).$$

(V.D.3) **Example:** If $n = 3$, then the procedure at the end of Section V.B embeds V in $H^1(J^2_{\mathrm{G}})^{\otimes 2} \otimes H^1(J^2_{\mathrm{W}})^\vee(-1)$; put differently, $V \subset H^3(\mathcal{A})$ where $\mathcal{A}$ is the Abelian variety $J_{\mathrm{W}}(V)^\vee \times J_{\mathrm{G}}(V)^{\times 2}$. This was noticed by Borcea [Bo] when V is H^3 of a CM-CY 3-fold.

Finally, for an SCMHS (V, φ, F, η), by definition a *Galois-polarization Q* is a polarization satisfying

$$Q(\omega_i, \overline{\omega_j}) = 0 \text{ if } i \neq j$$

with respect to the $\eta(F)$-eigenbasis ω:

- the SCMHS admitting a polarization are $\Xi(\widetilde{OCMF})$ (see Section III.C);
- the SCMHS admitting a Galois-polarization are $\Xi(OCMF)$;
- *any* polarization of an *irreducible* SCMpHS ($= \Xi$(primitive OCMF)) is a Galois-polarization.

Turning to the Mumford-Tate groups, we can calculate a dense subset of the rational points of $M_{\tilde{\varphi}}(\mathbb{Q})$ explicitly for a SCMpHS $V = V^n_{(F,\Pi)}$ arising from an $\widetilde{OCMF}$.

(V.D.4) PROPOSITION: $\eta(N_{\widetilde{\Pi}^{-1}}(F^{c,*})) = \eta(N_{\Pi'}(F^{',*})) \subset M_{\tilde{\varphi}}(\mathbb{Q}) \subset \eta(F)$ $\subset E_\varphi$. *More precisely, underlying $F^{',*} \xrightarrow{N_{\Pi'}} F^* \overset{\eta}{\hookrightarrow} \mathrm{GL}(V,\mathbb{Q})$ are morphisms (defined over $\mathbb{Q}$) of algebraic groups $\mathrm{Res}_{F'/\mathbb{Q}}\mathbb{G}_m \xrightarrow{\mathcal{N}_{\Pi'}} \mathrm{Res}_{F/\mathbb{Q}}\mathbb{G}_m \subset \mathrm{GL}(V)$ and $\mathrm{Im}(\mathcal{N}_{\Pi'}) = M_{\tilde{\varphi}}$.*

Here $\eta(F) \cong F$; and $\eta(F) = E_\varphi$ if, and only if, (V, φ) is irreducible. If a SCMpHS (V, φ) has a sub-Hodge structure then they will have the same Mumford-Tate group — this is because they will share the same irreducible SCMpHS (V_0, φ_0), and both Mumford-Tate groups will just be the diagonal embedding of $M_{\tilde{\varphi}_0}$. Finally, for V a more general CMpHS, say V decomposes into $V_1, \ldots, V_\ell$ non-isomorphic irreducibles $V_1^{\oplus m_1} \oplus \cdots \oplus V_\ell^{\oplus m_\ell}$, with $V_i = V^n_{(K_i, \Pi_i)}$. Then

$$E_\varphi \cong \mathrm{Mat}_{m_1}(K_1) \oplus \cdots \oplus \mathrm{Mat}_{m_\ell}(K_\ell) \supset K_1^* \times \cdots \times K_\ell^* \supset M_{\tilde{\varphi}}(\mathbb{Q}),$$

where the $K_i^* \subset \mathrm{Mat}_{m_i}(K_i)$ inclusion is given by $k \mapsto \mathrm{diag}\{k, \ldots, k\}$. On the level of algebraic groups, the last inclusion becomes

$$\mathrm{Res}_{K_1/\mathbb{Q}} \mathbb{G}_m \times \cdots \times \mathrm{Res}_{K_\ell/\mathbb{Q}} \mathbb{G}_m \supset \mathcal{N}_{\Pi_1'}(K_1^{',*}) \times \cdots \times \mathcal{N}_{\Pi_\ell'}(K_\ell^{',*}) \supset M_{\tilde{\varphi}}.$$

We can also fairly easily compute the rank of $M_{\tilde{\varphi}}$ (in the SCMpHS case), using the Kubota rank defined in Section V.A.

(V.D.5) PROPOSITION: $\dim(M_{\tilde{\varphi}^n_{(F,\Pi)}}) = \mathcal{R}(F, \Pi)$ *for* $(F, \Pi) \in \widetilde{OCMF}(n)$.

Proofs of Propositions V.D.4–V.D.5 will be sketched in Section V.F.

Define the class of *degenerate* SCMpHS to be Ξ(degenerate $\widetilde{\text{OCMF}}$'s); in light of Proposition V.D.5, an irreducible SCMpHS is then nondegenerate if the always satisfied inequality

$$\dim(M_{\tilde{\varphi}}) \leq \frac{1}{2} rk(V) + 1$$

$$\text{[equivalently, } \dim(M_\varphi) \leq \frac{1}{2} rk(V)\text{]}$$

is an equality. Of course, if $n = 1$ the corresponding abelian variety is called [non]degenerate, etc. For $(F, \Theta) \in \text{OCMF}(1)$ nondegenerate, it is known[7] [Mo1] that all powers of the abelian variety associated to $V^1_{(F,\Theta)}$ satisfy the Hodge conjecture.

(V.D.6) **Remark:** We recall from Section II.B that nondegeneracy for a Hodge structure means that Mumford-Tate is cut out by rational 1- and 2-tensors, or equivalently, is determined solely by which endomorphisms it centralizes. If a Hodge structure underlies a SCMpHS, this is exactly equivalent to nondegeneracy for the SCMpHS (regardless of irreducibility/primitivity). In particular, if

[7] From Murty and Hazama.

the Mumford-Tate group of an irreducible CM-Hodge structure is cut out solely by endomorphisms, then it is a torus with complex points of the form

$$\mathrm{diag}\{z_1, \ldots, z_g, z_1^{-1}, \ldots, z_g^{-1}\}$$

with respect to the Hodge basis ω; consequently one has $\dim(M_\varphi) = \frac{1}{2}\dim(V)$. (In Section VI we shall frequently speak of irreducible nondegenerate CM-Hodge structures, since they are so easy to get one's hands on.)

(V.D.7) **Example:** We are interested in the Hodge structures produced, via Ξ, from the examples in (V.A.5). For cases (i)–(iii) the Kubota rank is 3, while for (iv) and (v) it is 7. In particular this says that the SCMpHS $V^3_{(\mathbb{Q}(\zeta_{13}),\Pi)}$ arising from (v) is strongly nondegenerate, with nondegenerate (but reducible) G-Jacobian and strongly nondegenerate W-Jacobian. For instance the computation for (v) boils down to checking that the rank of

$$\begin{bmatrix} 1 & 1 & 1 & 1 & 1 & 1 & 1 & 1 & 1 & 1 & 1 & 1 \\ 3 & 2 & 2 & 1 & 2 & 2 & 1 & 1 & 2 & 1 & 1 & 0 \\ 1 & 3 & 1 & 2 & 2 & 2 & 1 & 1 & 1 & 2 & 0 & 2 \\ 2 & 2 & 3 & 1 & 2 & 2 & 1 & 1 & 2 & 0 & 1 & 1 \\ 1 & 1 & 1 & 3 & 1 & 1 & 2 & 2 & 0 & 2 & 2 & 2 \\ 1 & 2 & 1 & 2 & 3 & 2 & 1 & 0 & 1 & 2 & 1 & 2 \\ 1 & 2 & 1 & 2 & 2 & 3 & 0 & 1 & 1 & 2 & 1 & 2 \end{bmatrix}$$

is maximal, where the top row comes from the fact that $\tau(\sigma) + \tau(\rho\sigma) = (n, \ldots, n)$ for all σ in the Galois group.

The immediate question that this example brings up is: What is the relationship between $\mathcal{R}(F, \Pi)$ and $\mathcal{R}(F, \Theta)$ for $\Theta \in \Theta(\Pi)$? Suppose we know $V^n_{(F,\Pi)} \subset \otimes \left(V^1_{(F,\Theta_i)}\right)^{\otimes m_i}$ where the product is over those Θ_i in some subset of $\Theta(\Pi)$ (not the whole thing) — this is easy to obtain by the inductive half-twist procedure. Then the Mumford-Tate group of the tensor product is contained in $\times_i M_{\varphi^1_{(F,\Theta_i)}}$ and surjects onto each factor as well as onto $M_{\varphi^n_{(F,\Pi)}}$; consequently

$$\mathcal{R}(F, \Pi) - 1 \leq \sum_i (\mathcal{R}(F, \Theta_i) - 1).$$

In particular, for $V^3_{(F,\Pi)}$, this says that $\mathcal{R}(F, \Pi) + 1 \leq \mathcal{R}(F, \Theta^G_\Pi) + \mathcal{R}(F, \Theta^W_\Pi)$, which is satisfied above ($8 \leq 3 + 7$). This inequality could be a useful tool for SCMpHS of CY type in view of (V.C.2): if both W and G types are *very* degenerate or reducible, then a weight $n = 3$ SCMpHS of this sort will be degenerate and irreducible. We do not know if this ever happens, however.

(V.D.8) **Example:** For the examples (V.A.6), we have $\mathcal{R}(L,\Pi)-1=8$ but $\mathcal{R}(L,\Theta)-1=6$. Consequently the SCMpHS $V^3_{(L,\Pi)}$ is irreducible and nondegenerate, while $V'_{(L,\Theta)}$ is irreducible and *degenerate* (by (V.A.9)(ii)). So (in (V.D.2)) Abdulali's abelian variety need *not* be nondegenerate when the SCMpHS V is.

V.E AN EXTENDED EXAMPLE

To give the reader a more explicit feel for CM-Hodge structures and their Mumford-Tate groups, we will discuss Example (V.A.5) (i)–(ii) in detail. The first approach will use (V.D.5) ; the second will be more or less from scratch, and will only apply to (i).

Let $F=\mathbb{Q}(\zeta_5)$. To use (V.D.5) , we need to compute the group homomorphism from $F^*\to F^*$ given by $N_{\Theta'}$ (resp $N_{\Pi'}$) for $n=1$ (resp. $n=3$). We find that

$$\begin{aligned} &\text{(i)} \qquad N_{\Theta'}(f)=\sigma_1(f)\cdot\sigma_3(f);\\ &\text{(ii)} \qquad N_{\Pi'}(f)=\sigma_1(f)^3\cdot\sigma_3(f)^2\cdot\sigma_2(f),\end{aligned}$$

where we note that σ_1 is the identity in $\mathrm{Gal}(F/\mathbb{Q})$. Writing θ_i for σ_i viewed as an embedding, the composition with η, written with respect to the natural Hodge basis ω, is

$$\mathrm{diag}\{\theta_1(f)\theta_3(f),\,\theta_2(f)\theta_1(f),\,\theta_3(f)\theta_4(f),\,\theta_4(f)\theta_2(f)\}$$

for (i) and

$$\begin{aligned}\mathrm{diag}\{&\theta_1(f)^3\theta_3(f)^2\theta_2(f),\,\theta_2(f)^3\theta_1(f)^2\theta_4(f),\\ &\qquad\qquad\theta_3(f)^3\theta_4(f)^2\theta_1(f),\,\theta_4(f)^3\theta_2(f)^2\theta_3(f)\}\end{aligned}$$

for (ii). In each case these map elements $f\in F^*$ to rational points of $M_{\tilde{\varphi}}$.

Now, $M_{\tilde{\varphi}}$ must be a torus, and writing z_1,z_2,z_3,z_4 for the matrix entries, it is clear in both cases that it is the one cut out by the relation $z_1z_4=z_2z_3$. Hence, with respect to ω, for (i) and (ii)

$$\text{(V.E.1)} \qquad M_{\varphi}(\mathbb{C})=\left\{\left(\begin{array}{cccc} z_1 &&& \\ & z_2 && \\ && z_3 & \\ &&& z_4\end{array}\right)\middle|\, z_1z_4=z_2z_3=1\right\}.$$

To focus more closely on case (i) next, we shall begin with some generalities on abelian g-folds with endomorphisms by $\mathcal{O}_L$, for L a CM field of degree $2g$.

Henceforth we will *not* work up to isogeny; for clarity we shall begin from scratch.

Let $\{\alpha_i\} \subset \mathcal{O}_L$ be a $\mathbb{Z}$-basis; we have

$$\alpha_p \alpha_i = \sum_j \mu^p_{ij} \alpha_j \qquad (\mu^p_{ij} \in \mathbb{Z})$$

and we write $\mu^p := \{\mu^p_{ij}\}$, which is a $2g \times 2g$ matrix. For the embeddings $L \hookrightarrow \mathbb{C}$ write $\{\theta_1, \ldots, \theta_g; \overline{\theta}_1, \ldots, \overline{\theta}_g\}$ and $\theta_{g+i} := \overline{\theta}_i$; the choice of CM-type $\Theta = \{\theta_1, \ldots, \theta_g\}$ is implicit. Take A to be the abelian variety with period matrix

$$\mathfrak{P} := {}_e[\mathbb{I}]_\omega = \left\{ \int_{\gamma_i} \omega_j \right\}_{i,j=1,\ldots,2g} = \{\theta_j(\alpha_i)\}_{i,j}.$$

Here

- $\{\gamma_i\} \subset H_1(A, \mathbb{Z})$ is a basis with dual basis $e := \gamma^\vee$ for $H^1(A, \mathbb{Z}) =: V_\mathbb{Z}$;
- $\omega = \{\omega_1, \ldots, \omega_g; \overline{\omega}_1, \ldots, \overline{\omega}_g\}$ is *Hodge basis* for $H^1(A, \mathbb{C})$; and
- $\mathbb{I}$ is the identity transformation and ${}_e[\mathbb{I}]_\omega$ the change-of-basis matrix.

It is clear that

$$A = A^{(L,\Theta)}_{\mathcal{O}_L} = \mathbb{C}^g / \Theta(\mathcal{O}_L).$$

Now $\ell = \sum_p \ell_p \alpha_p \in \mathcal{O}_L$ acts on $V_\mathbb{C}$ by $\eta(\ell)\omega_i := \theta_i(\ell)\omega_i$; and writing $\boldsymbol{\theta}(\ell) := \mathrm{diag}\{\theta_1(\ell), \ldots, \theta_{2g}(\ell)\}$, $\mu(\ell) := \sum \ell_p \mu^p$,

$$\begin{aligned}
\sum_k \mathfrak{P}_{ik} (\boldsymbol{\theta}(\alpha_p))_{kj} &= \sum_k \theta_k(\alpha_i) \delta_{kj} \theta_j(\alpha_p) = \theta_j(\alpha_i \alpha_p) \\
&= \theta_j \left(\sum_k \mu^p_{ki} \alpha_k \right) = \sum \mu^p_{ki} \mathfrak{P}_{kj} \\
\Longrightarrow [\eta(\ell)]_e &= ({}_e[\mathbb{I}]_\omega)\, [\eta(\ell)]_\omega\, ({}_\omega[\mathbb{I}]_e) \\
&= \mathfrak{P}(\boldsymbol{\theta}(\ell))\mathfrak{P}^{-1} = {}^t\mu(\ell).
\end{aligned}$$

Since $\mu(\ell)$ is integral, $\eta(\ell)$ acts on $V_\mathbb{Z}$. It follows that if $(u_1, \ldots, u_g)$ $[\mathrm{mod}\, \Theta(\mathcal{O}_L)]$ denote coordinates on A in correspondence with the $\{\omega_j\}_{j=1}^g$, then $\ell \in \mathcal{O}_L$ acts on A by $(u_1, \ldots, u_g) \overset{\mathcal{M}(\ell)}{\mapsto} (\theta_1(\ell)u_1, \ldots, \theta_g(\ell)u_g)$, and $\eta(\ell) = \mathcal{M}(\ell)$.[8] Note that $\mathrm{trace}(\mu(\ell)) = \mathrm{Tr}_{L/\mathbb{Q}}(\ell)$, $\det(\mu(\ell)) = N_{L/\mathbb{Q}}(\ell)$, and $\det\{\mathfrak{P}^t\mathfrak{P}\} = \Delta_L$.

[8] Though $V^1_{(L,\Theta)}$ and $V^1_{(L,\overline{\Theta})}$ are integrally isomorphic as HS (by applying ρ to $\mathcal{O}_L$), this isomorphism is somewhat unnatural. We have tried to stick with the most natural identifications here.

Now we have a convenient formula (Section V.D) that constructs polarizations for CM HS; to adapt this to the present situation, we need to compare V with $V' := V^1_{(L,\overline{\Theta})}$ and A with the abelian variety A' having V' as its H^1. The first difference is that cohomology (not homology) is put into correspondence

$$\beta : \mathcal{O}_L \xrightarrow{\cong} V'_{\mathbb{Z}}$$

with the ring of integers, so that $e'_i = \beta(\alpha_i) \subset V'_{\mathbb{Z}}$ is a basis. Furthermore, ω' is defined by

$$e'_i = \sum_{j=1}^{2g} \overline{\theta_j}(\alpha_i)\omega'_j,$$

yielding period matrix $\mathfrak{P}' := {}_{e'}[\mathbb{I}]_{\omega'} = {}^t\overline{\mathfrak{P}}^{-1}$. Once again θ_L acts on A', by $\mathcal{M}'(\ell)(u'_1, \ldots, u'_g) = (\overline{\theta}_1(\ell)u'_1, \ldots, \overline{\theta}_g(\ell)u'_g)$, which means $[\mathcal{M}'(\ell)^*]_{e'} = \mu(\ell)$ and $[\mathcal{M}'(\ell)^*]_{\omega'} = \overline{\boldsymbol{\theta}}(\ell)$. Put $\gamma' := (e')^\vee$ and

$$\partial' := \{\partial/\partial\overline{u'_1}, \ldots, \partial/\partial\overline{u'_g}; \partial/\partial u'_1, \ldots, \partial/\partial u'_g\}$$

in $(V')^\vee = H_1(A')$; more precisely, $\{\partial/\partial\overline{u'_1}, \ldots, \partial/\partial\overline{u'_g}\}$ are a basis for

$$H_1(A')^{(0,-1)} = F^0H_1(A') = (V'/F^1V')^\vee,$$

and ∂' is the conjugate dual basis of ω'. We have the "period matrix" ${}_{\gamma'}[\mathbb{I}]_{\partial'} = \mathfrak{P}$, $[\mathcal{M}'(\ell)_*]_{\gamma'} = {}^t\mu(\ell)$, $[\mathcal{M}'(\ell)_*]_{\partial'} = \boldsymbol{\theta}(\ell)$, so that

$$V = (V')^\vee(-1) \qquad \text{(V.E.2)}$$

as $\mathbb{Z}$-Hodge structure (and $\mathcal{O}_L$-modules). It follows that $A' \cong A^\vee$.

(V.E.3) **Remark:** For an alternate perspective on the relationship between A and A', consider the complex linear transformation[9]

$$V'_{\mathbb{C}} \xrightarrow{T} V_{\mathbb{C}}$$

given by $\omega'_i \mapsto \omega_i$. One may choose $\{\alpha_i\} \subset \mathcal{O}_L$ so that $\rho(\alpha_j) = (-1)^j\alpha_j$ $(j = 1, \ldots, 2g)$. (We shall not do this outside this Remark!) With this choice,

$${}_e[T]_{e'} = ({}_e[\mathbb{I}]_\omega)_\omega[T]_{\omega'}({}_{\omega'}[\mathbb{I}]_{e'}) = \mathfrak{P}{}^t\overline{\mathfrak{P}}$$

has ij^{th} entry $(-1)^j\mathrm{Tr}_{L/\mathbb{Q}}(\alpha_i\alpha_j)$ and determinant $(-1)^g\Delta_L$. In particular, this matrix is integral, and thus T is a morphism of $\mathbb{Z}$-Hodge structure corresponding to a Δ_L-to-1 isogeny $A \twoheadrightarrow A'$.

[9] It only becomes an $\mathcal{O}_L$-module homomorphism if we replace $(\mathcal{M}')^*$ by $(\mathcal{M}')^* \circ \rho$.

Let $\delta \in L_0$ and $\xi = \sqrt{\delta} \in L$ be such that $\sqrt{-1}\theta_i(\xi) < 0$ $(i = 1, \ldots, g)$; in fact, we may take $\xi \in \mathcal{O}_L$. Then $Q'(\beta(\ell), \beta(\ell')) := \mathrm{Tr}_{L/\mathbb{Q}}(\xi \ell \rho(\ell'))$ yields an *integral* polarization of V' with

$$[Q']_{\omega'} = \begin{pmatrix} 0 & -\sqrt{-1}D \\ \sqrt{-1}D & 0 \end{pmatrix}$$

where $D = \mathrm{diag}\{-\sqrt{-1}\theta_i(\xi)\}$ has positive real entries, and

$$\begin{aligned} [Q']_{e'} = \overline{[Q']_{e'}} &= \overline{{}^t_{\omega'}[\mathbb{I}]_{e'}} \cdot \overline{[Q']_{\omega'}} \cdot \overline{{}_{\omega'}[\mathbb{I}]_{e'}} \\ &= \mathfrak{P} \begin{pmatrix} 0 & \sqrt{-1}D \\ -\sqrt{-1}D & 0 \end{pmatrix} {}^t\mathfrak{P}. \end{aligned}$$

By (V.E.2), $V' \cong H_1(A)(-1)$ as $\mathbb{Z}$-Hodge structure (with $\gamma \mapsto e$) and hence Q' is equivalent to a polarization

$$\mathcal{Q} : H_1(A, \mathbb{Z}) \times H_1(A, \mathbb{Z}) \to \mathbb{Z}$$

with

$$\text{(V.E.4)} \quad [\mathcal{Q}]_\gamma = [Q']_{e'} = \begin{pmatrix} \theta_1(\alpha_1) & \cdots & \theta_{2g}(\alpha_1) \\ \vdots & \ddots & \vdots \\ \theta_1(\alpha_{2g}) & \cdots & \theta_{2g}(\alpha_{2g}) \end{pmatrix}$$

$$\times \left(\begin{array}{c|c} \mathbb{O} & \begin{matrix} \theta_1(\xi) & & 0 \\ & \ddots & \\ 0 & & \theta_g(\xi) \end{matrix} \\ \hline \begin{matrix} \theta_{g+1}(\xi) & & 0 \\ & \ddots & \\ 0 & & \theta_{2g}(\xi) \end{matrix} & \mathbb{O} \end{array}\right) \begin{pmatrix} \theta_1(\alpha_1) & \cdots & \theta_1(\alpha_{2g}) \\ \vdots & \ddots & \vdots \\ \theta_{2g}(\alpha_1) & \cdots & \theta_{2g}(\alpha_{2g}) \end{pmatrix}.$$

The induced polarization Q on $V = H^1(A)$ (which may no longer be integral) has matrix $[Q]_e = -[\mathcal{Q}]_\gamma^{-1}$, and $\eta(\ell)^t = \eta(\rho(\ell))$ essentially as in the proof of (V.2).

With (V.E.4) in hand, we can turn our focus exclusively to A and V. There are some nice bijections between data for A and L:

$$\begin{aligned} \mathrm{Corr}(A) &\cong \mathrm{End}(V_{\mathbb{Q}}, \varphi) \cong L, \\ \mathrm{End}(A) &\cong \mathrm{End}(V_{\mathbb{Z}}, \varphi) \cong \mathcal{O}_L, \\ \mathrm{Aut}(A) &\cong \mathrm{End}(V_{\mathbb{Z}}, \varphi)^* \cong \mathcal{O}_L^*, \end{aligned}$$

where "$*$" means invertible (e.g., $\mathcal{O}_L^*$ is the group of units in $\mathcal{O}_L$). For correspondences preserving Q,

(V.E.5) $\quad \mathrm{Corr}(A,Q) = \mathrm{End}(V_{\mathbb{Q}},\varphi,Q) = \{\ell \in L \mid |\theta_i(\ell)| = 1 \;\text{ for all }\; i\}$

(we need $\theta_i(\ell)\theta_{i+g}(\ell) = 1$), and

(V.E.6) $$\mathrm{End}(A,Q) = \mathrm{Aut}(A,Q) = \omega_L,$$

where ω_L, the roots of unity in L, equals $\{\ell \in L \mid |\theta_i(\ell)| = 1 \text{ (for all } i)\} \cap \mathcal{O}_L^*$ by a theorem of Kronecker. We observe that Dirichlet's theorem now translates as $\frac{\mathrm{Aut}(A)}{\mathrm{Aut}(A,Q)} \cong \mathbb{Z}^{g-1}$. Note that

(V.E.7) $\quad \mathrm{Corr}(A)$ [resp. $\mathrm{Corr}(A,Q)$] contains $M_{\tilde{\varphi}}(\mathbb{Q})$ [resp. $M_{\varphi}(\mathbb{Q})$].

Now let A be the abelian *surface* with CM by the ring of integers of the 5th cyclotomic field

$$L = \frac{\mathbb{Q}[X]}{(1+X+X^2+X^3+X^4)}.$$

Obviously $[L:\mathbb{Q}] = 4$ and $g = 2$, and L has totally real subfield

$$L_0 = \frac{\mathbb{Q}[Z]}{(Z^2+Z-1)}$$

that embeds in L by $Z \mapsto X + X^4$. As before, we choose our two principal complex embeddings (the other two are conjugates) θ_1 and θ_2 to be induced by $X \mapsto \zeta$ resp. ζ^2, where $\zeta := e^{\frac{2\pi\sqrt{-1}}{5}}$; this yields for instance $\theta_1(Z) = \frac{-1+\sqrt{5}}{2}$, $\theta_2(Z) = \frac{-1-\sqrt{5}}{2}$. We write bases for the integers and units of L:

$$\mathcal{O}_L = \mathbb{Z}\left\langle 1, X, X^2, X^3 \right\rangle,$$

which is an additive group with identity 0,[10] and

$$\frac{\mathcal{O}_L^*}{\omega_L} = \mathbb{Z}\left\langle Z \right\rangle,$$

which is a multiplicative group with identity 1;[11] the roots of unity are

$$\omega_L = \{1, X, X^2, X^3, X^4, -1, -X, -X^2, -X^3, -X^4\}.$$

[10] Note $X^4 = -1 - X - X^2 - X^3$.

[11] The inverse of $Z = X + X^4$ is $-1 - Z = X^2 + X^3$, hence it still lies in $\mathcal{O}_L$.

The period matrix for A is

$$\mathfrak{P} = \begin{pmatrix} 1 & 1 & 1 & 1 \\ \zeta & \zeta^2 & \zeta^4 & \zeta^3 \\ \zeta^2 & \zeta^4 & \zeta^3 & \zeta \\ \zeta^3 & \zeta & \zeta^2 & \zeta^4 \end{pmatrix}.$$

For the polarization matrix, first observe that $\xi := (-2X + 2X^4 - X^3 + X^2)^{-1}$ yields

$$D = \operatorname{diag}\left\{ \frac{\sqrt{-1}}{2\zeta - 2\zeta^4 + \zeta^3 - \zeta^2}, \frac{\sqrt{-1}}{2\zeta^2 - 2\zeta^3 + \zeta - \zeta^4} \right\} > 0$$

and by (V.E.4)

$$[\mathcal{Q}]_\gamma = \begin{pmatrix} & 1 & & \\ -1 & & 1 & \\ & -1 & & 1 \\ & & -1 & \end{pmatrix}, [Q]_e = \begin{pmatrix} & & 1 & 1 \\ -1 & & & \\ & & & 1 \\ -1 & & -1 & \end{pmatrix}.$$

This choice of ξ is not in $\mathcal{O}_L$, but if we replace it by (say) $-\frac{1}{\xi} (\in \mathcal{O}_L)$, then $[\mathcal{Q}]_\gamma$ (while integral) is not nearly as nice.

For the Mumford-Tate group, here is how to find it the explicit computational way: for $\alpha = a + b\sqrt{-1} \in \mathbb{C}^*$, $[\tilde{\varphi}(\alpha)]_e = \mathfrak{P}.[\tilde{\varphi}(\alpha)]_\omega.\mathfrak{P}^{-1} =$

$$= \frac{\zeta^2}{(\zeta - 1)^3(\zeta+1)^2} \begin{pmatrix} 1 & 1 & 1 & 1 \\ \zeta & \zeta^2 & \zeta^4 & \zeta^3 \\ \zeta^2 & \zeta^4 & \zeta^3 & \zeta \\ \zeta^3 & \zeta & \zeta^2 & \zeta^4 \end{pmatrix} \begin{pmatrix} \alpha & & & \\ & \alpha & & \\ & & \bar{\alpha} & \\ & & & \bar{\alpha} \end{pmatrix}$$

$$\begin{pmatrix} -\zeta & -(\zeta + 1) & \zeta^2(\zeta + 1) & \zeta^2 \\ -\zeta(\zeta + 1) & \zeta^3 & -\zeta^2 & \zeta^3(\zeta + 1) \\ 1 & \zeta + 1 & -\zeta^3(\zeta + 1) & -\zeta^4 \\ \zeta^4(\zeta + 1) & -\zeta^3 & \zeta^4 & -\zeta^2(\zeta + 1) \end{pmatrix}.$$

Writing $S(m) = \sin(\frac{m}{20}2\pi)$, $C(s) = \cos(\frac{m}{20}2\pi)$, this equals

$$
\begin{aligned}
&\frac{\sqrt{-1}}{(\zeta-1)^3(\zeta+1)^2}\begin{pmatrix} {\scriptstyle (4C(3)+2C(1))a \atop +(4S(3)-2S(1))b} & (4S(3)+2)b & 2b & (4S(1)+2)b \\ -(4S(1)+2)b & {\scriptstyle (4C(3)+2C(1))a \atop -2S(1)b} & 2b & (-4S(3)+2)b \\ (4S(3)-2)b & -2b & {\scriptstyle (4C(3)+2C(1))a \atop +2S(1)b} & (4S(1)+2)b \\ -(4S(1)+2)b & -2b & -(4S(3)+2)b & {\scriptstyle (4C(3)+2C(1))a \atop -(4S(3)-2S(1))b} \end{pmatrix} \\
&= a\mathbb{I} + \frac{2\sqrt{-1}b}{(\zeta-1)^3(\zeta+1)^2}\begin{pmatrix} 2S(3)-S(1) & 2S(3)+1 & 1 & 2S(1)+1 \\ -2S(1)-1 & -S(1) & 1 & -2S(3)+1 \\ 2S(3)-1 & -1 & S(1) & 2S(1)+1 \\ -2S(1)-1 & -1 & -2S(3)-1 & -2S(3)+S(1) \end{pmatrix} \\
&= a\mathbb{I} + b\xi \begin{pmatrix} \frac{3+\sqrt{5}}{4} & \frac{3+\sqrt{5}}{2} & 1 & \frac{1+\sqrt{5}}{2} \\ -\left(\frac{1+\sqrt{5}}{2}\right) & \frac{1-\sqrt{5}}{2} & 1 & \frac{1-\sqrt{5}}{2} \\ \frac{-1+\sqrt{5}}{2} & -1 & \frac{-1+\sqrt{5}}{4} & \frac{1+\sqrt{5}}{2} \\ -\left(\frac{1+\sqrt{5}}{2}\right) & -1 & -\left(\frac{3+\sqrt{5}}{2}\right) & -\left(\frac{3+\sqrt{5}}{4}\right) \end{pmatrix} \\
&=: a\mathbb{I} + b\xi\mathsf{M},
\end{aligned}
$$

where $2\xi = \frac{\sqrt{5}}{5}\sqrt{\frac{8}{5+\sqrt{5}}}$. Writing $\tilde{\mathsf{M}}$ for the conjugate of M induced by $\sqrt{5} \mapsto -\sqrt{5}$, it can be checked explicitly that $\mathsf{M}^2 = -\frac{5}{8}(5+\sqrt{5})\mathbb{I}$ and $[\mathsf{M}, \tilde{\mathsf{M}}] = 0$. The fact that the $\mathbb{Q}$-closure of the image of $\tilde{\varphi}$ must contain in its real points all elements of the form

$$\left\{ (a_1\mathbb{I} + b_1\mathsf{M})(a_2\mathbb{I} + b_2\tilde{\mathsf{M}}) \middle| (a_i, b_i) \in \mathbb{R}^2 \setminus (0,0)\,,\, i = 1,2 \right\}$$

with respect to the e-basis, indicates that $\dim M_{\tilde{\varphi}} \geq 3$. This is enough to show equality in (V.E.7).

To identify the $\mathbb{Q}$-points of $M_{\tilde{\varphi}}$ [resp. M_{φ}] in the e-basis, first note that for $\ell = a + bX + cX^2 + dX^3$ $(a, b, c, d \in \mathbb{Q})$, we have

$$\mu(\ell) = \begin{pmatrix} a & b & c & d \\ -d & a-d & b-d & c-d \\ d-c & -c & a-c & b-c \\ c-b & d-b & -b & a-b \end{pmatrix}.$$

The condition $|\theta_1(\ell)|^{(2)} = |\theta_2(\ell)|^{(2)}$ gives $M_{\tilde{\varphi}}(\mathbb{Q})$, and is expressed by

$$ab + bc + cd = ac + bd + ad. \tag{V.E.8}$$

[PROOF: $|\theta_1(\ell)|^2 = r_1 + r_2\sqrt{5}$, $|\theta_2(\ell)|^2 = r_1 - r_2\sqrt{5}$, where $r_2 = \text{LHS(V.E.8)} - \text{RHS(V.E.8)}$.] For $M_{\varphi}(\mathbb{Q})$, we need $|\theta_1(\ell)| = |\theta_2(\ell)| = 1$ (cf. (V.E.5)), and this is equivalent to the Diophantine equation

$$ab + bc + cd = ac + bd + ad = a^2 + b^2 + c^2 + d^2 - 1. \tag{V.E.9}$$

A solution of (V.E.9) of infinite order is $(a, b, c, d) = \left(\frac{3}{11}, \frac{-3}{11}, \frac{9}{11}, \frac{7}{11}\right)$. More generally, taking elements of L of norm $(\prod_{i=1}^{4} \theta_i(\ell) =)\ 1$ and computing their image (in L) under $N_{\Theta'}$ would give further solutions to (V.E.9).

V.F PROOFS OF PROPOSITIONS V.D.4 AND V.D.5 IN THE GALOIS CASE

We now demonstrate that the Mumford-Tate group $M_{\tilde{\varphi}}$ of a SCMpHS $V = V^n_{(F,\Pi)}$ is computed by the reflex norm map $N_{\Pi'}$ from Section V.A. To eschew unenlightening notational and technical complication, we shall assume $F/\mathbb{Q}$ is Galois, of degree $r = 2g = \dim(V)$.

Begin with the diagram

$$\begin{array}{ccccc} F'^{,*} & \xrightarrow{\quad N_{\Pi'} \quad} & F^* & & \\ {\scriptstyle \eta|_{F'}} \downarrow \wr\| & & {\scriptstyle \eta} \downarrow \wr\| & \searrow^{f \mapsto \eta(f)} & \\ (\mathrm{Res}_{F'/\mathbb{Q}}(\mathbb{G}_m))(\mathbb{Q}) & \xrightarrow{\eta(N_{\Pi'})} & (\mathrm{Res}_{F/\mathbb{Q}}(\mathbb{G}_m))(\mathbb{Q}) & \subset & \mathrm{GL}(V, \mathbb{Q}) \end{array}$$

where, since all $\sigma \in \mathrm{Gal}(F/\mathbb{Q})$ define rational automorphisms of $\mathrm{Res}_{F/\mathbb{Q}}(\mathbb{G}_m)$, the underlying map

$$\mathcal{N}_{\Pi'} : \mathrm{Res}_{F'/\mathbb{Q}}(\mathbb{G}_m) \to \mathrm{Res}_{F/\mathbb{Q}}(\mathbb{G}_m)(\subset \mathrm{GL}(V))$$

is obviously defined over $\mathbb{Q}$. For (V.D.4) we must show that the algebraic group $\mathrm{Im}(\mathcal{N}_{\Pi'})$

(a) is contained in $M_{\tilde{\varphi}}$;

(b) contains $M_{\tilde{\varphi}}$.

For (b), it is enough to show that $(\mathrm{Im}(\mathcal{N}_{\Pi'}))(\mathbb{C}) \supset (\mathrm{Im}(\mu))(\mathbb{C})$, where $\mu : \mathbb{G}_m \to \mathrm{GL}(V)$ is defined (over $\mathbb{C}$) by $\mu(z)\big|_{V^{p,q}} = z^p\, \mathrm{id}_{V^{p,q}}$.

The $\{\theta_i\}$ define (over $F^{(\prime)}$) maps $\mathrm{Res}_{F'/\mathbb{Q}}(\mathbb{G}_m) \xrightarrow{\theta_i} \mathbb{G}_m$. Under the identification $(\mathrm{Res}_{F'/\mathbb{Q}}(\mathbb{G}_m))(\mathbb{Q}) \cong (F')^*$, a rational point f' is sent to $\theta_i(f') \in \mathbb{G}_m(F)$. Under the identification $(\mathrm{Res}_{F'/\mathbb{Q}}(\mathbb{G}_m))(\mathbb{C}) \cong (\mathbb{C}^*)^{[F':\mathbb{Q}]=:m}$, $(z_1, \ldots, z_m) \mapsto z_j$, where j is such that $\theta_i\big|_{F'} = \theta'_j$. We also have a map (defined over F') $\mathbb{G}_m \xrightarrow[\jmath_1]{} \mathrm{Res}_{F'/\mathbb{Q}}(\mathbb{G}_m)$ defined on $\mathbb{C}$-points by $z \mapsto (z, 1, \ldots, 1)$. The composition $\theta_i \circ \jmath_1$ is the identity if $\theta_i\big|_{F'} = \theta'_1$, and is otherwise trivial.

From Section V.A we have the formula (with $\mathcal{G}_0 := \mathrm{Gal}(F/F')$)

$$(\theta_i \circ \mathcal{N}_{\Pi'})(f') = \prod_{\mathcal{G}_0\sigma \in (\mathcal{G}_0\backslash\mathcal{G})} (\theta_i(\sigma^{-1}(f')))^{p(\sigma)}.$$

Now

$$(\theta_i\sigma^{-1})(\jmath_1(z)) = \begin{cases} z, & \theta_i\sigma^{-1}\big|_{F'} = \theta_1'(=\theta_1\big|_{F'}) \\ 1, & \text{otherwise,} \end{cases}$$

and by the definition of the reflex field,

$$\begin{aligned} \theta_i\sigma^{-1}\big|_{F'} = \theta_1\big|_{F'} \quad &\Longleftrightarrow \\ \theta_\sigma^{\{-1\}}\big|_{F'} = \theta_i^{\{-1\}}\big|_{F'} \quad &\Longleftrightarrow \\ \theta_\sigma, \theta_i \text{ are in the same } \Pi^{p,q} &\Longleftrightarrow \\ p(\sigma) = p. \end{aligned}$$

Hence

$$(\theta_i \circ \mathcal{N}_{\Pi'})(\jmath(z)) = z^p,$$

i.e., $\mathcal{N}_{\Pi'}(\jmath(z)) = \mu(z)$, proving (b).

For (a), we will show that any $\mathbb{Q}$-algebraic group $\mathfrak{M} \subset \mathrm{GL}(V)$ with $\mathfrak{M}(F)$ containing $\mu(F^{',*})$ (in particular, $M_{\tilde{\varphi}}$) has $\mathbb{Q}$-points containing $\eta(N_{\Pi'}(F^{',*}))$, hence contains $\mathrm{Im}(\mathcal{N}_{\Pi'})$.

If $\mathfrak{M}$ is defined over $\mathbb{Q}$ then $\mathfrak{M}(F)$ must also contain

$$\{{}^{\sigma}\mu(f') \mid \sigma \in \mathcal{G}, f' \in F^{',*}\}.$$

So we may define

$$\mathcal{N} := \prod_{\mathcal{G}_0\sigma \in (\mathcal{G}_0\backslash\mathcal{G})} ({}^{\sigma^{-1}}\mu(f')) : F^{',*} \to \mathfrak{M}(\mathbb{Q});$$

to check that this recovers $\mathcal{N}_{\Pi'}$ we compare their compositions with each θ_i. Now $\theta_i(\mathcal{N}(f'))$ is the eigenvalue of the action of $\mathcal{N}(f')$ on ω_i:

$$\begin{aligned} \mathcal{N}(f')\omega_i &= \prod_{\mathcal{G}_0\sigma} {}^{\sigma^{-1}}(\mu(f')\omega_{\sigma(i)}) \\ &= \left[\prod_{\mathcal{G}_0\sigma} (\sigma^{-1}(f))^{p(\sigma(i))}\right]\omega_i. \end{aligned}$$

Replacing σ by $\sigma\theta_i^{-1}$, this becomes

$$\prod_{\mathcal{G}_0\sigma} (\theta_i(\sigma^{-1}(f')))^{p(\sigma)} = \theta_i(\mathcal{N}_{\Pi'}(f'))$$

as desired. This completes the proof of Proposition V.D.4.

To compute the dimension of $M_{\tilde{\varphi}}$ in terms of the generalized Kubota rank (Prop. V.D.5), first note that τ from Section V.A is factored by the transformation from $\mathbb{Z}[\mathcal{G}] \to \mathbb{Z}[\mathcal{G}_0\backslash\mathcal{G}]$ given by

$$\tilde{\tau}([g]) = \sum_{\mathcal{G}_0 g' \in (\mathcal{G}_0\backslash\mathcal{G})} p(\mathcal{G}_0 g')[\mathcal{G}_0 g' g].$$

It is easy to see that this has the same rank as

$$\hat{\tau}([g'\mathcal{G}_0]) = \sum_{g\in\mathcal{G}} p(\mathcal{G}_0(g')^{-1}g)[g] : \mathbb{Z}[\mathcal{G}\backslash\mathcal{G}_0] \to \mathbb{Z}[\mathbb{G}].$$

As before we may define $\jmath_j : \mathbb{G}_m \to \mathrm{Res}_{F'/\mathbb{Q}}(\mathbb{G}_m)$ by

$$z \mapsto (1, \ldots, \underset{j^{\text{th}}\text{ entry}}{z}, \ldots, 1)$$

on complex points, and

$$\theta_i(\sigma^{-1}(\jmath_j(z))) = \begin{cases} z & \text{if } \theta_i\sigma^{-1}\big|_{F'} \overset{(*)}{=} \theta'_j \\ 1 & \text{otherwise.} \end{cases}$$

Writing $\theta'_j = \theta_1 g'_j \mathcal{G}_0$ and $\theta_i = \theta_1 g_i$, we find that $(*) \Rightarrow p(\sigma) = p((g'_j)^{-1}g_i)$, so that

$$\theta_i(\mathcal{N}_{\Pi'}(\jmath_j(z))) = z^{p(\mathcal{G}_0(g'_j)^{-1}g_i)}$$

and $\mathcal{N}_{\Pi'}$ may be described on complex points by

$$(z_1, \ldots, z_m) \mapsto \left(\prod_{j=1}^{m} z_j^{p(\mathcal{G}_0(g'_j)^{-1}g_1)}, \ldots, \prod_{j=1}^{m} z_j^{p(\mathcal{G}_0(g'_j)^{-1}g_{[F:\mathbb{Q}]})} \right).$$

Taking log of each entry on the RHS gives a linear map with the same rank as $\mathcal{N}_{\Pi'}$; its matrix is just the matrix of $\hat{\tau}$, hence has rank $\mathcal{R}(F, \Pi)$. This proves (V.D.5), and as a byproduct we get the following more precise statement.

(V.F.1) PROPOSITION: *In the Hodge-Galois basis $\{\omega_i\}$ of V, the Lie algebra $\mathfrak{m} \subset \mathrm{End}(V)$ of $M_{\tilde{\varphi}^n_{(F,\Pi)}}$ is given by the image of the matrix*

$$\{p(\mathcal{G}_0(g'_j)^{-1}g_i)\}_{\substack{i=1,\ldots,r \\ j=1,\ldots,m.}}$$

That is, writing $\{\widehat{\omega}_{kl}\}_{k,l=1,\ldots,r}$ for the basis of $\mathrm{End}(V_{\mathbb{C}})$ determined by $\widehat{\omega}_{kl}(\omega_i) := \delta_{il}\omega_k$, we have

$$\mathfrak{m} = \mathbb{C}\left\langle \left\{ \sum_{i=1}^{r} p((g'_j)^{-1}g_i)\widehat{\omega}_{ii} \right\}_{j=1,\ldots,m} \right\rangle.$$

PROOF. Take the Jacobian matrix of $\mathcal{N}_{\Pi'}$ at 1, and interpret $\partial/\partial z_i$ as $\widehat{\omega}_{ii}$. $\square$

(V.F.2) **Example:** For our standard $n = 3$, $r = 4$ case, $(L = L' = \mathbb{Q}(\zeta_5)$, $\Pi^{3,0} = \{\theta_1\}$, $\Pi^{2,1} = \{\theta_2\}$, $N_{\Pi'} = \sigma_1^3 \cdot \sigma_3^2 \cdot \sigma_2)$, $m = 4$ and the matrix is

$$\begin{pmatrix} 3 & 1 & -1 & -3 \\ -1 & 3 & -3 & 1 \\ 1 & -3 & 3 & -1 \\ -3 & -1 & 1 & 3 \end{pmatrix};$$

consequently

$$\mathfrak{m} = \mathbb{C}\left\langle \widehat{\omega}_{22} - \widehat{\omega}_{33}, \widehat{\omega}_{11} - \widehat{\omega}_{44} \right\rangle.$$

Chapter VI

Arithmetic Aspects of Mumford-Tate Domains

We start by remarking on a few points concerning the structure and construction of Mumford-Tate domains. Let φ be a Q-polarized Hodge structure on V with Mumford-Tate group

$$M = M_\varphi \subseteq G := \mathrm{Aut}(V, Q)^0,$$

and H_φ (resp. P_φ) the stabilizer of φ (resp. $F_\varphi^\bullet$) in G. The algebraic groups M_φ, H_φ, and P_φ are connected,[1] closed, and reductive.

(VI.1) LEMMA: *$M_\varphi \cap H_\varphi$ and $M_\varphi \cap P_\varphi$ are connected (as algebraic groups).*

PROOF. We do H_φ; proof for P_φ is nearly identical. Suppose otherwise, and write $M_\varphi \cap H_\varphi = Z_0 \amalg \coprod_{i=1}^{m} Z_i$, with $Z_1^k \subset Z_0$, and where Z_0 denotes the identity component. Any element $\alpha \in Z_1(\mathbb{C})$ is of the form $\alpha_{ss}\alpha_u$ by the Jordan decomposition, where α_{ss} and α_u are unique, commuting, and contained in both M_φ and H_φ. Both groups also contain $\exp(\mathbb{C} \log \alpha_u)$ so that $\alpha_u \in Z_0$; consequently $\alpha_{ss} \in Z_1$. Notice that, being in H_φ, α_{ss} diagonalizes with respect to a Hodge basis for φ.

Clearly $\varphi(\mathbb{U}(\mathbb{C})) \subset Z_0(\mathbb{C})$. Let $z \in \mathbb{U}(\mathbb{C})$ be generic and consider the element $\alpha_{ss}\varphi(z) =: \alpha'_{ss} \in Z_1(\mathbb{C})$, which now *only* diagonalizes with respect to a Hodge basis. Let $T \subset M_\varphi$ be a maximal torus containing α'_{ss}; then it also only diagonalizes with respect to a Hodge basis, and is therefore contained in $M_\varphi \cap H_\varphi$. But only the identity component Z_0 can contain a torus, and so we reach the contradiction $\alpha'_{ss} \in T(\mathbb{C}) \subset Z_0(\mathbb{C})$. □

(VI.2) COROLLARY: *Mumford-Tate domains respect the $\mathbb{R}$-direct-product structure of M: that is, if $M = M_1 \times M_2$, then $\check{D}_M := M(\mathbb{C})F_\varphi^\bullet \cong \check{D}_{M_1} \times \check{D}_{M_2}$ and $D_M := M(\mathbb{R}) \cdot \varphi \cong D_{M_1} \times D_{M_2}$.*

[1] For M_φ this is by virtue of being the $\mathbb{Q}$-closure of $\varphi(\mathbb{U}(\mathbb{C}))$. The connectedness is in the "absolute" sense: the complex points $M(\mathbb{C})$, and hence the orbit $\check{D}_M = M(\mathbb{C})F_\varphi^\bullet$, are connected, but $M(\mathbb{R})$ (and $D_M = M(\mathbb{R})\varphi$) need not be.

PROOF. We regard M_1, M_2 as subgroups of M defined over $\mathbb{R}$. The isomorphism

$$M_1(\mathbb{R}) \times M_2(\mathbb{R}) \xrightarrow[\mu_{\mathbb{R}}]{\cong} M(\mathbb{R})$$

induces

$$D_{M_1} \times D_{M_2} \cong \frac{M_1(\mathbb{R})}{(M_1 \cap H_\varphi)(\mathbb{R})} \times \frac{M_2(\mathbb{R})}{(M_2 \cap H_\varphi)(\mathbb{R})} \overset{(*)}{\twoheadrightarrow} \frac{M(\mathbb{R})}{(M \cap H_\varphi)(\mathbb{R})}.$$

To know that $(*)$ is an isomorphism, we must show that the map

$$(M_1 \cap H_\varphi)(\mathbb{R}) \times (M_2 \cap H_\varphi)(\mathbb{R}) \overset{(**)}{\hookrightarrow} (M \cap H_\varphi)(\mathbb{R})$$

induced by $\mu_{\mathbb{R}}$ is surjective. As

$$\begin{aligned} \mathrm{LHS}(**) &= ((M_1 \cap H_\varphi)(\mathbb{C}) \times (M_2 \cap H_\varphi)(\mathbb{C})) \cap \underbrace{(M_1(\mathbb{R}) \times M_2(\mathbb{R}))}_{\downarrow \cong} \\ \mathrm{RHS}(**) &= \qquad (M \cap H_\varphi)(\mathbb{C}) \qquad \cap \qquad \overbrace{M(\mathbb{R})}, \end{aligned}$$

it is enough to show that

$$(M_1 \cap H_\varphi)(\mathbb{C}) \times (M_2 \cap H_\varphi)(\mathbb{C}) \twoheadrightarrow (M \cap H_\varphi)(\mathbb{C}).$$

By connectedness of $M \cap H_\varphi$, this can be checked on the level of tangent spaces, where we do indeed have

$$T_\varphi(M \cap H_\varphi)_{\mathbb{C}} = \mathfrak{m}_{\mathbb{C}}^{(0,0)} = \bigoplus_{i=1}^{2} \mathfrak{m}_{i,\mathbb{C}}^{(0,0)} = \bigoplus_{i=1}^{2} T_\varphi(M_i \cap H_\varphi)_{\mathbb{C}}.$$

The statement for $\check{D}_M$ follows from a similar Lie algebra statement and connectedness of $M \cap P_\varphi$. □

So Mumford-Tate groups built up from well-understood real[2] factors are one source of easily described examples of Mumford-Tate domains.

Another source is Hodge structures with complex multiplication. Borcea [Bo] proved the equivalence of

- φ is a CM-HS,
- There exists a family of automorphisms of D, induced by elements $g \in G(\mathbb{Q})$, with φ as an isolated fixed point;

[2] What is more, the almost-direct simple factors of M, a priori defined over $\mathbb{C}$, are known to be defined over $\mathbb{R}$ [Mo1].

and indeed this family of automorphisms is provided by the action of $\mathbb{Q}$-points of M_φ, which is contained in H_φ for φ of CM type (Proposition (II.A.7)). So CM-Hodge structures give 1-point Mumford-Tate domains, henceforth called "CM points." In contrast to Borcea, we want to study automorphisms of D exchanging the finitely many points of $\overset{(\vee)}{\mathrm{NL}}_{M_\varphi}$ and their arithmetic meaning.

Particularly, if M_φ is a maximal torus in G, the irreducible nondegenerate CM case, with $\mathrm{End}(V_\mathbb{Q}, \varphi)$ isomorphic to a CM field L, we shall be interested in the action of its Weyl group $W_G(M_\varphi)$ on the Noether-Lefschetz locus. Furthermore, one can easily define interesting nondegenerate Mumford-Tate groups, and hence "Shimura domains," by replacing $\mathrm{End}(V_\mathbb{Q}, \varphi)$ with a proper subfield of L. The corresponding Noether-Lefschetz locus components are permuted by a subgroup of the Weyl group, and the relationship between such subgroups and Galois theory seems intriguing and unexplored for general period domains.

Recall of notation: S^1 will mean the unit circle in $\mathbb{C}$; recall that $\mathbb{U}(\mathbb{R}) \cong S^1$. We will also write $M(\mathbb{R})^0$ for the identity connected component of $M(\mathbb{R})$ (viz., for even n we have that $\mathrm{SO}(p,q;\mathbb{R})^+ = G(\mathbb{R})^0$ in $\mathrm{SO}(p,q;\mathbb{R}) = G(\mathbb{R})$), and $D_M^0 := M(\mathbb{R})^0\varphi$ for the connected component of D_M. Note that if $M(\mathbb{R})^0 \neq M(\mathbb{R})$, it is not the real points of an algebraic group.

Finally, $\check{D}$ is disconnected and $G(\mathbb{C})$ does not act transitively, precisely when the weight $n = 2m$ is even and $h^{m,m} = 0$; we exclude this case, as well as the case where $h^{m,m}$ is the *only* non-zero Hodge number.

VI.A GROUPS STABILIZING SUBSETS OF D

Let $\varphi \in D$ be a Q-polarized Hodge structure with Mumford-Tate group $M = M_\varphi$. We will be interested in subgroups of $G := \mathrm{Aut}(V, Q)^0$ stabilizing $\overset{(\vee)}{\mathrm{NL}}_M$ and $\overset{(\vee)}{D}_M$ in $\overset{(\vee)}{D}$. Here, $\overset{(\vee)}{D}$ stands for either D or $\check{D}$. Suppose

$$M \leq \mathcal{G} \leq G,$$

with $\mathcal{G}$ connected and defined over $\mathbb{Q}$ and $\mathcal{G}(\mathbb{R})D_M \subset D_M$. Then, in analogy to the geometric monodromy group, writing

$$\Pi_\mathcal{G} := \text{identity connected component of the } \mathbb{Q}\text{(-Zariski)-closure of } \mathcal{G}(\mathbb{Z}),$$

we have

(VI.A.1) LEMMA: $\Pi_\mathcal{G} \leq M$.

PROOF. By definition, and recalling that $T^{k,l} = T^{k,l}(V)$

$$\mathrm{Hg}_M^{k,l} = T^{k,l} \cap F^{\frac{(k-l)n}{2}} \text{ for generic } F^\bullet \in D_M^0,$$

with $\subseteq$ for *all* $F^\bullet \in D_M$. For $g \in \mathcal{G}(\mathbb{Q})$,

$$g\mathrm{Hg}_M^{k,l} \subset T^{k,l}(V_{\mathbb{Q}}) \cap gF^{\frac{(k-l)n}{2}} \quad \text{for all } F^\bullet \in D_M,$$

and so

$$g\mathrm{Hg}_M^{k,l} \subset \mathrm{Hg}_M^{k,l}.$$

Since $\mathcal{G}$ is connected and defined over $\mathbb{Q}$, $\mathcal{G}(\mathbb{Q})$ is Zariski-dense in $\mathcal{G}$; so $\mathcal{G}$ itself, and hence $\mathcal{G}(\mathbb{R})$ and $\mathcal{G}_{\mathbb{C}}$, stabilize each $\mathrm{Hg}_{M,\mathbb{C}}^{k,l}$ (but may not fix it pointwise unlike M).

Now on each $\mathrm{Hg}_M^{k,l}$, the polarization is positive definite, so that $\mathcal{G}(\mathbb{Z})$ acts on $\mathrm{Hg}_M^{\bullet,\bullet}$ through the integer points of an orthogonal group, and hence finitely. Consequently $\Pi_{\mathcal{G}}$ acts as the identity on each $\mathrm{Hg}_M^{k,l}$, which implies it is contained in $M' = M$. □

Next we look at groups stabilizing various Noether-Lefschetz loci:

$$\begin{aligned} \mathrm{NL}_M &:= \{Q\text{-polarized Hodge structures with Mumford-Tate group} \\ &\qquad \text{contained in } M\} \subseteq D \\ \breve{\mathrm{NL}}_M &:= \{\text{flags } F^\bullet \in \check{D} \text{ with Mumford-Tate group contained in } M\} \subseteq \check{D} \end{aligned}$$

where we recall that the Mumford-Tate group of a flag is the group stabilizing its Hodge tensors. Let $\mathfrak{Z} \subseteq \check{D}$ denote the non-Hodge structure locus, and $\check{D}^-$ the complement. Then $\check{D}^-$ are the Hodge structures *not necessarily polarized by* Q. Putting

$$\breve{\mathrm{NL}}_M^- := \breve{\mathrm{NL}}_M \cap \check{D}^-,$$

we note that the proof of (II.C.1) shows
(VI.A.2)

$$\begin{cases} \breve{\mathrm{NL}}_M^- \text{ is a disjoint union of finitely many Mumford-Tate domains,} \\ \text{i.e., } M(\mathbb{R})\text{-orbits, with } \breve{\mathrm{NL}}_M^- \cap D = \mathrm{NL}_M \text{ the subset} \\ \text{consisting only of the } Q\text{-polarized Mumford-Tate domains.} \end{cases}$$

This leaves open the question of whether (a) $\breve{\mathrm{NL}}_M$ is just the Zariski-closure of $\breve{\mathrm{NL}}_M^-$ and therefore is a union of compact duals of Mumford-Tate domains

($M(\mathbb{C})$-orbits of Hodge structure),[3] or (b) there are connected components of $\widecheck{\mathrm{NL}}_M$ contained entirely in $\mathfrak{Z}$. We address this in IV.B, as it is of some consequence to the arithmetic problems we consider later in this section.

(VI.A.3) THEOREM: (i) *The largest subgroup of* $G(\mathbb{R})$ *stabilizing* NL_M *is the normalizer* $N_G(M,\mathbb{R})$.[4] (ii) *The largest subgroup of* $G(\mathbb{C})$ *stabilizing* $\widecheck{\mathrm{NL}}_M$ *is the normalizer* $N_G(M,\mathbb{C})$.

PROOF. It suffices to prove the equality of the following algebraic groups, all defined over $\mathbb{Q}$:

- $N_G(M)$,
- $\mathfrak{M} :=$ (largest) subgroup of G stabilizing spaces of M-Hodge tensors,
- $\widetilde{\mathfrak{M}} :=$ (largest) subgroup of G stabilizing $\widecheck{\mathrm{NL}}_M$,

and it is enough to consider complex points. There are a few warnings here: some non-identity components of these groups may have complex (and $\overline{\mathbb{Q}}$-) points but no real points. Also, "$\mathfrak{M}$ stabilizes each $\mathrm{Hg}_M^{k,l}$" does not mean that an individual $g \in \mathfrak{M}(\mathbb{C})$ has $g\mathrm{Hg}_M^{k,l} \subset \mathrm{Hg}_M^{k,l}$, only that $g\mathrm{Hg}_{M,\mathbb{C}}^{k,l} \subset \mathrm{Hg}_{M,\mathbb{C}}^{k,l}$ ($:= \mathrm{Hg}_M^{k,l} \otimes \mathbb{C}$), which implies $g\mathrm{Hg}_{M,\mathbb{C}}^{k,l} = \mathrm{Hg}_{M,\mathbb{C}}^{k,l}$.

Step 1: To see that $\mathfrak{M} = N_G(M)$: $g \in \mathfrak{M}(\mathbb{C}) \iff g\mathrm{Hg}_{M,\mathbb{C}}^{k,l} \subset \mathrm{Hg}_{M,\mathbb{C}}^{k,l} \iff M$ fixes $g\mathrm{Hg}_{M,\mathbb{C}}^{k,l}$ pointwise $\iff g^{-1}Mg$ fixes $\mathrm{Hg}_{M,\mathbb{C}}^{k,l}$ pointwise $\iff g^{-1}Mg \subset M \iff g \in N_G(M,\mathbb{C})$.

Step 2: The inclusion $\mathfrak{M} \subseteq \widetilde{\mathfrak{M}}$ is also easy: Let $F^\bullet \in \widecheck{\mathrm{NL}}_M$. Then

$$g\mathrm{Hg}_{M,\mathbb{C}}^{k,l} = \mathrm{Hg}_{M,\mathbb{C}}^{k,l} \implies$$
$$gF^\bullet \text{ satisfies } (gF)^{\frac{(k-l)n}{2}}\left(T^{k,l}(V_\mathbb{C})\right) = g\left(F^{\frac{(k-l)n}{2}}T^{k,l}(V_\mathbb{C})\right)$$
$$\supseteq g\mathrm{Hg}_{M,\mathbb{C}}^{k,l} = \mathrm{Hg}_{M,\mathbb{C}}^{k,l},$$

hence also belongs to $\widecheck{\mathrm{NL}}_M$. Thus, $g \in \widetilde{\mathfrak{M}}(\mathbb{C})$.

Step 3: Next, we show that the identity component $\widetilde{\mathfrak{M}}^0 \subseteq \mathfrak{M}$: This group is connected and defined over $\mathbb{Q}$, and $\widetilde{\mathfrak{M}}^0(\mathbb{R})$ stabilizes M-Mumford-Tate domains. Therefore the 1st paragraph of the proof of (VI.A.1) shows $\widetilde{\mathfrak{M}}^0$, i.e., $\widetilde{\mathfrak{M}}^0(\mathbb{C})$, stabilizes spaces of Hodge tensors.

[3]The terminology "$M(\mathbb{C})$-orbits of Hodge structure" is better, since some of these may not intersect D.

[4]Recall that this notation means $N_{G(\mathbb{R})}(M(\mathbb{R}))$.

Step 4: Finally, we must show one point in each connected component of $\widetilde{\mathfrak{M}}$ belongs to $\mathfrak{M}$ (then Step 3 implies $\widetilde{\mathfrak{M}} \subseteq \mathfrak{M}$), and to do this we can work with points defined over a number field $\mathbb{F}$, where we may assume that $\mathbb{F}/\mathbb{Q}$ is Galois. Moreover, given $g \in \widetilde{\mathfrak{M}}(\mathbb{F})$ and $\sigma \in \mathrm{Gal}(\mathbb{F}/\mathbb{Q})$, ${}^{\sigma}g \in \widetilde{\mathfrak{M}}(\mathbb{F})$ since $\widetilde{\mathfrak{M}}$ is defined over $\mathbb{Q}$; also obviously ${}^{\sigma}g^{-1} \in \widetilde{\mathfrak{M}}(\mathbb{F})$.

Pick $F^{\bullet} \in \mathrm{NL}_M$ with $M_{F^{\bullet}} = M$, so that $\mathrm{Hg}_{F^{\bullet}} = \mathrm{Hg}_M$. Then for all $\sigma \in \mathrm{Gal}(\mathbb{F}/\mathbb{Q})$ and $k, l \in \mathbb{Z}$ we have

$$ {}^{\sigma}g^{-1} F^{\frac{(k-l)n}{2}} T^{k,l}(V_{\mathbb{C}}) \supseteq \mathrm{Hg}_M^{k,l}, $$

hence

$$ {}^{\sigma}g\mathrm{Hg}_{M,\mathbb{C}}^{k,l} \subset F^{\frac{(k-l)n}{2}} T^{k,l}(V_{\mathbb{C}}). $$

Thus, the *rationally generated* vector space $\sum_{\sigma \in \mathrm{Gal}(\mathbb{F}/\mathbb{Q})} {}^{\sigma}g\mathrm{Hg}_{M,\mathbb{C}}^{k,l}$ is contained in $F^{\frac{(k-l)n}{2}} T^{k,l}(V_{\mathbb{C}})$, whose largest $\mathbb{Q}$-generated subspace is

$$ \left(F^{\frac{(k-l)n}{2}} T^{k,l}(V_{\mathbb{C}}) \cap T^{k,l}(V_{\mathbb{Q}})\right)_{\mathbb{C}} = \mathrm{Hg}_{F^{\bullet},\mathbb{C}}^{k,l} = \mathrm{Hg}_{M,\mathbb{C}}^{k,l}. $$

Thus

$$ \left(g\mathrm{Hg}_{M,\mathbb{C}}^{k,l} \subseteq\right) \sum_{\sigma} {}^{\sigma}g\mathrm{Hg}_{M,\mathbb{C}}^{k,l} \subseteq \mathrm{Hg}_{M,\mathbb{C}}^{k,l} $$

and $g \in \mathfrak{M}(\mathbb{F})$. □

(VI.A.4) **Remark:** It is clear now that in (VI.A.1) we also have $\mathcal{G} \subseteq N_G(M)$.

As a simple application of the foregoing, we compute the Mumford-Tate group of a generic Hodge structure in a connected component D^0 of D.

(VI.A.5) PROPOSITION: *If φ is general in D^0, $M_{\varphi} = G$.*

PROOF. We give two arguments, the first of which works for all odd weight but only some even weight cases.

Since $D^0_{M_{\varphi}}$ is defined over $\overline{\mathbb{Q}}$ and φ is general, we have $D^0_{M_{\varphi}} = D^0$. Obviously G acts on D, and so by (VI.A.1) $\Pi_G \leq M_{\varphi} \leq G$; all we have to show is that $\Pi_G = G$. Here $G_{\mathbb{R}}$ is a symplectic group if n is odd, and some $\mathrm{SO}(p,q)$ if n is even. $\mathrm{Sp}(2k,\mathbb{Z})$ is $(\mathbb{Q})$-Zariski-dense in $\mathrm{Sp}(2k)$, proving the proposition for odd n. (The corresponding density statement for $\mathrm{SO}(p,q;\mathbb{Z})$ is conditional; cf. [PS, Examples 10.19].)

For the second proof, we use Remark (VI.A.4) to see that $M_{\varphi} \trianglelefteq G$. Since $\mathrm{Sp}(2k,\mathbb{Z})$ and $\mathrm{SO}(p,q)$, excluding the case $\mathrm{SO}(1,1)$, which does not occur as G, are simple, $M_{\varphi} = G$ or $\{e\}$. The latter is impossible unless n is even and all Hodge numbers but $h^{m,m}$ are zero, a case we have excluded. □

VI.B DECOMPOSITION OF NOETHER-LEFSCHETZ INTO HODGE ORIENTATIONS

In this subsection we will show how to index connected components of NL_M, $\breve{\mathrm{NL}}_{\overline{M}}$ and $\breve{\mathrm{NL}}_M$ by real and complex Hodge orientations. In order that the result for $\breve{\mathrm{NL}}_M$ be complete, we need to know:

(VI.B.1) THEOREM: *Let M be the Mumford-Tate group of a Q-polarized Hodge structure $\varphi \in D$. Then no connected component of $\breve{\mathrm{NL}}_M$ is contained in $\mathfrak{Z}$. More precisely, $\breve{\mathrm{NL}}_M$ is a disjoint union of finitely many $M(\mathbb{C})$-orbits of Hodge structures.*

This will be a consequence of the following:

(VI.B.2) Let $\varepsilon \in S^1$ be general. For $\varphi \in D$, the fixed points in $\check{D}$ of $\varphi(\varepsilon)$ acting on flags in $\check{D}$ belong to $\check{D} \setminus \mathfrak{Z}$. That is, they are Hodge structures, but are not necessarily polarized by Q.

PROOF. Let $F^\bullet = \varphi(\varepsilon) \cdot F^\bullet$ be a fixed point. From the basic principle that subspaces of a vector space stabilized by a linear transformation are direct sums of (generalized) eigenspaces, it follows that $F^\bullet$ may be written in terms of the $\mathbb{C}$-spans of subsets of a φ-Hodge basis, starting with

(VI.B.3)
$$F^n = \mathbb{C}\left\langle \omega_1^{(n,0)}, \ldots, \omega_{k_0}^{(n,0)}; \omega_1^{(n-1,1)}, \ldots, \omega_{k_1}^{(n-1,1)}; \ldots ; \omega_1^{(0,n)}, \ldots, \omega_{k_n}^{(0,n)} \right\rangle .$$

Here the superscript denotes the φ-Hodge-type of a basis element.

By the first Riemann bilinear relations on $F^\bullet$, $Q(F^n, F^1) = \{0\}$. If we take any

$$\xi = \sum_{i_0=1}^{k_0} \alpha_{0,i_0} \overline{\omega_{i_0}^{n,0}} + \sum_{i_1=1}^{k_1} \alpha_{1,i_1} \overline{\omega_{i_1}^{n-1,1}} + \cdots + \sum_{i_n=1}^{k_n} \alpha_{n,i_n} \overline{\omega^{0,n}} \in \overline{F^n}$$

and suppose $\xi \in F^1$, then for each $j \in \{0, \ldots, n\}$

$$\begin{aligned} 0 &= Q\left(\sum_{i_j=1}^{k_j} \overline{\alpha_{j,i_j}} \omega_{i_j}^{n-j,j}, \xi \right) \\ &= \left(\sqrt{-1}\right)^{2j-n} \sum_{i_j=1}^{k_i} |\alpha_{j,i_j}|^2 \left\{ \left(\sqrt{-1}\right)^{n-2j} Q\left(\omega_{i_j}^{n-j,j}, \overline{\omega_{i_j}^{n-j,j}} \right) \right\} . \end{aligned}$$

By the second Riemann bilinear relation for φ, the quantities in curly brackets are all positive, so that every $\alpha_{\bullet,\bullet}$ must be zero. Hence $\xi = 0$, i.e., $\overline{F^n} \cap F^1 = \{0\}$, meaning that the basis elements "ω" in (VI.B.3) satisfy: $\bar{\omega}$ span F^0/F^1, and $\{\omega, \bar{\omega}\}$ is a linearly independent set.

Now write

$$\begin{aligned} F^{n-1} &= \mathbb{C}\Big\langle \omega_1^{(n,0)}, \ldots, \omega_{k_0}^{(n,0)}; \grave{\omega}_1^{(n,0)}, \ldots, \grave{\omega}_{k_0'}^{(n,0)}; \text{etc.}\Big\rangle \\ &= \operatorname{span}\{\omega, \grave{\omega}\}. \end{aligned}$$

A similar argument (using $\{0\} = Q(F^{n-1}, F^2))$ shows $\overline{F^{n-1}} \cap F^2 = \{0\}$, so that $\{\omega, \grave{\omega}, \grave{\bar{\omega}}, \bar{\omega}\}$ is linearly independent and $\{\grave{\bar{\omega}}, \bar{\omega}\}$ spans F^0/F^2. We also need that $\grave{\bar{\omega}}$ belong to F^1; i.e., that under Q the $\grave{\omega}$ annihilate the $\bar{\omega}$. But on each $V_\varphi^{n-j,j}$, $\sqrt{-1}^{n-2j}Q(\cdot, \bar{\cdot})$ is positive-definite, meaning that we may choose each subset $\grave{\omega}^{(n-j,j)}$ in $(V_\varphi^{n-j,j})^{\perp \omega^{(n-j,j)}}$ to accomplish this purpose.

Continuing on in this fashion until $F^{\lfloor \frac{n+1}{2} \rfloor}$ we reach a basis $\{\omega, \grave{\omega}, \grave{\grave{\omega}}, \ldots; \ldots, \grave{\grave{\bar{\omega}}}, \grave{\bar{\omega}}, \bar{\omega}\}$ subordinate to both $F^\bullet$ and φ and closed under complex conjugation. For even weight $n = 2m$, of course, some basis elements in $F^m \cap \overline{F^m} \cap V_\varphi^{m,m}$ may be fixed rather than exchanged by conjugation. This is obviously a Hodge structure. □

(VI.B.4) COROLLARY: *Let $\varphi \in D$ and $\mathfrak{F}^\bullet \in \mathfrak{Z} \subset \check{D}$.*

(i) $\lim_{z\to\infty} \varphi(z)\mathfrak{F}^\bullet \in \check{D}\backslash\mathfrak{Z}$. *Here $\varphi(z)$, $z \in \mathbb{C}^*$, uses the identification of $\mathbb{C}^*$ with $\mathbb{U}(\mathbb{C})$.*

(ii) *If $\varphi \in \mathrm{NL}_M$ and $\mathfrak{F}^\bullet \in \check{\mathrm{NL}}_M \cap \mathfrak{Z}$, then $\lim_{z\to\infty} \varphi(z)\mathfrak{F}^\bullet$ lies in $\check{\mathrm{NL}}_M^-$, in the same connected component of $\check{\mathrm{NL}}_M$ as $\mathfrak{F}^\bullet$.*

PROOF. (i) The map $\mathbb{C}^* \to \check{D}$ given by $z \mapsto \varphi(z)\mathfrak{F}^\bullet$ is algebraic and $\check{D}$ is complete. So this map extends to $\mathbb{P}^1 \to \check{D}$, defining the limit. Obviously $\varphi(\varepsilon) \cdot \varphi(z)\mathfrak{F}^\bullet = \varphi(\varepsilon z)\mathfrak{F}^\bullet$ and $\{\infty\} \in \mathbb{P}^1$ is a fixed point of multiplication by $\varepsilon \in S^1$. So $\mathfrak{F}^\bullet_\infty := \lim_{z\to\infty} \varphi(z)\mathfrak{F}^\bullet$ is a fixed point of $\varphi(\varepsilon)$. Now we may apply (VI.B.2).

(ii) $\varphi \in \mathrm{NL}_M \implies M_\varphi \leq M \implies \varphi(\mathbb{C}^*) \subset M \implies \varphi(\mathbb{C}^*)\mathfrak{F}^\bullet \subseteq M(\mathbb{C})\mathfrak{F}^\bullet \implies \mathfrak{F}^\bullet_\infty \in M(\mathbb{C})\mathfrak{F}^\bullet$, using compactness of $M(\mathbb{C})$ and its orbits. By (i), $\mathfrak{F}^\bullet_\infty \notin \mathfrak{Z}$. □

PROOF OF (VI.B.1). Clearly the first statement follows immediately from (VI.B.4)(ii). The $M(\mathbb{C})$-orbit-closure of $\check{\mathrm{NL}}_M^-$ is a finite union of (necessarily disjoint and irreducible) $M(\mathbb{C})$-orbits, each of which contains Hodge structures. If the algebraic variety $\check{\mathrm{NL}}_M$ is bigger than this, then the complement is closed under $M(\mathbb{C})$ and contained in $\mathfrak{Z}$, contradicting the proof of (VI.B.4). □

Before going further we shall need a couple of lemmas, in which we continue our running assumption that $M = M_\varphi$ for $\varphi \in D$.

(VI.B.5) LEMMA: *The maximal compact real tori of M are all conjugate, are maximal tori, and diagonalize with respect to a Q-quasi-unitary*[5] *self-complex-conjugate basis with at most one self-conjugate member. This basis is always a Hodge basis for some $\varphi' \in D$.*

PROOF. Let $T \supset \varphi(\mathbb{U}(\mathbb{R}))$ be a maximal torus of $M_{\mathbb{R}}$, and $\widetilde{T} \subset G_{\mathbb{R}}$ be a maximal torus containing T; clearly $T = (\widetilde{T} \cap M)^0$. Since it centralizes $\varphi(\mathbb{U}(\mathbb{R}))$, $\widetilde{T}$ diagonalizes with respect to a Hodge basis $\{\omega_i\}$. As $\widetilde{T}$ is defined over $\mathbb{R}$, we must have $\overline{\{\omega_i\}} = \{\omega_i\}$, so that the $\bar{\omega}_i$ are eigenvectors with conjugate eigenvalues for $\widetilde{T}(\mathbb{R})$; set things up so that $\bar{\omega}_i = \omega_{r-i}$ where r is $\operatorname{rank}(V)$. Then for any $t \in \widetilde{T}(\mathbb{C})$,

$$Q(\omega_i, \bar{\omega}_j) = Q(t\omega_i, t\bar{\omega}_j) = t_i t_{r-j} Q(\omega_i, \bar{\omega}_j).$$

If $i = j$ then by the 2nd Hodge-Riemann bilinear relation for φ, $Q(\omega_i, \bar{\omega}_i) \neq 0$ so that $t_{r-i} = \frac{1}{t_i}$. (For even weight and odd rank, $\omega_i = \bar{\omega}_i$ and[6] $t_1 = 1$ for $i = \frac{r+1}{2}$; if this happened for any other i, $\widetilde{T}$ would not be maximal). By maximality of $\widetilde{T}$, there can be no further relations on the eigenvalues t_i. Hence, the ω_i satisfy $Q(\omega_i, \bar{\omega}_j) = \xi_i \delta_{ij}$, and we can scale to have that the ξ_i are $\pm\sqrt{-1}$ (odd weight) or ± 1 (even weight) — that is, the basis $\{\omega_i\}$ is *quasi-unitary*.

The real points of the torus

$$\begin{pmatrix} t_1 & & & & & & \\ & \ddots & & & & \bigcirc & \\ & & t_k & & & & \\ & & & (1) & & & \\ & & & & t_k^{-1} & & \\ & \bigcirc & & & & \ddots & \\ & & & & & & t_1^{-1} \end{pmatrix} \cong (\mathbb{C}^*)^{\lfloor \frac{r}{2} \rfloor},$$

where the 1 occurs only for even weight odd rank, are of the form

$$\begin{pmatrix} e^{i\theta_1} & & & & & & \\ & \ddots & & & & \bigcirc & \\ & & e^{i\theta_k} & & & & \\ & & & (1) & & & \\ & & & & e^{-i\theta_k} & & \\ & \bigcirc & & & & \ddots & \\ & & & & & & e^{-i\theta_1} \end{pmatrix} \cong (S^1)^{\lfloor \frac{r}{2} \rfloor}.$$

So the real torus $\widetilde{T}(\mathbb{R})$ is compact, and therefore, noting that M closed $\Longrightarrow$ $M(\mathbb{R})$ closed, so is $T(\mathbb{R})$. Since (by the theory of real Lie groups) all maximal-compact real tori are conjugate by $M(\mathbb{R})^{(0)}$, they are all *maximal real tori* and

[5] Defined in proof.

[6] Note that t_i cannot be ± 1 since $\widetilde{T}$ is connected.

all diagonalize with respect to a Q-quasi-unitary self-conjugate basis. Since the sign-multiplicities of $\{\sqrt{-1}^n \cdot Q(\omega_i, \bar{\omega}_i)\}$ are preserved, the last statement is clear. □

One question this does raise, is: What rôle do the noncompact maximal real tori play? This will be addressed next.

(VI.B.6) LEMMA: *Let $T \subset M(\mathbb{R})$ be a maximal compact (i.e., anisotropic) real torus, and $\chi \in X_*(T_{\mathbb{C}})$ a complex co-character. Then the composition*

$$\begin{array}{ccccc} \mathbb{U}(\mathbb{C}) & \stackrel{\mu}{\to} & \mathbb{G}_m(\mathbb{C}) & \stackrel{\chi}{\to} & T(\mathbb{C}) \subset M(\mathbb{C}) \\ \| \wr & & \| \wr & & \\ \mathbb{C}^* & \stackrel{\text{id.}}{\longrightarrow} & \mathbb{C}^* & & \end{array}$$

is defined over $\mathbb{R}$, hence defines a Hodge structure $\varphi \in \check{\mathrm{NL}}_M^-$.

PROOF. We should first remark that μ is *not* defined over $\mathbb{R}$, since $\mathbb{U}(\mathbb{R})$ is represented by $S^1 \subset \mathbb{C}^*$ and $\mathbb{G}_m(\mathbb{R})$ by $\mathbb{R}^*$; "complex conjugation" on $\mathbb{U}(\mathbb{C})$ is given by $z \mapsto 1/\bar{z}$. Writing

$$w \stackrel{\chi}{\mapsto} \begin{pmatrix} w^{\chi_1} & & & & & & \\ & \ddots & & & & \bigcirc & \\ & & w^{\chi_k} & & & & \\ & & & (1) & & & \\ & & & & w^{-\chi_k} & & \\ & \bigcirc & & & & \ddots & \\ & & & & & & w^{-\chi_1} \end{pmatrix}$$

in terms of a Q-quasi-unitary self-conjugate basis,[7] we have

$$\overline{(\chi \circ \mu)(z)} = \overline{(\chi \circ \mu)(1/\bar{z})} = \overline{\chi(1/\bar{z})}$$

$$= \overline{\begin{pmatrix} \bar{z}^{-\chi_1} & & & & & & \\ & \ddots & & & & \bigcirc & \\ & & \bar{z}^{-\chi_k} & & & & \\ & & & (1) & & & \\ & & & & \bar{z}^{\chi_k} & & \\ & \bigcirc & & & & \ddots & \\ & & & & & & \bar{z}^{\chi_1} \end{pmatrix}}$$

[7]Here M will impose further conditions on the $\{\chi_i\}$.

$$= \begin{pmatrix} z^{\chi_1} & & & & & & \\ & \ddots & & & & \bigcirc & \\ & & z^{\chi_k} & & & & \\ & & & (1) & & & \\ & & & & z^{-\chi_k} & & \\ & \bigcirc & & & & \ddots & \\ & & & & & & z^{-\chi_1} \end{pmatrix}$$

$$= (\chi \circ \mu)(z),$$

where, at the end, complex conjugation acts on the linear transformation, not just the eigenvalues. □

Now we are ready to describe the components of the Noether-Lefschetz locus in terms of Lie theory. The first step is to define a "co-character bundle" over

$$\mathcal{B}_{\mathbb{R}}^{M} := \text{set of maximal compact real tori in } M_{\mathbb{R}}^{(0)},$$

keeping (VI.B.6) in mind. Given $T \in \mathcal{B}_{\mathbb{R}}^{M}$, one usually considers the weights of the fixed Lie algebra representation $\mathfrak{m} \to \mathfrak{gl}(V)$ with respect to the Cartan sub-algebra $\mathfrak{t} = \mathrm{Lie}(T) \subseteq \mathfrak{m}$ as a *subset* $\mathcal{X}_{\mathfrak{m}}^{*}(\mathfrak{t})(\subset X^{*}(T_{\mathbb{C}})) \subset \mathfrak{t}^{*}$. We shall consider instead the *zero-cycle* $\mathcal{Z}_{\mathfrak{m}}^{*}(\mathfrak{t})$, with support on $\mathcal{X}_{\mathfrak{m}}^{*}(\mathfrak{t})$, which keeps track of multiplicities (i.e., dimensions of $T_{\mathbb{C}}$-eigenspaces of $V_{\mathbb{C}}$). Any co-character $\chi \in X_{*}(T_{\mathbb{C}})$ induces a projection $\mathrm{pr}^{\chi} : \mathfrak{t}^{*} \to \mathbb{C}$ sending $X^{*}(T_{\mathbb{C}}) \to \mathbb{Z}$; in fact, the two notions are equivalent. Writing

$$\mathcal{Z}_{\mathrm{hdg}} := \sum_{p+q=n} h^{p,q}[p-q]$$

for the 0-cycle on $\mathbb{Z}$ that records the Hodge type of the Hodge structures classified by $\check{D}\backslash\mathfrak{Z}$, we define the bundle

$$\Xi_{\mathbb{R}}^{M} \xrightarrow{\pi_{\mathbb{R}}^{M}} \mathcal{B}_{\mathbb{R}}^{M}$$

by

$$\text{(VI.B.7)} \qquad \pi^{-1}(T) := \{\chi \in X_{*}(T_{\mathbb{C}}) \mid \mathrm{pr}_{*}^{\chi}\, \mathcal{Z}_{\mathfrak{m}}^{*}(\mathfrak{t}) = \mathcal{Z}_{\mathrm{hdg}}\}.$$

Similarly, if we introduce

$$\widetilde{\mathcal{B}}_{\mathbb{R}}^{M} := \text{set of maximal real tori in } M_{\mathbb{R}}$$
$$\mathcal{B}_{\mathbb{C}}^{M} := \text{set of maximal tori in } M_{\mathbb{C}},$$

then exactly the same formula (VI.B.7) gives $\mathbb{Z}$-projection/co-character bundles over each of these. The bundles fit into a diagram

$$\begin{array}{ccccc} \Xi_{\mathbb{R}}^{M} & \subset & \widetilde{\Xi}_{\mathbb{R}}^{M} & \subset & \Xi_{\mathbb{C}}^{M} \\ \downarrow \pi_{\mathbb{R}}^{M} & & \downarrow \tilde{\pi}_{\mathbb{R}}^{M} & & \downarrow \pi_{\mathbb{C}}^{M} \\ \mathcal{B}_{\mathbb{R}}^{M} & \subset & \widetilde{\mathcal{B}}_{\mathbb{C}}^{M} & \subset & \mathcal{B}_{\mathbb{C}}^{M} \end{array}$$

with all three π's finite and étale. $M(\mathbb{R})$ acts compatibly on the entire diagram by conjugation ($T \mapsto mTm^{-1}, \chi(z) \mapsto m \circ \chi(z) \circ m^{-1}$), with transitive action on $\mathcal{B}_{\mathbb{R}}^{M}$; while $M(\mathbb{C})$ acts on $\Xi_{\mathbb{C}}^{M} \to \mathcal{B}_{\mathbb{C}}^{M}$ with transitive action on $\mathcal{B}_{\mathbb{C}}^{M}$.

Given $(T,\chi) \in \Xi_{\mathbb{C}}^{G}$, we define a filtration on $V_{\mathbb{C}}$ in terms of eigenspaces of $T(\mathbb{C})$, or of $\chi(z)$ for general $z \in \mathbb{C}^{*}$, by

$$\text{(VI.B.8)} \qquad \mathfrak{F}(T,\chi)^{p} := \sum_{a \geq p} \sum_{\xi \in (\mathrm{pr}^{\chi})^{-1}(2a-n)} E_{\xi}^{T(\mathbb{C})}(V_{\mathbb{C}}) = \sum_{a \geq p} E_{z^{2a-n}}^{\chi(z)}(V_{\mathbb{C}}).$$

One easily shows that since $\chi(z) \in G(\mathbb{C})$, $\mathfrak{F}(T,\chi)^{\bullet}$ must satisfy the 1st Hodge-Riemann bilinear relation; hence $\mathfrak{F}$ gives a map $\Xi_{\mathbb{C}}^{G} \to \check{D}$.

(VI.B.9) THEOREM: *Restricting the definition of $\mathfrak{F}$ induces an $M(\mathbb{R})$-equivariant diagram, with the right-hand column $M(\mathbb{C})$-equivariant,*

$$\begin{array}{ccccc} \breve{\mathrm{NL}}_{M}^{-} & \subset & \breve{\mathrm{NL}}_{M} & = & \breve{\mathrm{NL}}_{M} \\ \upuparrows \mathfrak{F} & & \uparrow \mathfrak{F} & & \upuparrows \mathfrak{F} \\ \Xi_{\mathbb{R}}^{M} & \subset & \widetilde{\Xi}_{\mathbb{R}}^{M} & \subset & \Xi_{\mathbb{C}}^{M} \end{array}$$

in which there are one-to-one correspondences between connected components and $M(\mathbb{R})^{0}$- resp. $M(\mathbb{C})$-orbits of $\Xi_{\mathbb{R}}^{M}$ and $\breve{\mathrm{NL}}_{M}^{-}$, resp. $\Xi_{\mathbb{C}}^{M}$ and $\breve{\mathrm{NL}}_{M}$.

PROOF. If $\tau \in T^{k,l}(V_{\mathbb{C}})$ is fixed by M, then it is fixed by $T^{k,l}(\chi(z))$. Since (VI.B.8) is compatible with tensors, $\tau \in (T^{k,l}\mathfrak{F}(T,\chi))^{\frac{(k-l)n}{2}}$ follows easily, and $\mathfrak{F}(\Xi_{\mathbb{C}}^{M}) \subset \breve{\mathrm{NL}}_{M}$.

Now let $(T,\chi) \in \Xi_{\mathbb{C}}^{M}$. Writing φ for the composition indicated in (VI.B.6), we have $\mathfrak{F}(T,\chi) = F_{\varphi(T,\chi)}^{\bullet}$. Moreover, if $T \in \mathcal{B}_{\mathbb{R}}^{M}$, then $\varphi(T,\chi)$ is, by (VI.B.6), defined over $\mathbb{R}$; this implies $\mathfrak{F}(\Xi_{\mathbb{R}}^{M})$ produces a Hodge structure and hence lies in $\breve{\mathrm{NL}}_{M}^{-}$.

To see the surjectivity on the left, take $\varphi \in \widecheck{\mathrm{NL}}_M^-$. Then $\varphi(\mathbb{U}(\mathbb{R})) \subset M(\mathbb{R})$ and we have a maximal compact real torus $T \subset M(\mathbb{R})$ containing $\varphi(\mathbb{U}(\mathbb{R}))$. So $F_\varphi^\bullet = \mathfrak{F}(T, \varphi \circ \mu^{-1})^\bullet$.

The compatibility with $M(\mathbb{R})$ and $M(\mathbb{C})$ is just

$$\mathfrak{F}(mTm^{-1}, m \circ \chi(\cdot) \circ m^{-1})^\bullet = m\mathfrak{F}(T, \chi)^\bullet.$$

With this, we can see the right-surjectivity; let $F^\bullet \in \widecheck{\mathrm{NL}}_M$. Then by (VI.B.1) $F^\bullet = m\widetilde{F}^\bullet$ for $m \in M(\mathbb{C})$ and $\widetilde{F}^\bullet \in \widecheck{\mathrm{NL}}_M^-$. By left-surjectivity $\widetilde{F}^\bullet = \mathfrak{F}(T, \chi)$, and so $F^\bullet = \mathfrak{F}(mTm^{-1}, m \circ \chi(\cdot) \circ m^{-1})$.

By left resp. right surjectivity and $M(\mathbb{R})^0$- resp. $M(\mathbb{C})$-equivariance, we get that connected components of $\Xi_{\mathbb{R}}^M$ resp. $\Xi_{\mathbb{C}}^M$ surject onto connected components of $\widecheck{\mathrm{NL}}_M^-$ resp. $\widecheck{\mathrm{NL}}_M$.

To get a one-to-one correspondence, start on the left:[8] Let (T, φ) and (T', φ') $\in \Xi_{\mathbb{R}}^M$ yield the same point in $D_M \subset D$, i.e. $\varphi = \varphi'$. Then $T, T' \subset (M \cap H_\varphi)_{\mathbb{R}}^0$ are maximal compact real tori, hence are conjugate by $m \in M(\mathbb{R})^0 \cap H_\varphi(\mathbb{R})$, which fixes φ. So (T, φ) and (T', φ) are in the same connected component of $\Xi_{\mathbb{R}}^M$.

Next, consider $(T, \varphi), (T', \varphi') \in \Xi_{\mathbb{C}}^M$ with $F_\varphi^\bullet = F_{\varphi'}$, so that $T \subset (M \cap H_\varphi)_{\mathbb{C}}$ and $T' \subset (M \cap P_\varphi)_{\mathbb{C}}$. So there exists $m \in M(\mathbb{C}) \cap P_\varphi(\mathbb{C})$ with $mTm^{-1} = T$; writing $\varphi'' = m\varphi' m^{-1}$, $F_{\varphi''} = F_\varphi$ and $F_{\varphi''} = F_\varphi^\bullet$ and φ, φ'' factor through T. But then φ, φ'' are the same "co-character," or else they would not induce the same filtration. Again, (T, φ) and (T', φ') are shown to be in the same component of $\Xi_{\mathbb{C}}^M$.

Finally, we observe that this induces one-to-one correspondences between $M(\mathbb{R})$-orbits on $\Xi_{\mathbb{R}}^M$ and $M(\mathbb{R})$-orbits on $\widecheck{\mathrm{NL}}_M^-$ — i.e., Mumford-Tate-domains — even when these consist of more than one component. □

(VI.B.10) **Remark:** We expect that the middle vertical arrow is also surjective, giving a stratification of $\widecheck{\mathrm{NL}}_M \cap \mathfrak{Z}$ by the $\mathbb{R}$-rank of non-anisotropic maximal real tori (or more precisely, by the images of connected components of $\widetilde{\mathfrak{B}}_{\mathbb{R}}^M$). To see this in case $M = G$, consider the (unique) semi-Hodge decomposition $V_{\mathbb{C}} = \bigoplus_{i,j} V_j^i$ associated to an *arbitrary* $F^\bullet \in \check{D}$, where

- $(F^p \cap \overline{F^q})V_{\mathbb{C}} = \bigoplus_{\substack{i \geq p \\ j \geq q}} V_j^i$ think of $V_j^i \cong \frac{F^i \cap \overline{F}^j}{F^{i+1} \cap \overline{F}^j + F^i \cap \overline{F}^{j+1}}$;
- $\overline{V_j^i} = V_i^j$;

[8] In the following we shall frequently write (T, φ) instead of (T, χ), where $\varphi = \chi \circ \mu$.

- $Q(V^i_j, V^k_l) = 0$ unless $i + k = j + l = n$;
- $Q : V^i_j \times V^{n-i}_{n-j} \to \mathbb{C}$ is nondegenerate.

All of this follows from just the 1st Hodge-Riemann bilinear relation; if $F^\bullet \in \check{D}\backslash\mathfrak{Z}$ then only the $V^i_{n-i} = V^{i,n-i}$ are non-zero. Those $g \in G(\mathbb{C})$ multiplying V^i_j by $z_{ij} \in \mathbb{C}^*$ with $z_{ij}z_{n-i,n-j} = 1$ are the complex points of a real torus in G. (It is real because *closed* under complex conjugation; the maximal real torus T containing it is non-anisotropic unless $F^\bullet \in \check{D}\backslash\mathfrak{Z}$, so (VI.B.6) will not apply.) A co-character of this torus is the χ defined by: $\chi(z)$ multiplies V^i_j by z^{2i-n}. Clearly $\mathfrak{F}(T,\chi)$ recovers $F^\bullet$, so that $\widetilde{\Xi}^G_{\mathbb{R}} \twoheadrightarrow \check{D}$. □

Another "corollary" of (VI.B.9) is the next statement, which may shed some light on the nature of the components of an individual $D_M := M(\mathbb{R})\varphi$ ($\varphi \in \mathrm{NL}_M$).

(VI.B.11) PROPOSITION: *$D_M = \check{D}_M \cap D$, where*[9] *$\check{D}_M = M(\mathbb{C})\varphi$.*

PROOF. Referring to the figure,

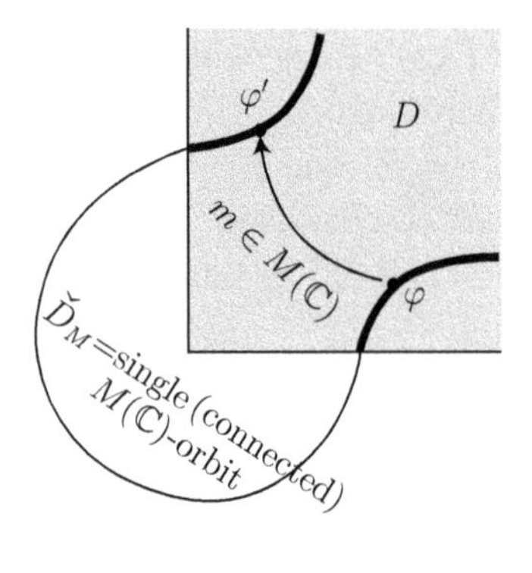

we want to prove $M(\mathbb{R})\varphi$ contains φ'. By (VI.B.9), $M(\mathbb{R})^0\varphi$ and $M(\mathbb{R})^0\varphi'$ contain φ's factoring through every maximal compact real torus in $M_{\mathbb{R}}$. Hence, we may assume φ and φ' both factor through a single such torus defined over $\mathbb{Q}$, so they are CM-HS. Clearly then $m \in N_M(T,\mathbb{C})$.

Both φ's have their corresponding Hodge decompositions built out of eigenspaces E of T in $V_{\mathbb{C}}$ to which a quasi-unitary Hodge basis Ω of φ must be subordinate. Here $E \subset V^{p,n-p}_{\mathbb{C}}$ for some p, and we write $p =: p(E)$. To preserve Q, and the 1st Hodge-Riemann bilinear relation

- m must permute E's (of the same dimension);

[9]Though non-traditional, it is legitimate to let $M(\mathbb{C})$ act on φ by conjugation (and equivalent to action on $F^\bullet_\varphi$) in light of (VI.B.9)*ff*.

- if $m(E) = E'$ then $m(\overline{E}) = \overline{E'}$ (this does not mean, or assume, that m is in $M(\mathbb{R})$!);
- if $\{\omega_i\}$ resp. $\{\omega'_i\}$ is a basis of E resp. E', which is part of Ω, then setting ${}_{\omega'}[m|_E]_\omega =: A$, ${}_{\bar{\omega}'}[m|_{\overline{E}}]_{\bar{\omega}} =: B$, we have

$$\text{(i) } B = {}^tA^{-1} \text{ if } p(E) - p(E') \overset{(2)}{\equiv} 0;$$
$$\text{(ii) } B = -{}^tA^{-1} \text{ if } p(E) - p(E') \overset{(2)}{\equiv} 1.$$

One should remark that case (ii) for even one $E \Rightarrow m \notin M(\mathbb{R})$. Now for φ, φ' both polarized by Q, we are in case (i) for every E. Furthermore, $N_M(T)$ is defined over $\mathbb{Q}$ and $m \in N_M(T, \mathbb{C}) \Rightarrow \overline{m} \in N_M(T, \mathbb{C})$. Since $\overline{m}$ will execute the same permutations of E's, $m\overline{m}^{-1} \in Z_m(T, \mathbb{C}) = T(\mathbb{C})$ (by maximality); and $(m\overline{m}^{-1})^{-1} = \overline{(m\overline{m}^{-1})}$ implies

$$\left.\begin{array}{l} (\overline{B}^{-1}A =) m\overline{m}^{-1}|_E = \gamma I, \\ (\overline{A}^{-1}B =) m\overline{m}^{-1}|_E = \overline{\gamma}^{-1} I, \end{array}\right. \gamma \in \mathbb{C}^*.$$

From this and (i) we have ${}^t\overline{A}A = \gamma I \Rightarrow \gamma \in \mathbb{R}_{>0} \Rightarrow B = \frac{1}{\gamma}\overline{A}$ (and $\overline{B} = \gamma A$).

Now, reality ($\overline{m} = m$) for m means: $B = \overline{A}$. Modifying m by precomposition with $t \in T(\mathbb{C})$, which multiplies E by $\frac{1}{\sqrt{\gamma}}$ and $\overline{E}$ by $\sqrt{\gamma}$ (that the multipliers be inverse is the only constraint on t), we are in this case: $m \in M(\mathbb{R})$. □

In the one-to-one correspondence between components of $\Xi^M_{\mathbb{R}}$ and $\breve{\mathrm{NL}}^-_M$, let $\Xi^{M,+}_{\mathbb{R}} \subset \Xi^M_{\mathbb{R}}$ be the components corresponding to NL_M. Fixing a maximal compact real torus $T \subset M$, and noting that $(\pi^M_{\mathbb{R}})^{-1}(T) = (\pi^M_{\mathbb{C}})^{-1}(T)$, by virtue of how Lemma (VI.B.6) allowed us to do the construction of (VI.B.9), we will just write $\pi^{-1}(T)$ (cf. (VI.B.7)). Define the Weyl groups of T

$$\begin{array}{ccccc} W_M(T,\mathbb{R})^0 & \subset & W_M(T,\mathbb{R}) & \subset & W_M(T,\mathbb{C}) \\ \ddot{\|} & & \ddot{\|} & & \ddot{\|} \\ \frac{N_{M(\mathbb{R})^0}(T)}{T(\mathbb{R})} & & \frac{N_{M(\mathbb{R})}(T)}{T(\mathbb{R})} & & \frac{N_{M(\mathbb{C})}(T)}{T(\mathbb{C})} \end{array}$$

and note that these act on $\pi^{-1}(T)$ in the obvious way. We shall use the resulting Weyl-orbits to index Noether-Lefschetz components.

(VI.B.12) THEOREM: (i) *We have equivalencies of sets*

$$\frac{\pi^{-1}(T)}{W_M(T,\mathbb{R})^0} = \pi_0(\breve{\mathrm{NL}}^-_M), \quad \frac{\pi^{-1}(T)}{W_M(T,\mathbb{R})} = \left\{ \begin{array}{c} M(\mathbb{R})\text{-orbits} \\ \text{in } \breve{\mathrm{NL}}^-_M \end{array} \right\},$$

$$\frac{\pi^{-1}(T)}{W_M(T,\mathbb{C})} = \pi_0(\breve{\mathrm{NL}}_M).$$

(ii) *Fix (equivalently)* $\varphi_0 : U \to T/\mathbb{R}$, $\chi_0 : \mathbb{G}_m \to T/\mathbb{C}$, *or* $\mathrm{pr}_0 : \mathfrak{t}^* \to \mathbb{C}$ *so that* $(T, \chi_0) \in \Xi_{\mathbb{R}}^{M,+}$. *Define* $\pi^{-1}(T)^+$ *to be those* $\varphi/\chi/\mathrm{pr}$ *such that* $\frac{\chi}{\chi_0}$ *is the fourth power of a co-character of* $T_{\mathbb{C}}$, *or equivalently* $(\mathrm{pr} - \mathrm{pr}_0)(X^*(T_{\mathbb{C}})) \subset 4\mathbb{Z}$. *Then*

$$\frac{\pi^{-1}(T)^+}{W_M(T,\mathbb{R})^0} = \pi_0(\mathrm{NL}_M), \quad \frac{\pi^{-1}(T)^+}{W_M(T,\mathbb{R})} = \left\{ \begin{array}{c} M(\mathbb{R})\text{-orbits} \\ \text{in } \mathrm{NL}_M \end{array} \right\},$$

and we have the diagram

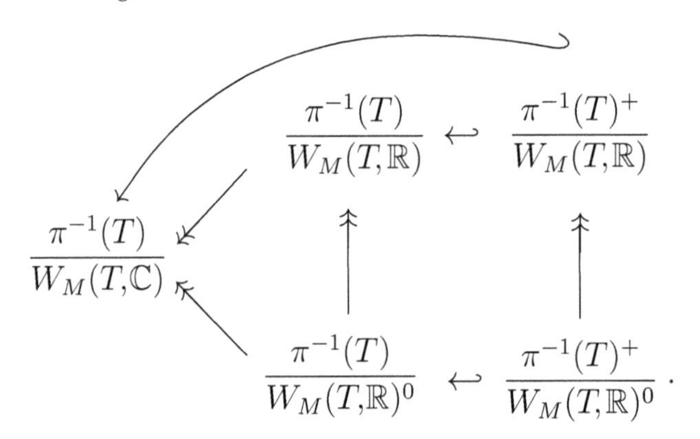

PROOF. (i) By virtue of connectedness of $\mathcal{B}_{\mathbb{R}}^M$ and $\mathcal{B}_{\mathbb{C}}^M$, T occurs at least once in every $M(\mathbb{R})^0$, $M(\mathbb{R})$, or $M(\mathbb{C})$ orbit concerned, and $\pi^{-1}(T)$ indexes the sheets of $\pi_{\mathbb{R}}^M$ or $\pi_{\mathbb{C}}^M$. Two sheets lie in the same orbit or component precisely when an element of the group sends one point of $\pi^{-1}(T)$ to the other, and hence normalizes T.

(ii) It is always true that two projections satisfying (VI.B.7) have

$$(\mathrm{pr} - \mathrm{pr}_0)(X^*(T)) \subset 2\mathbb{Z},$$

as distinct co-characters simply assign different Hodge types to each T-eigenspace and $(p_1 - q_1) - (p_2 - q_2) \underset{(2)}{\equiv} 0$. In order that both φ and φ_0 be polarized by Q, this must be $(p_1 - q_1) - (p_2 - q_2) \underset{(4)}{\equiv} 0$. Finally, injectivity of the top arrow in the diagram is by (VI.B.11). □

(VI.B.13) **Remark:** In practice, the way to use this theorem will be: first, write down the weights of the Lie algebra representation of $\mathfrak{m}$ on V with respect to a Cartan sub-algebra $\mathfrak{t}$ (corresponding to a maximal *compact* real torus). Then determine the (finitely many) integral projections that yield the right Hodge type, as in (VI.B.7). Finally, identify the orbits of the Weyl groups. We refer to these orbits as $M_{\mathbb{C}}$- (resp. $M_{\mathbb{R}}$-, $M_{\mathbb{R}}^{\circ}$-) *Hodge orientations*, as they all boil down to (CM) n-orientations when M is a torus, in which case the above Weyl groups are all trivial. It makes sense to speak of Q-polarized $M_{\mathbb{R}}^{(0)}$- (but not $M_{\mathbb{C}}$-) Hodge orientations.

Before demonstrating the CM case in an example, we want to emphasize — although we have not attempted it in this paper — that it may be possible to use this technique to obtain a complete classification of Calabi-Yau Hodge structures. This is because the Mumford-Tate groups are sufficiently constrained that one may be able to identify all the possibilities. For example, if $\mathrm{End}_{\mathrm{HS}}(V) \cong \mathbb{Q}$ and V is of Calabi-Yau type ($h^{n,0} = 1$), then M is semi-simple, and any φ is nontrivial in every $\mathbb{R}$-simple factor M_i of M. These factors turn out to be absolutely simple [Mo1, §1], and the representation $\rho : \mathfrak{m} \to \mathfrak{gl}(V)$ decomposes as $\rho_1 \boxtimes \cdots \boxtimes \rho_k$, $\rho_i : \mathfrak{m} \to \mathfrak{gl}(V_i)$ with highest weights λ_i of length l_i [Mo1, §3] and $V = \bigotimes_i V_i$. Then we have $l_i \geq 1$ and $\Sigma l_i \leq n$ (in particular, $k \leq n$), and $\prod_i \dim(\rho_i) = \dim(V)$ — substantial restrictions if say $n = 3$ and $h^{2,1}$ is small. Finally, there have to exist $\mathbb{Z}$-projections of the weight spaces that map the "weight cycle" to $[n] + h^{n-1,1}[n-2] + \cdots + h^{1,n-1}[2-n] + [-n]$. □

In the next example we will contrast the "non-classical" situation with the "classical" one, which is partially summed up by the following. Here D^0 denotes the connected component of D, of which there are two if n is even and one if n is odd; and T, M as usual are M_φ for some $\varphi \in D$ and $T \subset M_{\mathbb{R}}$ maximal compact.

(VI.B.14) PROPOSITION:

$$\begin{aligned} n = 1 \quad &\Rightarrow \quad \left|\pi^{-1}(T)^+\right| = 1 \\ \left.\begin{array}{r} n = 2 \\ \text{and} \\ h^{2,0} = 1 \end{array}\right\} \quad &\Rightarrow \quad \left|\pi^{-1}(T)^+\right| = 2. \end{aligned}$$

In particular, $\mathrm{NL}_M \cap D^0$ *is connected*, i.e., $\mathrm{NL}_M \cap D$ is a single $M(\mathbb{R})$-orbit.

PROOF. For $n = 1$, since $V_\varphi^+ = V_\varphi^{1,0}$ and $V_\varphi^- = V_\varphi^{0,1}$, there is no other way than φ to build a Q-polarized Hodge structure out of T-eigenspaces. For $n = 2$, the only other possibility is $\overline{\varphi}$ (complex conjugate), and no $m \in M(\mathbb{R})^0$ takes $\varphi \mapsto \overline{\varphi}$. □

(VI.B.15) **Example:** Let $r = \mathrm{rank}(V)$ be even, so that the maximal compact real tori in G take the form ($r = 2s$)

$$\begin{pmatrix} z_1 & & & & & \\ & \ddots & & & \bigcirc & \\ & & z_s & & & \\ & & & z_1^{-1} & & \\ & \bigcirc & & & \ddots & \\ & & & & & z_s^{-1} \end{pmatrix}$$

with respect to a self-conjugate basis. By essentially the same argument as in [M2, §3], there exists a torus M of this form defined over $\mathbb{Q}$. Taking a co-character χ of M of $\check{D}$-type (cf. (VI.B.7)), $\varphi := \chi \circ \mu$ gives a CM-Hodge structure with Mumford-Tate group $M_\varphi \leq M$. In what follows, we assume $M_\varphi = M$, or equivalently that the CM-HS is irreducible and nondegenerate (see Chapter V).

Writing $X^*(M_\mathbb{C}) = \mathbb{Z}\langle \hat{e}_1, \ldots, \hat{e}_s \rangle$, we obviously have for the weights of $\mathfrak{g}$ on V $\mathcal{X}^*_\mathfrak{g}(\mathfrak{m}) = \{\hat{e}_1, -\hat{e}_1, \ldots, \hat{e}_s, -\hat{e}_s\}$. Now, suppose

$$\{h^{n,0}, h^{n-1,1}, \ldots\} = \{\dim(F^n), \dim(F^{n-1}/F^n), \ldots\}$$

is the type of flag classified by $\check{D}$. Then the point components of $\check{\mathrm{NL}}_M$ $(= \check{\mathrm{NL}}_M^-$ by (VI.B.11)) correspond to the projections on $\mathfrak{m}^*$ induced by the Euclidean product with vectors of the form required by (VI.B.7). Namely, we consider the vector

(VI.B.16)

$$\left(\underbrace{n, \ldots, n}_{h^{n,0} \text{ times}}; \underbrace{n-2, \ldots, n-2}_{h^{n-1,1} \text{ times}}; \ldots; \underbrace{1, \ldots, 1}_{\substack{h^{\frac{n+1}{2}, \frac{n-1}{2}} \text{ times} \\ (n \text{ odd})}} \text{ OR } \underbrace{0, \ldots, 0}_{\substack{\frac{1}{2} h^{n/2, n/2} \text{times} \\ (n \text{ even})}} \right),$$

which we assume corresponds to a type φ polarized by Q. Then components of $\check{\mathrm{NL}}_M$ correspond bijectively to

(VI.B.17) arbitrary permutations of the entries of (VI.B.16) and arbitrary signs placed on these entries,

resulting in precisely

$$2^s \cdot \frac{s!}{(h^{n,0})!(h^{n-1,1})! \ldots \begin{cases} (h^{\frac{n+1}{2}, \frac{n-1}{2}})! \\ \text{OR} \\ \left(\frac{h^{n/2, n/2}}{2} \right)! \end{cases}}$$

components. Again, these all give *distinct* Hodge structures because the relevant Weyl group is $W_M(M)$, which is trivial.

For rank 4 ($s = 2$), we contrast the weight 1 (Hodge type $(2, 2)$) and weight 3 (Hodge type $(1, 1, 1, 1)$) cases. For weight 1, (VI.B.17) yields

(VI.B.18)
$$\boxed{\begin{pmatrix}1\\1\end{pmatrix}}, \begin{pmatrix}1\\-1\end{pmatrix}, \begin{pmatrix}-1\\1\end{pmatrix}, \begin{pmatrix}-1\\-1\end{pmatrix},$$

while in weight 3 we get

(VI.B.19)
$$\boxed{\begin{pmatrix}3\\1\end{pmatrix}}, \begin{pmatrix}-3\\1\end{pmatrix}, \begin{pmatrix}3\\-1\end{pmatrix}, \begin{pmatrix}-3\\-1\end{pmatrix}\begin{pmatrix}1\\3\end{pmatrix}, \begin{pmatrix}-1\\3\end{pmatrix}, \begin{pmatrix}1\\-3\end{pmatrix}, \boxed{\begin{pmatrix}-1\\-3\end{pmatrix}}.$$

Here the box means "polarized by Q", and we note $\binom{3}{1} - \binom{-1}{-3} = \binom{4}{4} \in (4\mathbb{Z})^2$, see (VI.B.12)(ii). So we see that $\mathrm{NL}_M \cap D$ *is* **not** *a single* $M(\mathbb{R})$*-orbit in general in the* **non-classical case** (compare (VI.B.14)).

Consider finally the weight 1 rank 12 case corresponding to the CM-HS generated by $\mathbb{Q}(\zeta_{13})$ in Example (V.D.7). If we take the vector $\begin{pmatrix}1\\1\\1\\1\\1\\1\end{pmatrix}$, hence φ, to correspond to the primitive CM type $\{\theta_{12}, \theta_2, \theta_3, \theta_5, \theta_6, \theta_9\}$, then (since $\overline{\theta_{12}} = \theta_1$) $\begin{pmatrix}-1\\1\\1\\1\\1\\1\end{pmatrix}$ corresponds to the *non*-primitive type $\{\theta_1, \theta_2, \theta_3, \theta_5, \theta_6, \theta_9\}$. This demonstrates that, for a given $\underset{\smile}{\varphi} \in D$ with Mumford-Tate group M, it can happen that some *components* of $\check{\mathrm{NL}}_M$ have a strictly smaller Mumford-Tate group; we shall simply call such components *degenerate*.

VI.C WEYL GROUPS AND PERMUTATIONS OF HODGE ORIENTATIONS

In (VI.A.4) we have seen that the largest subgroups of G stabilizing Noether-Lefschetz-loci of a Mumford-Tate group M are the normalizers of M. In this subsection and the next we will look at the orbits of these normalizers on the Noether-Lefschetz locus, especially the induced permutations of components. First we shall consider the related issue of the action of the Weyl group $W_M(T)$ when T is a CM-MT group itself, starting from the following simple statement:

(VI.C.1) LEMMA: *Let $M = M_\varphi$ for $\varphi \in D$. Then there exist in every component of $\overline{\check{\mathrm{NL}}}_M$, CM points whose Mumford-Tate groups are contained in a common $\mathbb{Q}$-torus.*

PROOF. This is a consequence of the fact that $\mathcal{B}_{\mathbb{R}}^M$ (being a complete conjugacy class of real tori, cf. [M2, §3]) contains a torus T defined over $\mathbb{Q}$, together with the one-to-one correspondence in (VI.B.9). □

Leaving aside the question of whether this CM point $\mathfrak{F}(T,\chi_0) = F^\bullet_{\varphi_0}$ has $M_{\varphi_0} < T$ or $M_{\varphi_0} = T$, we can still define the finite point set $\breve{\text{NL}}_T$ (containing φ_0) and observe that $W_M(T,\mathbb{C})$ (resp. $W_M(T,\mathbb{R})^0$) acts on it (via $N_M(T,\mathbb{C}$ (resp. $\mathbb{R}$)). This yields all permutations of CM-HS with Mumford-Tate group $\leq T$, which can be achieved while stabilizing components of $\breve{\text{NL}}_M$ (resp. $\breve{\text{NL}}_{\bar{M}}$). The first obvious question is, when $M = G$, whether this action is transitive on $\breve{\text{NL}}_T$.

(VI.C.2) **Example:** We continue the rank 4 examples from (VI.B.15), writing $\omega = \{\omega_1, \omega_2, \bar{\omega}_2 = \omega_3, \bar{\omega}_1 = \omega_4\}$ for the eigenvectors (in $V_{\mathbb{C}}$) of

$$T(\mathbb{C}) = \left\{ \left. \begin{pmatrix} z_1 & & & \\ & z_2 & & \\ & & z_2^{-1} & \\ & & & z_1^{-1} \end{pmatrix} \right| z_1, z_2 \in \mathbb{C}^* \right\}.$$

We may assume that ω is a quasi-unitary basis, viz.

$$\sqrt{-1}Q(\omega_i, \bar{\omega}_j) = \begin{cases} \delta_{ij}, & n = 1 \\ (-1)^i \delta_{ij}, & n = 3 \end{cases}$$

for $i = 1, 2$. In these cases T is of course a Mumford-Tate group for CM-HS derived from $\mathbb{Q}(\zeta_5)$.

One way to treat the permutation issue would be to compute the action of $W_G(T)$ on (VI.B.18), (VI.B.19). Or we can argue in a less direct but more Hodge-theoretically motivated manner, which we do now. To normalize T, $g \in G(\mathbb{C})$ must be a product of permutation and scalings of the basis vectors ω. To preserve Q, a permutation sending $\omega_i \mapsto \alpha\omega_j$ must send $\bar{\omega}_i \mapsto \beta\bar{\omega}_j$, with

$$\alpha\beta = (-1)^{j-1} \text{ for } n = 3$$

and

$$\alpha\beta = \begin{cases} 1 & \text{if } i,j \in \{1,2\} \text{ or } i,j \in \{3,4\} \\ -1 & \text{otherwise} \end{cases} \quad \text{for } n = 1.$$

For $\alpha\beta = 1$ (resp. $\alpha\beta = -1$) we can take $\alpha = \beta = 1$ (resp. $\alpha = \beta = \sqrt{-1}$). Finally, if we want $g \in G(\mathbb{R})$ then we must also require $\beta = \overline{\alpha}$. Since $\alpha\beta = -1$ together with $\beta = \overline{\alpha}$ is impossible, this effectively restricts us to permuting elements of ω belonging to the same $V_{\mathbb{C}}^+$ or $V_{\mathbb{C}}^-$.

The conclusion is that the following permutation group emerges in both cases, ignoring scalings:
(VI.C.3)

$$W_{\mathbb{C}} = \{e, (14), (23), (14)(23), (12)(34), (13)(24), (1243), (1342)\}$$

are the permutations of ω achievable by means of $W_G(T, \mathbb{C})$. The "real Weyl groups" $W_G(T, \mathbb{R})$ identify with

$$W_{1,\mathbb{R}} = \{e, (12)(34)\}$$

for $n = 1$ and

$$W_{3,\mathbb{R}} = \{e, (13)(24)\}$$

for $n = 3$.

Now referring to (VI.B.18), (VI.B.19), here is how the "projection vectors" there correspond to Hodge structures: the first (resp. second) entry says what $p - q$ is for ω_1 (resp. ω_2) $\in V^{p,q}$. So "$\binom{3}{-1}$" means $V^{3,0} = \mathbb{C}\langle\omega_1\rangle, V^{1,2} = \mathbb{C}\langle\omega_2\rangle$, etc. When we compute the action of the group $W_{\mathbb{C}}$ on each set of projection vectors: on (VI.B.18) (numbered 1–4) we obtain:

(VI.C.4) $$\{\boxed{e}, (13)(24), (12)(34), (14)(23), \boxed{(23)}, (14), (1342), (1243)\}$$

and on (VI.B.19) (numbered 1–8)

(VI.C.5) $$\{\boxed{e}, (12)(34)(56)(78), (13)(24)(57)(68), (14)(23)(58)(67),$$
$$(15)(27)(36)(48), \boxed{(18)(26)(37)(45)}, (1647)(2835), (1746)(2538)\}$$

where we have boxed the permutations induced by $W_{n,\mathbb{R}}$. Obviously both actions of $W_{\mathbb{C}}$ are transitive, and this is just the action of $W_G(T, \mathbb{C})$ on $\breve{\mathrm{NL}}_T$. $\square$

By exactly the same reasoning as used in the above calculation we have the following general statement, which includes the case where T is the Mumford-Tate group of an irreducible nondegenerate CM-HS. Alternatively, as the reader may have guessed, the following is a corollary of (VI.B.12)(i) with $M = G$:

(VI.C.6) Proposition: *Let $T \leq G$ be a maximal anisotropic real torus (rank $= \lfloor \frac{\dim V}{2} \rfloor$) that is defined over $\mathbb{Q}$. Then $W_G(T, \mathbb{C})$ (resp. $W_G(T, \mathbb{R})$) acts transitively on $\breve{\mathrm{NL}}_T$ (resp. NL_T.)*

If we want to study permutations of components of a more general $\overset{(\vee)}{NL}_M$, a priori the groups of interest, over $\mathbb{R}$ or $\mathbb{C}$, are $N_G(M)/M$. To refine this, note that by (VI.B.9) and (VI.C.1), we have $T/\mathbb{Q}$ in $\mathcal{B}_{\mathbb{R}}^{M}$ and can modify by the action of M any $g \in N_G(M)$ so that it sends $(T, \varphi) \mapsto (T, \varphi')$, and hence normalizes T. So it suffices to consider the action of

$$\text{(VI.C.7)} \qquad W_G(M) := \frac{N_G(M) \cap N_G(T)}{N_M(T)} \leq \frac{N_G(M)}{M}$$

on components of $\overset{(\vee)}{\mathrm{NL}}_M$, which is facilitated by viewing it as a subquotient of $W_G(T)$ via

$$W_G(M) \twoheadleftarrow \frac{N_G(M) \cap N_G(T)}{T} \hookrightarrow \frac{N_G(T)}{T} = W_G(T)$$

and looking at the action on $\overset{(\vee)}{\mathrm{NL}}_T$ ($\subseteq \overset{(\vee)}{\mathrm{NL}}_M$). Note that $T, N_M(T) \trianglelefteq N_G(M) \cap N_G(T)$. Transitivity of the action of $W_G(M)$ on the set of components of $\overset{(\vee)}{\mathrm{NL}}_M$ is by no means obvious, and we look at some key examples in the next subsection.

VI.D GALOIS GROUPS AND FIELDS OF DEFINITION

One approach to finding an upper bound on the field of definition of a component $\check{D}_M$ of $\widecheck{\mathrm{NL}}_M$ in $\check{D}$, uses a density argument: first, by (VI.C.1) we have a CM point $F_0^\bullet \in \check{D}_M$, which via Chapter V is defined over the compositum K of the totally imaginary fields involved in its construction. Conjugating by $M(\mathbb{Q})^0$ yields a Zariski-dense set of points in $\check{D}_M$, also defined over K. Hence, for any Galois conjugation σ fixing K, $(\check{D}_M)^\sigma \cap \check{D}_M$ contains a set of points dense in both $\check{D}_M$ and $(\check{D}_M)^\sigma$, which implies $\check{D}_M = (\check{D}_M)^\sigma$ so that $\check{D}_M$ is defined over K in this sense. Typically $\check{D}_M$ will contain CM-points defined over different fields K', so one could easily refine this approach. We will not do that here, but one should note that this argument shows:

(VI.D.1) the degree of the field of definition of any $\check{D}_M$ is bounded above by $2^{\lfloor r/2 \rfloor}$ ($r = \dim(V)$),

since, via Chapter V, the worst-case scenario is for $F_0^\bullet$ to be built out of rank 2 CM-HS with "algebraically independent" CM fields.[10] It would be *very* interesting if (VI.D.1) could be used to produce an upper bound on the degrees of Hodge tensors required to define the Mumford-Tate groups of Hodge structures arising in $\overset{(\vee)}{D}$, since it is not known if such a bound exists.

In this subsection we will concern ourselves with questions about:

[10] If we knew that $F_0^\bullet$ can be chosen irreducible, the bound improves to r.

- transitivity of the action of $W_G(M)$ on $\breve{\mathrm{NL}}_M$;
- field of definition of components of, and Galois action on, $\breve{\mathrm{NL}}_M$,

and the relationship between them, when M is is cut out by degree-2 Hodge tensors (i.e., endomorphisms) — the nondegenerate case. Here is a key example, which will essentially be our setup for the remainder of the section: let $\varphi_0 \in D$ be a nondegenerate, irreducible CM-HS of odd weight. We have a CM field L with

$$L \xrightarrow[\cong]{\beta} \mathrm{End}(V_{\mathbb{Q}}, \varphi_0)$$

and by nondegeneracy the Mumford-Tate group of φ_0 is just its commutator in G,

$$T = G^{\beta(L)}.$$

The easiest way to produce nondegenerate Mumford-Tate groups containing T is to let $K \subseteq L$ be a subfield and set

(VI.D.2) $$M := G^{\beta(K)};$$

clearly then the "weak CM locus"

$$\breve{\mathrm{NL}}_M = \bigcup_{F^\bullet \in \breve{\mathrm{NL}}_T} M(\mathbb{R})^0 F^\bullet$$

and this has the property that any two $M(\mathbb{R})^0 F^\bullet$, $M(\mathbb{R})^0 \widetilde{F}^\bullet$ are either equal or disjoint.

(VI.D.3) **Example:** Continuing (VI.C.2) (both $n = 1$ and $n = 3$) we have $L = \mathbb{Q}(\zeta_5)$ and take K to be the totally real subfield $\mathbb{Q}(\sqrt{5})$. In the basis $\omega = \{\omega_1, \omega_2, \omega_3 = \bar{\omega}_2, \omega_4 = \bar{\omega}_1\}$ we have

$$[\beta(\ell)]_\omega = \begin{pmatrix} \sigma_1(\ell) & & & \\ & \sigma_2(\ell) & & \\ & & \sigma_3(\ell) & \\ & & & \sigma_4(\ell) \end{pmatrix}$$

for degree-2 Hodge tensors of T, hence for those of M all endomorphisms of the form

(VI.D.4) $$\begin{pmatrix} \mu & & & \\ & \tilde{\mu} & & \\ & & \tilde{\mu} & \\ & & & \mu \end{pmatrix}, \qquad \mu \in K$$

where $\widetilde{a + b\sqrt{5}} := a - b\sqrt{5}$. For $g \in G(\mathbb{C})$ to normalize M is obviously equivalent to conjugation by g stabilizing the set (VI.D.4), which all the permutations in (VI.C.3) do. Indeed, we have $N_G(T) \leq N_G(M)$ in this case. Since the action of $W_{\mathbb{C}}$ on (CM) points of $\breve{\mathrm{NL}}_T$ is transitive and each component of $\breve{\mathrm{NL}}_M$ contains at least one, its action — hence that of $N_G(M)$ — on components of $\breve{\mathrm{NL}}_M$ is transitive.

The subgroup

$$\text{(VI.D.5)} \qquad \{e, (14), (23), (14)(23)\} \subseteq W_{\mathbb{C}}$$

actually fixes the set (VI.D.3) pointwise, and so comes from the action of $W_M(T, \mathbb{C})$. That is, $N_M(T, \mathbb{C})$ induces actions on $\breve{\mathrm{NL}}_T$ described by the subgroup

$$\{e, (13)(24), (12)(34), (14)(23)\}$$

of (VI.C.4) for $n = 1$, and the subgroup

$$\{e, (12)(34)(56)(78), (13)(24)(57)(68), (14)(23)(58)(67)\}$$

of (VI.C.5) for $n = 3$. From this we deduce that for $n = 1$, $\breve{\mathrm{NL}}_M$ has one component containing all four points of $\breve{\mathrm{NL}}_T$ (and defined over $\mathbb{Q}$); while for $n = 3$, there are 2 components (both intersecting D), each of which contains half of the eight points of $\breve{\mathrm{NL}}_T$. It is also easy to see that for $n = 1$, $\dim \check{D} = 3$ and $\dim \breve{\mathrm{NL}}_n = 1$; while for $n = 3$, $\dim \check{D} = 4$ and $\dim \breve{\mathrm{NL}}_M = 2$. □

More generally, a CM field L has embeddings $\sigma_1, \ldots, \sigma_{2s}$ with $\overline{\sigma_i} = \sigma_{i+s}$ (indices mod $2s$) and an element $\rho \in \mathrm{Gal}(L/\mathbb{Q})$ with $\sigma_i \circ \rho = \sigma_{i+s}$. Its fixed field L_0 ($[L : L_0] = 2$) is totally real and we consider the case $M = L_0$ in (VI.D.2).

(VI.D.6) PROPOSITION: *$N_G(M)$ acts transitively on the components of the "totally-real-weak-CM" locus $\breve{\mathrm{NL}}_M$.*

PROOF. Let $\omega = \{\omega_1, \ldots, \omega_{2s}\}$ ($\omega_{i+s} = \overline{\omega_i}$) be a Hodge basis corresponding to the embeddings of L; then $W_G(T)$ permutes ω and hence the embeddings. These permutations do not all come from the Galois group but, by virtue of preserving Q, satisfy $\pi(i + x) \overset{(2s)}{\equiv} \pi(i + s)$. This ensures that they stabilize the sub-algebra $\beta(L_0) \subseteq \beta(L)$ which, in the ω-basis, consists precisely of those diagonal matrices $\mathrm{diag}(\alpha_1, \ldots, \alpha_{2s})$ satisfying $\alpha_i = \alpha_{i+s}$ (for all i). Hence $N_G(M)$ contains $N_G(T)$, and we can apply (VI.C.6) to the CM-points lying in every component of $\breve{\mathrm{NL}}_M$. □

Retaining the notation from the proof, suppose $\xi \in \mathrm{Gal}(L/\mathbb{R})$ takes $\sigma_i \mapsto \sigma_{\pi(i)}$, and let $\Pi_\xi \in \mathrm{Aut}(V_\mathbb{C})$ be the transformation sending $\omega_i \mapsto \omega_{\pi(i)}$. Furthermore, let Δ_ξ be the linear transformation that sends $\omega_j \mapsto \epsilon_{\pi,j} \cdot \omega_j$, where

$$\epsilon_{\pi,j} := \begin{cases} +1, & \text{if } \{\omega_j, \omega_{\pi^{-1}(j)}\} \subseteq V^+_{\varphi_0} \text{ or } V^-_{\varphi_0} \\ +\sqrt{-1}, & \text{otherwise.} \end{cases}.$$

Then we can define

$$\text{(VI.D.7)} \qquad \widetilde{\Pi} : \mathrm{Gal}(L/\mathbb{Q}) \hookrightarrow W_G(T, \mathbb{C})$$

by

$$\xi \mapsto \Delta_\xi \Pi_\xi =: \widetilde{\Pi}_\xi$$

since $\Delta_\xi \Pi_\xi$ preserves Q and normalizes T.

(VI.D.8) **Example:** For $L = \mathbb{Q}(\zeta_5)$, the resulting subgroup corresponds to

$$\{e, (14)(23), (1243), (1342)\}$$

in (VI.C.3), inducing $\{e, (14)(23), (1342), (1234)\}$ in (VI.C.4) and $\{e, (14)(23)(58)(67), (1647)(2835), (1746)(2538)\}$ in (VI.C.5). Hence it acts on $\breve{\mathrm{NL}}_T$ transitively for $n = 1$ but *not* for $n = 3$. □

Now let $K \subseteq L$ be arbitrary, but assume L is Galois over $\mathbb{Q}$. Considering $\beta(L)$ as tensors in $T^{1,1}$, the key formula is now

(VI.D.9)

$$\beta(\ell) \mapsto \widetilde{\Pi}_\xi \circ \beta(\ell) \circ \widetilde{\Pi}_\xi^{-1} = \beta(\xi(\ell)) \qquad \text{for all } \ell \in L, \xi \in \mathrm{Gal}(L/\mathbb{Q}).$$

In particular, this says that if ξ fixes the subfield K pointwise — that is, $\xi \in \mathrm{Gal}(L/K)$ — then $\widetilde{\Pi}_\xi$ fixes $\beta(K)$ and hence belongs to $W_M(T)$. This leads to a first refinement of (VI.D.7):

$$\text{(VI.D.10)} \qquad \mathrm{Gal}(K^c/\mathbb{Q}) \cong \frac{\mathrm{Gal}(L/\mathbb{Q})}{\mathrm{Gal}(L/K^c)} \hookrightarrow \frac{W_G(T, \mathbb{C})}{W_M(T, \mathbb{C})},$$

where K^c is the normal closure of K and $\mathrm{Gal}(L/K^c) \cong \mathrm{Gal}(L/K)$. Moreover, if ξ stabilizes K, then $\widetilde{\Pi}_\xi$ stabilizes $\beta(K)$, hence normalizes M (so belongs to $N_G(M) \cap N_G(T)$). Since all $\xi \in \mathrm{Gal}(L/\mathbb{Q})$ stabilize K^c, *if* $K = K^c$ we get the much nicer embedding

$$\text{(VI.D.11)} \qquad \widehat{\Pi} : \mathrm{Gal}(K/\mathbb{Q}) \hookrightarrow W_G(M);$$

cf. (VI.C.7).

Now, $\xi \in \mathrm{Gal}(L/\mathbb{Q})$ acts directly by Galois conjugation on the points of $\breve{\mathrm{NL}}_T$, and quite clearly $\tilde{\Pi}_\xi$ duplicates this action *in the context of a continuous automorphism of* $\check{D}$. Since $\breve{\mathrm{NL}}_M$ is defined over $\mathbb{Q}$, conjugation by ξ permutes *its* components as well, in a manner consistent with its action on $\breve{\mathrm{NL}}_T$. This leads to the following partial, but we feel very nice, transitivity result:

(VI.D.12) THEOREM: *Let $K \subseteq L$ be a* **normal** *subfield and define M by* (VI.D.2). *Then via $\hat{\Pi}$, the permutation action of $W_G(M)$ (or $N_G(M)$) on components of $\breve{\mathrm{NL}}_M$ reproduces (at least) that of* $\mathrm{Gal}(\mathbb{C}/\mathbb{Q})$.

PROOF. L gives an upper bound on the field of definition of components of $\breve{\mathrm{NL}}_M$. So we need only consider $\mathrm{Gal}(L/\mathbb{Q})$, and $\tilde{\Pi}$ maps $\mathrm{Gal}(L/K)$ into M, hence stabilizes components. □

This has obvious implications for fields of definition:

(VI.D.13) COROLLARY: *In the situation of* (VI.D.12),

(i) *The orbits of $W_G(M)$ on $\breve{\mathrm{NL}}_M$ (unions of components) are defined over $\mathbb{Q}$;*

(ii) *Individual components of $\breve{\mathrm{NL}}_M$ are defined over K (and in the non-normal case, over K^c).*

PROOF OF (II). By the reasoning preceding (VI.D.10) and (VI.D.12), K^c is the fixed field of a set of Galois conjugations that fix components of $\breve{\mathrm{NL}}_M$. □

(VI.D.14) **Example** (continuing (VI.D.3)): For $n = 3$, (VI.D.13)(ii) $\Longrightarrow$ the two components of $\breve{\mathrm{NL}}_M$ are defined over $\mathbb{Q}(\sqrt{5})$, and exchanged by $W_G(M)$ $\underset{\text{(VI.D.5)}}{\overset{\text{(VI.C.3)}}{=}} \hat{\Pi}(\mathrm{Gal}(\mathbb{Q}(\sqrt{5})/\mathbb{Q})) \cong \mathbb{Z}/2$. □

In general, (VI.D.13)(ii) will not be the best result. For the L-CM points themselves (which are components of a NL-locus), the precise field of definition is the reflex field L' (as defined in (V.A.4)) — which is generally smaller than L, the field given by (VI.D.13).

Though our preference has been to keep things simple here, we remark in conclusion that essentially all of the foregoing can be easily extended to direct sums of the irreducible nondegenerate CM-HS we have been considering.

APPENDIX TO CHAPTER VI: CM POINTS IN UNITARY MUMFORD-TATE DOMAINS

We felt this was an important example to include (even if peripheral to the main discussion), because of the centrality of unitary domains in the work of Carayol, and the fact that it illustrates the relationship between the fields of definition of $\check{D}_M$ and CM points in a nontrivial case.

Recall the construction of unitary groups: let $\psi_{\mathbb{C}}$ be a nondegenerate Hermitian form on an m-dimensional $\mathbb{C}$-vector space $W_{\mathbb{C}}$, and set $V_{\mathbb{C}} := W_{\mathbb{C}} \oplus \overline{W_{\mathbb{C}}}$. Define an alternating form on $V_{\mathbb{C}}$ having $W_{\mathbb{C}}, \overline{W_{\mathbb{C}}}$ as isotropic subspaces, by $Q_{\mathbb{C}}(u, \overline{v}) := i\psi_{\mathbb{C}}(u, v)$ (for $u \in W_{\mathbb{C}}$, $\overline{v} \in \overline{W_{\mathbb{C}}}$). One can view these as the complexification of a pair $V_{\mathbb{R}}, Q_{\mathbb{R}}$, and $W_{\mathbb{C}}, \overline{W_{\mathbb{C}}} \subset V_{\mathbb{C}}$ as eigenspaces of a real transformation $\eta_{\mathbb{R}}$. Setting $M := \mathrm{Aut}(V_{\mathbb{R}}, Q_{\mathbb{R}}, \eta_{\mathbb{R}})$, we have $M(\mathbb{R}) \cong \mathcal{U}(p, q)$ for some p, q $(p + q = n)$ determined by ψ. M is defined over $\mathbb{R}$.

In order for this construction to yield a $\mathbb{Q}$-algebraic group, we must have V, Q, η rational. For this to happen, $W \oplus \overline{W}$ must be a complete set of $\mathrm{Gal}(\mathbb{C}/\mathbb{Q})$-conjugates of W; hence W has to be defined over an imaginary quadratic field $\mathbb{F} = \mathbb{Q}(\sqrt{-d})$. Further, we must take $Q(u, \overline{v}) := \sqrt{-d}\psi(u, v)$, and replace $\eta_{\mathbb{R}}$ by $\eta : \mathbb{F} \hookrightarrow \mathrm{End}(V)$ such that $\eta(f)_{\mathbb{F}}$ has eigenvalues $f, \overline{f}$ on $W, \overline{W}$. Then $M := \mathrm{Aut}(V, Q, \eta)$ is defined over $\mathbb{Q}$, with $M(\mathbb{R})$ as before.

If M arises as the Mumford-Tate group of a Hodge structure φ on V polarized by Q, we must have $\eta(F) \subset \mathrm{End}(V, \varphi)$ (since M commutes with $\eta(\mathbb{F})$). If $D_M = M(\mathbb{R}) \cdot \varphi$ contains an irreducible CM Hodge structure φ_0, then $\mathbb{F} \cong \eta(\mathbb{F}) \subset \mathrm{End}(V, \varphi) \subset \mathrm{End}(V, \varphi_0) =: \mathbb{L}$. In this way we see that

> *for the CM Hodge structures in a unitary Mumford-Tate domain, the CM fields all contain a common $\mathbb{Q}(\sqrt{-d})$ (for some fixed squarefree $d \in \mathbb{N}$).*

Chapter VII

Classification of Mumford-Tate Subdomains

In this chapter we will produce an algorithm for deducing all Mumford-Tate subdomains of a given period domain — or equivalently, all possible Mumford-Tate groups of polarized Hodge structure with given Hodge numbers. We will then apply it in the case of rank 4 Hodge structures of weight 1 (Hodge numbers $(2, 2)$) and weight 3 (Hodge numbers $(1, 1, 1, 1)$), working out as much data as possible via the techniques of Chapter VI — Hodge tensors, endomorphism algebra, number of components in the Mumford-Tate Noether-Lefschetz loci (in D resp. $\check{D}$), and so forth.

VII.A A GENERAL ALGORITHM

The key motivating idea for our approach is that every Mumford-Tate domain contains a CM Hodge structure (VI.C.1). So it makes good sense to look in the tangent space to D at each CM point φ and ask which subspaces are the tangents to Mumford-Tate subdomains through φ. The main general result, with which we start, gives necessary and sufficient criteria for a subspace of $T_\varphi D$ to be of this type, which in principle could be fed to a computer if one knows the Galois group associated to φ. More precisely, it gives criteria for a subspace of $\mathrm{End}(V, Q)$ to be the Lie algebra of a Mumford-Tate group M with domain $D_M \ni \varphi$.

Let D be a period domain, excluding the same cases as in Chapter VI, and let $\varphi \in D$ be a CM Hodge structure. We have by Section V.B

(VII.A.1) $$V_\varphi = \underset{j}{\oplus} \left(V^n_{(L_j, \Pi_j)} \right)^{\oplus m_j};$$

write L for the compositum of the $\{L_j\}$, L^c for its normal closure, and $\mathcal{G} := \mathrm{Gal}(L^c/\mathbb{Q})$. The eigenspaces $\mathfrak{g}^{(i,-i)} \subset \mathfrak{g}$ for the action of $\mathrm{Ad}\,\varphi$ are defined over L^c, and $\mathcal{G}$ acts on $\mathfrak{g}$.

(VII.A.2) **Classification Theorem:** *Consider sub-vector-spaces $\mathcal{V}^i \subset \mathfrak{g}^{(i,-i)}$ defined over L^c with $\mathcal{V}^{-i} = \overline{\mathcal{V}^i} (= {}^\rho\mathcal{V}^i)$. Assume that*

- $\underset{i}{\oplus} \mathcal{V}^i =: \mathcal{V}$ *is closed under the action of $\mathcal{G}$;*

- *$\mathcal{V}$ is closed under the Lie bracket;*
- *$\mathcal{V}^{(0)}$ contains Lie (M_φ);*
- *among subspaces of $\mathfrak{g}$ with these properties and the same $\mathcal{V}^i$ (for all $i \neq 0$), $\mathcal{V}$ has the smallest $\mathcal{V}^0$.*

Then $\mathcal{V}$ is the Lie algebra of a Mumford-Tate group for M for some (set of) Hodge structures in D, and $\mathcal{V}^i = \mathfrak{m}^{(i,-i)}$. Conversely, the Lie algebra of any Mumford-Tate group whose Noether-Lefschetz locus passes through φ, is obtained in this way.

This result leads to the following "algorithm":

- classify CM fields of rank up to $r := \Sigma h^{p,q}$ by the Galois group over $\mathbb{Q}$, and use this to refine (VII.A.1);
- for each "Galois class" of φ, compute all the $\mathcal{V} \subset \mathrm{Lie}(G)$ as described in (VII.A.2);
- for each such $(\varphi, \mathcal{V})$, find effective generators for the algebra of Hodge tensors killed by $\mathcal{V}$ (and try to interpret these).

The latter tensors then cut out the Mumford-Tate Noether-Lefschetz locus (and corresponding Mumford-Tate group) whose component through φ has tangent space identifying with $\mathcal{V}^- = \underset{i<0}{\oplus} \mathcal{V}^i$.

PROOF OF CLASSIFICATION THEOREM. Let $\mathcal{V}$ be as in the theorem. Then $\mathcal{V} = \mathfrak{m}_{\mathbb{C}}$ for $\mathfrak{m}$ a $\mathbb{Q}$-sub-Hodge structure (with respect to $\mathrm{Ad}\,\varphi$) and sub-Lie algebra of $\mathfrak{g}$, with $\mathfrak{m}^{(0,0)} \supset \mathfrak{m}_\varphi$. The analytic closure $M_{\mathbb{C}}$ of $\exp(\mathfrak{m}_{\mathbb{C}})$ is $M(\mathbb{C})$ for a (connected) $\mathbb{Q}$-algebraic group M. Taking $\mu \in M(\mathbb{R})$ to be generic, and $\widetilde{\varphi} := \mu\varphi\mu^{-1}$, we will show that

$$M_{\widetilde{\varphi}} = M. \tag{VII.A.3}$$

Suppose otherwise, i.e.,[1] $\widetilde{M} := M_{\widetilde{\varphi}} \subsetneq M$. By genericity of μ, $\widetilde{M} \supseteq M_\varphi$; in particular, $\widetilde{M} \supset \varphi(\mathbb{U})$ so that $\widetilde{\mathfrak{m}} = \mathrm{Lie}(\widetilde{M})$ is closed under $\mathrm{Ad}\,\varphi$. We therefore have at the Lie algebra level

$$\mathfrak{m}_\varphi \subseteq \widetilde{\mathfrak{m}} \subsetneq \mathfrak{m}$$

[1] Warning: (for this proof only) the tildes do not refer to the "full Mumford-Tate group" as in Chapter II.

where *all three* are (Ad φ-) $\mathbb{Q}$ Hodge structures and Lie algebras. Moreover, if $\phi := \varphi'(1) \in \mathrm{Lie}(\varphi(\mathbb{U}))$, and $m = \Sigma m_j$ ($m_j \in \mathfrak{m}^{(j,-j)}$) is any vector in $\mathfrak{m}_{\mathbb{C}}$, then

(VII.A.4) $$(\mathrm{ad}(m))\phi = -(\mathrm{ad}\,\phi)m = -2\sum_j j \cdot m_j;$$

and so (by genericity of μ) $(\mathrm{Ad}\,\mu)\phi$ is equal to ϕ plus a generic element of $(\sum_{j\neq 0} \mathfrak{m}^{(j,-j)})_{\mathbb{R}}$. Since $\widetilde{\mathfrak{m}}$ contains this element and is defined over $\mathbb{Q}$,

$$\widetilde{\mathfrak{m}} \supseteq \widetilde{\mathfrak{m}}^{(0,0)} + \sum_{j\neq 0} \mathfrak{m}^{(j,-j)}.$$

That is, $\widetilde{\mathfrak{m}}^{(i,-i)} = \mathfrak{m}^{(i,-i)}$ ($\forall i \neq 0$) and $\widetilde{\mathfrak{m}}^{(0,0)} \subsetneq \mathfrak{m}^{(0,0}$. This contradicts the assumed minimality of $\mathcal{V}^0 = \mathfrak{m}^{(0,0)}$, establishing (VII.A.3). That is, for the generic $\mu \in M(\mathbb{R})$, $M_{\mu\varphi\mu^{-1}} = M$ exactly; furthermore, $M(\mathbb{R})\varphi$ is a Mumford-Tate domain through φ.

Conversely, given a Mumford-Tate domain through φ (i.e., an orbit $D_M := M(\mathbb{R})\varphi$ with the property $M = M_{\widetilde{\varphi}}$ for generic $\widetilde{\varphi} \in D_M$) we will show that the properties in the theorem obtain for $\mathfrak{m} := \mathrm{Lie}(M)$. Clearly $M_\varphi \subseteq M$ since $\varphi \in D_M$, and so $\mathfrak{m}_\varphi \subset \mathfrak{m}$. Also, $\mathfrak{m}$ is a $\mathbb{Q}$-Hodge structure with respect to Ad φ because it is defined over $\mathbb{Q}$ (as M is a Mumford-Tate group) and closed under Ad_φ (as φ factors through M); obviously it is a sub-Lie algebra of φ. To see the minimality property, suppose there is a sub-$\mathbb{Q}$ Hodge structure/Lie algebra $\widehat{\mathfrak{m}} \subsetneq \mathfrak{m}$ with

$$\begin{cases} \widehat{\mathfrak{m}}^{(i,-i)} = \mathfrak{m}^{(i,-i)} & (\forall i \neq 0) \\ \widehat{\mathfrak{m}}^{(0,0)} \supset \mathfrak{m}_\varphi. \end{cases}$$

Take $\widehat{M} \subsetneq M$ to be the $\mathbb{Q}$-algebraic group associated to $\widehat{\mathfrak{m}}$. By the calculation (VII.A.4), we see that

$$(\mathrm{ad}\,\widehat{\mathfrak{m}})\phi = (\mathrm{ad}\,\mathfrak{m})\phi,$$

which implies

$$\widehat{M}(\mathbb{R})\varphi = M(\mathbb{R})\varphi.$$

In particular, $\widetilde{\varphi} \in \widehat{M}(\mathbb{R})\varphi$ and $\widehat{M} \supset \varphi(\mathbb{U})$, so that $\widehat{M}$ contains $\widetilde{\varphi}(\mathbb{U})$ (and thus $M_{\widetilde{\varphi}}$), a contradiction. □

(VII.A.5) **Remarks.** (a) The reader will note that Theorem (V.F.1) gives a method for computing $\mathrm{Lie}(M_\varphi)$, at least when $L = L^c$ and $V_\varphi = V^n_{(L,\Pi)}$ (is a SCMpHS) in (VII.A.1). (This is part of the first step of the algorithm.)

(b) That $\mathcal{V}^i \subset \mathfrak{g}^{(i,-i)}$ be defined over L^c is a necessary condition for $\mathcal{V}$ to be defined over $\mathbb{Q}$, since the $\mathfrak{g}^{(i,-i)}$ are defined over L^c.

VII.B CLASSIFICATION OF SOME CM-HODGE STRUCTURES

In the remainder of this chapter we shall apply VII.A to weight 3 Hodge structure of type $(1,1,1,1)$ (and weight 1 Hodge structure as a "control"). (Note that in the *reducible* case the possible Mumford-Tate groups were found by the student group of J. de Jong in [BEKPW].) This section may be viewed as including the first step in the "algorithm" described in VII.A.

(VII.B.1) LEMMA: *Let L be a CM field, with totally real index-2 subfield*[2] *$L^{\rho} =: L_0$. If $L_0/\mathbb{Q}$ is Galois, then $L/\mathbb{Q}$ is Galois.*

PROOF. Since $L = L_0(\sqrt{\delta})$ for some $\delta \in L_0$, L/L_0 is a splitting-field extension. Therefore we have a short-exact sequence

$$1 \to \mathrm{Gal}(L/L_0) \to \mathrm{Stab}_{\mathrm{Gal}(L/\mathbb{Q})}(L_0) \to \mathrm{Gal}(L_0/\mathbb{Q}) \to 1,$$

where the middle term is a subgroup of $\mathrm{Gal}(L/\mathbb{Q})$ and $\mathrm{Gal}(L/L_0) = \langle\rho\rangle \cong \mathbb{Z}/2\mathbb{Z}$. If $L_0/\mathbb{Q}$ is Galois, then $|\mathrm{Gal}(L_0/\mathbb{Q})| = [L_0 : \mathbb{Q}] \Rightarrow$

$$|\,\mathrm{Stab}_{\mathrm{Gal}(L/\mathbb{Q})}(L_0)| = 2[L_0 : \mathbb{Q}],$$

which forces

$$|\mathrm{Gal}(L/\mathbb{Q})| = 2[L_0 : \mathbb{Q}] \quad (= [L : \mathbb{Q}]). \qquad \square$$

If the embeddings of L are $\{\theta_i, \overline{\theta}_i\}_{i=1}^{g}$ then δ must satisfy $\theta_i(\delta) < 0$ for all i. Write $\xi := \sqrt{\delta}$.

(VII.B.2) COROLLARY: *The CM fields L of degree 4 are all Galois. It therefore makes sense to write $\tilde{\xi}$ for the element of L with $\theta_1(\tilde{\xi}) = \theta_2(\xi)$. We have either*

(I) $\mathbb{Z}_4 \cong \mathcal{G} = \{1, \sigma, \rho, \sigma\rho\}$ *where $\sigma^2 = \rho$, and σ sends* $\xi \overset{\sigma}{\to} \tilde{\xi} \overset{\sigma}{\to} \overline{\xi} \overset{\sigma}{\to} \overline{\tilde{\xi}} \overset{\sigma}{\to} \xi$.

(II) $\mathbb{Z}_2 \times \mathbb{Z}_2 \cong \mathcal{G} = \{1, \eta, \rho, \eta\rho\}$ *and η sends* $\xi \overset{\eta}{\to} \tilde{\xi} \overset{\eta}{\to} \xi, \overline{\xi} \overset{\eta}{\to} \overline{\tilde{\xi}} \overset{\eta}{\to} \overline{\xi}$.

(VII.B.3) **Examples:**

(I) $\quad L = \mathbb{Q}(\zeta_5) \qquad (L_0 = \mathbb{Q}(\zeta_5 + \overline{\zeta}_5));$

(II) $\quad L = \mathbb{Q}\Big(\sqrt{-\frac{3+\sqrt{5}}{2}}\Big) \qquad (L_0 = \mathbb{Q}(\sqrt{5}));$

$\quad L = \mathbb{Q}(\sqrt{-2}, \sqrt{-3}) \qquad (L_0 = \mathbb{Q}(\sqrt{6})).$

(VII.B.4) LEMMA: *Given a CM field L of degree 4. If Θ (resp. Π) is a 1- (resp. 3-) orientation*[3] *for L, then we may arrange the labeling so that* $\Theta =$

[2] *Recall that $\rho \in Z(\mathrm{Gal}(L/\mathbb{Q}))$ corresponds to complex conjugation under every embedding.*

[3] *For the 3-orientation: with $|\Pi^{3,0}| = |\Pi^{2,1}| = 1$.*

$\{\theta_1, \theta_2\}$ *(resp.* $\Pi^{3,0} = \{\theta_1\}$, $\Pi^{2,1} = \{\theta_2\}$*). In case (I), we have Kubota ranks* $\mathcal{R}(L,\Theta)$ *and* $\mathcal{R}(L,\Pi)$ *both* 3. *In case (II),* $\mathcal{R}(L,\Pi) = 3$ *while* $\mathcal{R}(L,\Theta) = 2$*; in particular,* (L,Θ) *is degenerate.*

It turns out that in rank/degree 4, the Kubota ranks suffice to calculate $\mathrm{Lie}(M_\varphi)$, so that we need not apply the more general result (V.F.1). Their calculation will be incorporated into the proof of the following classification of CM-Hodge structures by endomorphism algebra E_φ.

(VII.B.5) THEOREM: *Let* (V,φ) *be a CM Hodge structure of rank* 4, $E_\varphi = \mathrm{End}(V_{\mathbb{Q}}, \varphi)$.

In weight 1, the possible cases are:

(I) $V = V^1_{(L,\Theta)}$ *for L CM of degree* 4 *with* $\mathcal{G} \cong \mathbb{Z}_4$

$$\Rightarrow V \text{ irreducible}, \quad E_\varphi \cong L, \quad \dim M_\varphi = 2.$$

(II) $V = (V^1_{(K,\psi)})^{\oplus 2}$ *for K imaginary quadratic*

$$\Rightarrow E_\varphi \cong M_2(K), \quad \dim M_\varphi = 1 \quad [\psi = \text{ embedding of } K].$$

(III) $V = V^1_{(K_1,\psi_1)} \oplus V^1_{(K_2,\psi_2)}$ *for* K_i **distinct** *imaginary quadratic*

$$\Rightarrow E_\varphi \cong K_1 \times K_2, \quad \dim M_\varphi = 2 \quad [\psi_i \text{ embeddings of } K_i].$$

In weight 3 ($h^{3,0} = h^{2,1} = 1$)*, the possible cases are:*

(I) $V = V^3_{(L,\Pi)}$ *for L CM degree* 4 *with* $\mathcal{G} \cong \mathbb{Z}_4$

$$\Rightarrow V \text{ irreducible}, \quad E_\varphi \cong L, \quad \dim M_\varphi = 2.$$

(II) $V = V^3_{(L,\Pi)}$ *for L CM degree* 4 *with* $\mathcal{G} \cong \mathbb{Z}_2 \times \mathbb{Z}_2$

$$\Rightarrow V \text{ irreducible}, \quad E_\varphi \cong L, \quad \dim M_\varphi = 2.$$

(Here $\Pi_1 = \{\Pi_1^{3,0}, \Pi_1^{0,3}\}$, $\Pi_2 = \{\Pi_2^{2,1}, \Pi_2^{1,2}\}$ *where* $\Pi_1^{3,0} = \{\psi_1\}$, $\Pi_2^{2,1} = \{\psi_2\}$*.)*

(III) $V = V^3_{(K_1,\Pi_1)} \oplus V^3_{(K_2,\Pi_2)} = (V^{3,0} \oplus V^{0,3}) \oplus (V^{2,1} \oplus V^{1,2})$ *for* K_i **distinct** *imaginary quadratic*

$$\Rightarrow E_\varphi \cong K_1 \times K_2, \quad \dim M_\varphi = 2.$$

(IV) $V = V^3_{(K_1,\Pi_1)} \oplus V^3_{(K_1,\Pi_2)} = (V^{3,0} \oplus V^{0,3}) \oplus (V^{2,1} \oplus V^{1,2})$ *for* K *imaginary quadratic*

$$\Rightarrow E_\varphi \cong K \times K, \ \ \dim M_\varphi = 1.$$

(Here $\Pi_1^{3,0} = \{\psi\} = \Pi_2^{2,1}$*.)*

Remarks: The fields of definition of the standard Hodge bases (cf. Section V.C) in each case are (weight 1) L, K, K_1, K_2; (weight 3) L, L, K_1K_2, K. The compositum K_1K_2 is a CM field with Galois group $\mathbb{Z}_2 \times \mathbb{Z}_2$; we shall write $L = K_1K_2$ in that case. We also note that the *only* case (among the seven above) in which V is degenerate as a Hodge structure (V.D.6), is case (IV wt. 3). In all other cases, M_φ is cut out of $\mathrm{Aut}(V, \mathbb{Q})$ by E_φ.

PROOF OF (VII.B.4)–(VII.B.5). From the discussion after (V.D.4), in either weight we have $V = V^n_{(L,\Pi)}$ or $V^n_{(K_1,\Pi_1)} \oplus V^n_{(K_2,\Pi_2)}$. Beyond this (and Corollary (VII.B.2)), the content of the classification is that $V^n_{(L,\Pi)}$ is reducible when $n = 1$ and $\mathcal{G} \cong \mathbb{Z}_2 \times \mathbb{Z}_2$ and otherwise irreducible. This (and the claimed $\dim M_\varphi$'s) follows from computing the (generalized) Kubota ranks and applying (V.A.9) and (V.D.5).

$$\left.\begin{array}{r} n=1 \\ \mathcal{G} \cong \mathbb{Z}_4 \end{array}\right\} \Rightarrow \mathfrak{R} = \mathrm{rank}\begin{pmatrix} 1&0&1&0\\1&1&0&0\\0&0&1&1\\0&1&0&1 \end{pmatrix} = 3, \tag{1}$$

$$\left.\begin{array}{r} n=1 \\ \mathcal{G} \cong \mathbb{Z}_2 \times \mathbb{Z}_2 \end{array}\right\} \Rightarrow \mathfrak{R} = \mathrm{rank}\begin{pmatrix} 1&1&0&0\\1&1&0&0\\0&0&1&1\\0&0&1&1 \end{pmatrix} = 2, \tag{2}$$

$$\left.\begin{array}{r} n=3 \\ \mathcal{G} \cong \mathbb{Z}_4 \end{array}\right\} \Rightarrow \mathfrak{R} = \mathrm{rank}\begin{pmatrix} 3&1&2&0\\2&3&0&1\\1&0&3&2\\0&2&1&3 \end{pmatrix} = 3, \tag{3}$$

$$\left.\begin{array}{r} n=3 \\ \mathcal{G} \cong \mathbb{Z}_2 \times \mathbb{Z}_2 \end{array}\right\} \Rightarrow \mathfrak{R} = \mathrm{rank}\begin{pmatrix} 3&2&1&0\\2&3&0&1\\1&0&3&2\\0&1&2&3 \end{pmatrix} = 3. \tag{4}$$

In particular, in case (2) we have $V^1_{(L,\Theta)} = \left(V^1_{(K,\psi)}\right)^{\oplus 2}$ where $K = L^\eta = \mathbb{Q}(\xi + \tilde{\xi})$.

Now we treat the reducible cases, where $V = V_1 \oplus V_2$. Let $z \in \mathbb{C}^*$ be general, and choose lifts of all Galois actions to $\mathbb{C}/\mathbb{Q}$ that fix z. With respect to the Hodge basis,

$$[\varphi(z)]_\omega = \left(\begin{array}{cc|cc} z & & & \\ & z^{-1} & & \\ \hline & & z & \\ & & & z^{-1} \end{array}\right) \quad \text{for weight 1}$$

and

$$[\varphi(z)]_\omega = \left(\begin{array}{cc|cc} z^3 & & & \\ & z^{-3} & & \\ \hline & & z & \\ & & & z^{-1} \end{array}\right) \quad \text{for weight 3.}$$

In cases (II, wt. 1) and (IV, wt. 3), ω is defined over K and the only Galois action is ρ; $[{}^\rho\varphi(z)]_\omega$ is just $[\varphi(z^{-1})]_\omega$ and so $\dim(M_\varphi) = 1$. In cases (III), writing $L = K_1K_2$ with $\mathcal{G} = \{1, \rho_1, \rho_2, \rho_1\rho_2\}$, we get

$$[{}^{\rho_1}\varphi(z)]_\omega = \left(\begin{array}{cc|cc} z^{-1} & & & \\ & z & & \\ \hline & & z & \\ & & & z^{-1} \end{array}\right) \quad \text{in weight 1}$$

and

$$[{}^{\rho_1}\varphi(z)]_\omega = \left(\begin{array}{cc|cc} z^{-3} & & & \\ & z^{3} & & \\ \hline & & z & \\ & & & z^{-1} \end{array}\right) \quad \text{in weight 3,}$$

so that $\dim(M_\varphi) = 2$ (the maximum). □

To describe more precisely the Mumford-Tate groups involved in this classification, recall that for a SCMpHS $V^n_{(F,\Pi)}$, $M_{\tilde{\varphi}} = \mathrm{im}(\mathcal{N}_{\Pi'})$ and $M_\varphi = \mathrm{im}(\mathcal{N}_{\Pi'}) \cap \ker(\mathcal{N}_{F/\mathbb{Q}})$, where $\mathcal{N}_{F/\mathbb{Q}} : \mathrm{Res}_{F/\mathbb{Q}}\,\mathbb{G}_m \to \mathbb{G}_m$ is the algebraic-group analog of the norm map $N_{F/\mathbb{Q}} : F^* \to \mathbb{Q}^*$. If F is imaginary quadratic, $\mathrm{im}(\mathcal{N}_{\Pi'})$ is just $\mathrm{Res}_{F/\mathbb{Q}}\,\mathbb{G}_m$ itself.

(VII.B.6) COROLLARY: *In the above classification, the Mumford-Tate groups are:*

- $\mathrm{im}(\mathcal{N}_{\Pi'}) \cap \ker(\mathcal{N}_{L/\mathbb{Q}})$ *in cases* (I; $n = 1, 3$) *and* (II; $n = 3$)*;*
- $\ker(\mathcal{N}_{K_1/\mathbb{Q}}) \times \ker(\mathcal{N}_{K_2/\mathbb{Q}})$ *in case* (III; $n = 1, 3$)*;*
- *diagonal (resp.* $\xi \mapsto (\xi^3, \xi)$*) embedding of* $\ker(\mathcal{N}_{K/\mathbb{Q}})$ *in cases* (II; $n=1$) *and* (IV; $n = 3$).

VII.C DETERMINATION OF SUB-HODGE-LIE-ALGEBRAS

Now we shall carry out "step 2" of the algorithm; i.e., determine all[4] $\mathcal{V}$ as described in (VII.A.2) for each "endomorphism class" of φ as in (VII.B.5). We

[4] *Notation:* $\mathcal{V}$ is a subspace of $\mathfrak{g}$; V is the underlying $\mathbb{Q}$-vector space of (V, φ) — not to be confused!

will first do as much as we can for weights 1 and 3 together, though later the treatments have to be separated.

In all cases we use the notation

$$\begin{aligned} \mathfrak{e} &= \{e_1, e_2, e_3, e_4\} && (\mathbb{Q}\text{-bases of } V) \\ \boldsymbol{\omega} &= \{\omega_1, \omega_2, \omega_3 = \overline{\omega}_2, \omega_4 = \overline{\omega}_1\} && (\text{Hodge basis of } V_{\mathbb{C}}) \end{aligned}$$

though we shall rarely use $\mathfrak{e}$. For $n = 1$, $F^1 V_{\mathbb{C}} = \mathbb{C}\langle \omega_1, \omega_2 \rangle$; for $n = 3$, ω_1 resp. ω_2 are pure of type $(3, 0)$ resp. $(2, 1)$.

In the cases (I; $n = 1, 3$) and (II; $n = 3$) we write $\beta : L \xrightarrow{\cong} V_{(\mathbb{Q})}$, and recall that $\beta(\ell) = \sum_{i=1}^{4} \theta_i(\ell)\omega_i$. If $\sigma \in \mathrm{Gal}(L/\mathbb{Q})$ (= abelian in all cases!) then we have

$$\beta(\ell) = {}^{\sigma}(\beta(\ell)) = \sum_i (\theta_i \circ \sigma)(\ell)\, {}^{\sigma}\omega_i =: \sum \theta_{\sigma(i)}(\ell)\, {}^{\sigma}\omega_i,$$

and then also

$$\text{(VII.C.1)} \qquad {}^{\sigma}\omega_i = \omega_{\sigma(i)}.$$

The notation (VII.C.1) is in fact good for *all* cases; note that ${}^{\sigma}\overline{\omega_i} = \overline{\omega_{\sigma(i)}}$ since $\overline{\omega_i} = \omega_{\rho(i)}$ and $\rho \in Z(\mathcal{G})$.

It will be useful to write $\widehat{\omega}_{ij} \in \mathrm{End}(V_{\mathbb{C}})$ for the transformation sending $\omega_j \longrightarrow \omega_i$ and all other $\omega_k \longrightarrow 0$. We may assume in all cases that the polarization is of the form

$$[Q]_{\boldsymbol{\omega}} = \sqrt{-1} \begin{pmatrix} 0 & 0 & 0 & Q_1 \\ 0 & 0 & Q_2 & 0 \\ 0 & Q_3 & 0 & 0 \\ Q_4 & 0 & 0 & 0 \end{pmatrix},$$

where

$$\text{(VII.C.2)} \qquad Q_i = -\sqrt{-1}Q(\omega_i, \overline{\omega_i}) \in \mathbb{R}$$

satisfies $Q_3 = -Q_2$, $Q_4 = -Q_1$; put

$$\text{(VII.C.3)} \qquad \Xi := \frac{-Q_1}{Q_2} =: \frac{-1}{\Xi^*}$$

(which is positive for $n = 3$ and negative for $n = 1$). A Hodge basis of $\mathfrak{g}_{\mathbb{C}} = \{\mathcal{T} \in \mathrm{End}(V_{\mathbb{C}}) \mid {}^t\mathcal{T}Q + Q\mathcal{T} = 0\}$ is then given by

$$\text{(VII.C.4)}\qquad \boxed{n=1}\left\{\begin{array}{ll}
\mathfrak{g}^{1,-1}\left\{\begin{array}{l} \\ \\ \\ \end{array}\right. & \begin{array}{ll}
\varepsilon_9 := \widehat{\omega}_{14} & \}\,\mathfrak{g}^{3,-3} \\
\varepsilon_7 := \widehat{\omega}_{24} + \Xi^*\widehat{\omega}_{13} & \}\,\mathfrak{g}^{2,-2} \\
\varepsilon_5 := \widehat{\omega}_{23} & \multirow{2}{*}{$\Big\}\,\mathfrak{g}^{1,-1}$} \\
\end{array} \\
\mathfrak{g}^{0,0}\left\{\begin{array}{l} \\ \\ \\ \\ \end{array}\right. & \begin{array}{ll}
\varepsilon_3 := \widehat{\omega}_{12} + \Xi\widehat{\omega}_{34} & \\
\varepsilon_1 := \widehat{\omega}_{11} - \widehat{\omega}_{44} & \Big\}\,\mathfrak{g}^{0,0} \\
\varepsilon_2 := \widehat{\omega}_{22} - \widehat{\omega}_{33} & \\
\varepsilon_4 := \widehat{\omega}_{43} + \Xi\widehat{\omega}_{21} & \Big\}\,\mathfrak{g}^{-1,1} \\
\end{array} \\
\mathfrak{g}^{-1,1}\left\{\begin{array}{l} \\ \\ \\ \end{array}\right. & \begin{array}{ll}
\varepsilon_6 := \widehat{\omega}_{32} & \\
\varepsilon_8 := \widehat{\omega}_{31} + \Xi^*\widehat{\omega}_{42} & \}\,\mathfrak{g}^{-2,2} \\
\varepsilon_{10} := \widehat{\omega}_{41} & \}\,\mathfrak{g}^{-3,3} \\
\end{array}
\end{array}\right\}\boxed{n=3}\,.$$

(Here $\varepsilon_5, \varepsilon_3$ span $\mathfrak{g}^{1,-1}$; $\varepsilon_1, \varepsilon_2$ span $\mathfrak{g}^{0,0}$; $\varepsilon_4, \varepsilon_6$ span $\mathfrak{g}^{-1,1}$ for $n=3$.)

The Galois action on (VII.C.4) can be read off from

$$\text{(VII.C.5)}\qquad {}^{\sigma}\widehat{\omega}_{ij} = \widehat{\omega}_{\sigma(i)\sigma(j)}$$

(which follows from (VII.C.1)) and

$$\text{(VII.C.6)}\qquad {}^{\sigma}\Xi^* = \frac{Q_{\sigma(2)}}{Q_{\sigma(1)}}$$

(follows from (VII.C.1)–(VII.C.3)). We also have the

(VII.C.7) Table of Lie brackets: $\left(\left[\begin{array}{cc}\text{left} & \text{top}\\ \text{entry,} & \text{entry}\end{array}\right] = ?\right)$

	ε_9	ε_7	ε_5	ε_3	ε_1	ε_2	ε_4	ε_6	ε_8	ε_{10}
ε_9	0	0	0	0	$-2\varepsilon_9$	0	$-\Xi\varepsilon_7$	0	$\Xi^*\varepsilon_3$	ε_1
ε_7	0	0	0	$-2\varepsilon_9$	$-\varepsilon_7$	$-\varepsilon_7$	$2\varepsilon_5$	$\Xi^*\varepsilon_3$	$\Xi^*(\varepsilon_1+\varepsilon_2)$	$-\Xi^*\varepsilon_4$
ε_5	0	0	0	$\Xi\varepsilon_7$	0	$-2\varepsilon_5$	0	ε_2	$-\Xi^*\varepsilon_4$	0
ε_3	0	$2\varepsilon_9$	$-\Xi\varepsilon_7$	0	$-\varepsilon_3$	ε_3	$\Xi(\varepsilon_1-\varepsilon_2)$	0	$-2\varepsilon_6$	$\Xi\varepsilon_8$
ε_1	$2\varepsilon_9$	ε_7	0	ε_3	0	0	$-\varepsilon_4$	0	$-\varepsilon_8$	$-2\varepsilon_{10}$
ε_2	0	ε_7	$2\varepsilon_5$	$-\varepsilon_3$	0	0	ε_4	$-2\varepsilon_6$	$-\varepsilon_8$	0
ε_4	$\Xi\varepsilon_7$	$-2\varepsilon_5$	0	$\Xi(\varepsilon_2-\varepsilon_1)$	ε_4	$-\varepsilon_4$	0	$-\Xi\varepsilon_8$	$2\varepsilon_{10}$	0
ε_6	0	$-\Xi^*\varepsilon_3$	$-\varepsilon_2$	0	0	$2\varepsilon_6$	$\Xi\varepsilon_8$	0	0	0
ε_8	$-\Xi^*\varepsilon_3$	$-\Xi^*(\varepsilon_1+\varepsilon_2)$	$\Xi^*\varepsilon_4$	$2\varepsilon_6$	ε_8	ε_8	$-2\varepsilon_{10}$	0	0	0
ε_{10}	$-\varepsilon_1$	$\Xi^*\varepsilon_4$	0	$-\Xi\varepsilon_8$	$2\varepsilon_{10}$	0	0	0	0	0

Specializing to weight $n = 3$,[5] we have

(VII.C.8) PROPOSITION: *For all cases* (I)–(IV), *the list of non-zero* $\mathcal{V} \subset \mathfrak{g}$ *which are*

- *a direct sum of* $\mathcal{V}^i \subset \mathfrak{g}^{(i,-i)}$ *with* $\mathcal{V}^{-i} = \overline{\mathcal{V}^i}$
- *closed under the Lie bracket,*

is (writing $\langle\ \rangle$ *for* $\mathbb{C}$*-span):*

$$\begin{array}{ll}
(1) & \langle A\varepsilon_1 + B\varepsilon_2\rangle \qquad ([A:B]\in\mathbb{P}^1(\mathbb{R}))\\
(2) & \langle\varepsilon_1,\varepsilon_2\rangle\\
(3) & \langle\varepsilon_9,\varepsilon_{10},\varepsilon_1\rangle\\
(4) & \langle\varepsilon_9,\varepsilon_{10},\varepsilon_1,\varepsilon_2\rangle\\
(5) & \langle\varepsilon_8,\varepsilon_7,\varepsilon_1+\varepsilon_2\rangle\\
(6) & \langle\varepsilon_8,\varepsilon_7,\varepsilon_1,\varepsilon_2\rangle\\
(7) & \langle\varepsilon_5,\varepsilon_6,\varepsilon_2\rangle\\
(8) & \langle\varepsilon_5,\varepsilon_6,\varepsilon_1,\varepsilon_2\rangle\\
(9) & \langle\varepsilon_3,\varepsilon_4,\varepsilon_1-\varepsilon_2\rangle\\
(10) & \langle\varepsilon_3,\varepsilon_4,\varepsilon_1,\varepsilon_2\rangle\\
(11) & \left.\begin{array}{l}\left\langle A\varepsilon_4+B\varepsilon_6,\overline{A}\varepsilon_3+\overline{B}\varepsilon_5,3\varepsilon_1+\varepsilon_2\right\rangle\\ \left\langle A\varepsilon_4+B\varepsilon_6,\overline{A}\varepsilon_3+\overline{B}\varepsilon_5,\varepsilon_1,\varepsilon_2\right\rangle\end{array}\right\} A,B\in\mathbb{C}\text{ with }\frac{3|B|^2}{4|A|^2}=\Xi.\\
(12) & \\
(13) & \langle\varepsilon_5,\varepsilon_6,\varepsilon_9,\varepsilon_{10},\varepsilon_1,\varepsilon_2\rangle\\
(14) & \mathfrak{g}.
\end{array}$$

PROOF. This is shown by taking each of $\mathfrak{g}^{(i,-i)} \oplus \mathfrak{g}^{(-i,i)}$ and computing closure under $[\ ,\ ]$, then considering pairs of the resulting spaces (the closure of their sum under $[\ ,\ ]$), etc. The one interesting bit is looking at 1-dimensional subspaces of $\mathfrak{g}^{(1,-1)}$, where one has to solve for A, B such that $\langle A\varepsilon_4 + B\varepsilon_6, \overline{A}\varepsilon_3 + \overline{B}\varepsilon_5\rangle$ does not bracket-generate all of $\mathfrak{g}$: $[1:0]$, $[0:1]$, or $[A:B]$ with $4|A|^2\Xi = 3|B|^2$. □

The next step is to check $\mathcal{G}$-stability of the subspaces (1)–(14). (It is not necessary to consider the action of ρ, since $\mathcal{V}^{-i} = \overline{\mathcal{V}^i}$ already ensures closure under complex conjugation.) A key observation is that (VII.C.1) gives a homomorphism

$$\text{(VII.C.9)}\qquad \begin{array}{ccc}\mathcal{G} & \to & \mathfrak{S}_4\\ \sigma & \longrightarrow & \text{“}\sigma(\cdot)\text{”}.\end{array}$$

We work through the cases (I)–(IV) of (VII.B.5):

[5] Weight 1 will be taken up again in Section VII.F.

Case (I). $\mathcal{G} \cong \mathbb{Z}_4$, (VII.C.9) sends the generator σ to (1243) by (VII.B.2)(I). Hence under σ we have therefore

$$\Xi^* \rightleftarrows \Xi$$

and

$$\text{(VII.C.10)} \qquad \begin{cases} \varepsilon_9 & \mapsto & \varepsilon_5 & \mapsto & \varepsilon_{10} & \mapsto & \varepsilon_6 & (\mapsto \varepsilon_9) \\ \varepsilon_7 & \mapsto & \varepsilon_4 & \mapsto & \varepsilon_8 & \mapsto & \varepsilon_3 & (\mapsto \varepsilon_7) \\ \varepsilon_1 & \mapsto & \varepsilon_2 & \mapsto & -\varepsilon_1 & \mapsto & -\varepsilon_2 & (\mapsto \varepsilon_1). \end{cases}$$

Also, $\mathrm{Lie}(M_\varphi) = \langle \varepsilon_1, \varepsilon_2 \rangle$ (as $\dim M_\varphi = 2$), eliminating (1), (3), (5), (7), (9), (11). The $\mathcal{G}$-stable requirement (in light of (VII.C.10)) eliminates (9), (6), (8), (10), (12). This leaves us with:

I(a)	$\langle \varepsilon_1, \varepsilon_2 \rangle$
I(b)	$\langle \varepsilon_5, \varepsilon_6, \varepsilon_9, \varepsilon_{10}, \varepsilon_1, \varepsilon_2 \rangle$.

Case (II). $\mathcal{G} \cong \mathbb{Z}_2 \times \mathbb{Z}_2$, (VII.C.9) sends $\zeta \longrightarrow (12)(34)$ by (VII.B.2)(II). So under ζ,

$$\begin{array}{lll} \Xi^* & \leftrightarrows & -\Xi \\ \varepsilon_9 & \leftrightarrows & \varepsilon_5, \\ \varepsilon_7 & \leftrightarrows & -\Xi\varepsilon_7, \\ \varepsilon_4 & \leftrightarrows & -\Xi^*\varepsilon_3, \end{array} \qquad \begin{array}{lll} \\ \varepsilon_{10} & \leftrightarrows & \varepsilon_6 \\ \varepsilon_8 & \leftrightarrows & -\Xi\varepsilon_8 \\ \varepsilon_1 & \leftrightarrows & \varepsilon_2. \end{array}$$

Once again $\dim M_\varphi = 2$ eliminates (1), (3), (5), (7), (9), (11); and $\mathcal{G}$-stability kills (4), (8), (12). We are left with:

II(a)	$\langle \varepsilon_1, \varepsilon_2 \rangle$
II(b)	$\langle \varepsilon_7, \varepsilon_8, \varepsilon_1, \varepsilon_2 \rangle$
II(c)	$\langle \varepsilon_3, \varepsilon_4, \varepsilon_1, \varepsilon_2 \rangle$
II(d)	$\langle \varepsilon_5, \varepsilon_6, \varepsilon_9, \varepsilon_{10}, \varepsilon_1, \varepsilon_2 \rangle$.

Case (III). $\mathcal{G} = \{1, \rho_1, \rho_2, \rho_1\rho_2 = \rho\} \cong \mathbb{Z}_2 \times \mathbb{Z}_2$, where ρ_i is complex conjugation on K_i fixing K_{2-i}; (VII.C.9) sends $\rho_1 \longrightarrow (14)$ and $\rho_2 \longrightarrow (23)$. Both ρ_i send

$$\Xi \leftrightarrows -\Xi, \qquad \Xi^* \leftrightarrows -\Xi^*.$$

Under ρ_1 we therefore have

$$\varepsilon_9 \leftrightarrows \varepsilon_{10}, \qquad \varepsilon_7 \leftrightarrows -\Xi^*\varepsilon_4,$$

$\varepsilon_2, \varepsilon_5, \varepsilon_6$ fixed,

$$\varepsilon_3 \leftrightarrows -\Xi\varepsilon_8, \qquad \varepsilon_1 \leftrightarrows -\varepsilon_1.$$

That $\mathrm{Lie}(M_\varphi) = \langle \varepsilon_1, \varepsilon_2 \rangle$ again kills (1)–(11) odd, and $\mathcal{G}$-stability (6), (10), (12) — we have (2), (4), (8), (13) left:

III(a)	$\langle \varepsilon_1, \varepsilon_2 \rangle$
III(b)	$\langle \varepsilon_9, \varepsilon_{10}, \varepsilon_1, \varepsilon_2 \rangle$
III(c)	$\langle \varepsilon_5, \varepsilon_6, \varepsilon_1, \varepsilon_2 \rangle$
III(d)	$\langle \varepsilon_5, \varepsilon_6, \varepsilon_9, \varepsilon_{10}, \varepsilon_1, \varepsilon_2 \rangle .$

Case (IV). $\mathcal{G} \cong \mathbb{Z}_2$, so the action on the ε_i's (by ρ) gives no $\mathcal{G}$-stability constraints not already covered by the $\mathcal{V}^{-i} = \overline{\mathcal{V}^i}$ requirement in (VII.C.8). The *only* constraint is that, since

$$\mathrm{Lie}(M_\varphi) = \langle 3\varepsilon_1 + \varepsilon_2 \rangle ,$$

$\mathcal{V}^0$ must contain this and should *equal this* (if possible, fixing all other $\mathcal{V}^i$'s) — which eliminates (2), (3), (5), (7), (9), and (12). We have left:

IV(a)	$\langle 3\varepsilon_1 + \varepsilon_2 \rangle$	
IV(b)	$\langle \varepsilon_9, \varepsilon_{10}, \varepsilon_1, \varepsilon_2 \rangle$	
IV(c)	$\langle \varepsilon_8, \varepsilon_7, \varepsilon_1, \varepsilon_2 \rangle$	
IV(d)	$\langle \varepsilon_5, \varepsilon_6, \varepsilon_1, \varepsilon_2 \rangle$	
IV(e)	$\langle \varepsilon_3, \varepsilon_4, \varepsilon_1, \varepsilon_2 \rangle$	
IV(f)	$\langle A\varepsilon_4 + B\varepsilon_6, \overline{A}\varepsilon_3 + \overline{B}\varepsilon_5, \zeta\varepsilon_1 + \varepsilon_2 \rangle ,$	$\frac{3\lvert B\rvert^2}{4\lvert A\rvert^2} = \Xi$
IV(g)	$\langle \varepsilon_5, \varepsilon_6, \varepsilon_9, \varepsilon_{10}, \varepsilon_{11}, \varepsilon_2 \rangle .$	

The four boxed sets of possibilities for $\mathcal{V}$ (for ϕ of each type (I)–(IV)) *almost* complete the second step of the "algorithm" in weight 3. What we have not considered, and what does not enter except for IV(f), is the requirement that the $\mathcal{V}^i$ be defined over L^c ($= K$ in this case).

VII.D EXISTENCE OF DOMAINS OF TYPE IV(F)

In the next two sections (continuing with $n = 3$) we address whether A and B can be chosen to satisfy this constraint, and "complete the algorithm" by finding the Hodge tensor(s) for this type. The underlying philosophy here is that there are different polarizations Q that can be put on the *reducible* CM Hodge structure φ. The choice of Q does not affect $M_\varphi(\cong T^1)$, but does determine the period domains D in which we are trying to "deform" φ. More precisely, it does not affect a certain 4-tensor $\mathfrak{t}^4$ (to be determined in Section VII.E) associated to φ, but *does* affect whether a nontrivial "deformation" of φ exists that simultaneously preserves Q and $\mathfrak{t}^4$ (as Hodge tensors).

Without loss of generality we can rewrite

IV(f) $\langle \varepsilon_4 + \mu\varepsilon_6, \varepsilon_3 + \overline{\mu}\varepsilon_5, 3\varepsilon_1 + \varepsilon_2 \rangle$

where $\mu = a + b\sqrt{-d} \in \mathbb{Q}(\sqrt{-d}) = K$ satisfies

(VII.D.1) $\frac{3}{4}\mu\overline{\mu} = \Xi \ (\in \mathbb{Q}).$

It is not always the case that such a μ exists! We first treat well-definedness of the question.

(VII.D.2) LEMMA: *Let $V = V^n_{(L,\Pi)}$ be a SCMpHS (say, of odd weight, and rank $2g$), and fix an ordering of the $\{\theta_i\}$ inside each $\Pi^{p,q}$. Then the polarization invariants $Q_i := \sqrt{-1}Q(\omega_i, \overline{\omega}_i)$ depend only on the choice of $\beta(1)$, and so are well-defined up to a transformation of the form*

$$(Q_1, \ldots, Q_g) \to \left(|\theta_1(\ell)|^2 Q_1, \ldots, |\theta_g(\ell)|^2 Q_g\right), \qquad \ell \in L^*.$$

PROOF/CLARIFICATION. The point is that $\eta : L \hookrightarrow \mathrm{End}(V, \varphi)$ and $Q : V \times V \to \mathbb{Q}$ are fixed, and so the choice of $\beta(1)$ determines $\beta : L \xrightarrow{\cong} V$, which in turn determines the $\{\omega_i\}$ via $\beta(\ell) = \sum_{i=1}^{2g} \theta_i(\ell)\omega_i$. Recall that Q is determined by a choice of $\xi = \sqrt{\delta} \in L$ and $\delta \in L^p$ such that $\sqrt{-1}^{\,p_i - q_i}\theta_i(\xi) > 0$ for all i, with $Q(\bullet, \bullet) = \mathrm{Tr}_{L/\mathbb{Q}}\Big(\xi \cdot \beta^{-1}(\bullet) \cdot \rho(\beta^{-1}(\bullet))\Big)$; in fact, $Q(\omega_i, \overline{\omega_i}) = \theta_i(\xi)$. But once such a Q is *fixed*, if the choice of $\beta(1)$ is changed (to the former $\beta(\ell)$), then the Q_i undergo the transformation described in the lemma. □

In case (IV), we wish to think of Ξ as an invariant of the direct sum of the two $g = 1$ SCMpHS $V^3_{(K,\Pi_1)} \oplus V^3_{(K,\Pi_2)}$. Clearly, Q_1 and Q_2 are each well-defined up to multiplication by norms of *independent* $k_1, k_2 \in K = \mathbb{Q}(\sqrt{-d})$.

(VII.D.3) COROLLARY: *In case IV(f), (a) Ξ is well-defined up to multiplication by $k\overline{k}$, for $k \in K$; and (b) The question of the existence of $\mu \in K$ satisfying* (VII.D.1) *is well-defined.*

In fact, we can give a necessary and sufficient criterion for μ's existence, and hence for the existence of a type IV(f) Mumford-Tate domain through φ, in terms of the ideal class group of K:

(VII.D.4) PROPOSITION: *Write $\frac{4}{3}\Xi = \frac{1}{D^2}\prod p_j$, where $D \in \mathbb{Q}_{>0}$ and $\{p_j\}$ are* **distinct** *primes in $\mathbb{N}$. Then there exists a solution $\mu \in K$ to $\mu\overline{\mu} = \frac{4}{3}\Xi \iff$ none of the (p_j) are inert*[6] *in $\mathcal{O}_K$ and, writing $(p_j) = \mathcal{P}_j\overline{\mathcal{P}_j}$, $\prod[\mathcal{P}_j]$ is a square in $C\ell(K)$.*

---

[6] *In practice, what this means is: $-d$ is a square mod p_j for each odd p_j; and if 2 is among the $\{p_j\}$, then also $-d \underset{(8)}{\not\equiv} 5$. (In particular, for the odd p_j, if $-d \underset{(p_j)}{\equiv} m^2$ then $\mathcal{P}_j = (p_j, m \pm \sqrt{-d})$.)*

PROOF. ($\Leftarrow$): $\prod[\mathcal{P}_j] = [\mathcal{Q}]^2$ ($\mathcal{Q} \subset \mathcal{O}_K$ ideal) $\Rightarrow$ setting $\mathcal{J} = \prod \mathcal{P}_j =: (\mu_0)\mathcal{Q}^2$ ($\mu_0 \in K$), we have $\prod(p_j) = \mathcal{J}\overline{\mathcal{J}} = (\mu_0)(\overline{\mu_0})\mathcal{Q}^2\overline{\mathcal{Q}}^2 = (\mu_0)(\overline{\mu_0})(\mathfrak{z})^2$ ($\mathfrak{z} \in \mathbb{Z}$) hence $\prod p_j = \mu_0\overline{\mu_0}\mathfrak{z}^2$ (since the only units in $\mathbb{Z}$ are ± 1 and both sides are > 0). Putting $\mu \in \mu_0\mathfrak{z}/D$, we are done.

($\Rightarrow$): Write $D = \prod e_m / \prod f_r = \frac{\prod e'_m \prod e''_p}{\prod f'_s \prod f''_t}$ and $\prod p_j = \prod'_k p'_k \prod p''_f$ where p'', e'', f'' are inert in $\mathcal{O}_k$ while p', e', f' ramify or split. Expanding $(\mu)(\overline{\mu})\prod(e_m) = \prod(p_j)\prod(f_r)$ into irreducibles (except $(\mu), (\overline{\mu})$),

$$(\mu)(\overline{\mu})\prod \mathcal{E}_n^2\overline{\mathcal{E}}_n^2 \prod (e''_p)^2 = \prod \mathcal{P}_k\overline{\mathcal{P}_k}\prod(p''_\ell)\prod \mathcal{F}_s^2\overline{\mathcal{F}_s}^2\prod(f''_t)^2.$$

By unique factorization of ideals, we find that (μ) must be a product of $\prod_k\{\mathcal{P}_k$ or $\overline{\mathcal{P}_k}\}$ by a square of a fractional ideal by some $\mathcal{J}\overline{\mathcal{J}} = (q)$, $q \in \mathbb{Q}$; and there can be no (p''_ℓ)'s. So in $C\ell(k)$ we have $1 = [(\mu)] = \prod\{[\mathcal{P}_k]$ or $[\overline{\mathcal{P}_k}]\} \cdot (\text{square}) = \prod\{[\mathcal{P}_k]$ or $[\mathcal{P}_k]^{-1}\} \cdot (\text{square}) = (\prod[\mathcal{P}_k]) \cdot (\text{square})$. □

VII.E CHARACTERIZATION OF DOMAINS OF TYPE IV(A) AND IV(F)

The next step is to compute sets of Hodge tensors cutting out each of the Mumford-Tate groups/Lie algebras I(a)–IV(g). Before doing this we recast some basic ideas from earlier on. Let

$$G \overset{\rho}{\hookrightarrow} \mathrm{GL}(V)$$

be a faithful representation and

$$\mathfrak{g} \overset{d\rho}{\hookrightarrow} \mathrm{End}(V)$$

the corresponding Lie algebra representation; they give rise to tensor representations

$$G \overset{\rho^{m,n}}{\hookrightarrow} \mathrm{GL}(T^{m,n}V)$$
$$\mathfrak{g} \overset{d\rho^{m,n}}{\hookrightarrow} \mathrm{End}(T^{m,n}V).$$

Writing $f : V^{\otimes n} \to V^{\otimes m}$, the latter are given by

$$\left(\rho^{m,n}(\gamma)\right)(f) = \rho(\gamma)^{\otimes m} \circ f \circ \left(\rho(\gamma)^{-1}\right)^{\otimes n}$$

and

$$\text{(VII.E.1)} \qquad \left(d\rho^{m,n}(g)\right)(f) = \sum_{j=1}^{n} \iota_j(d\rho(g)) \circ f - \sum_{k=1}^{m} f \circ \iota_k(d\rho(g))$$

where $\iota_j(g)$ means "operation on the jth factor."

In our setup,[7] $G = \mathrm{Aut}(V, Q)$ and so $\boldsymbol{\rho}$ is understood for G and all its subgroups M. Henceforth, the left-hand side of (VII.E.1) will just be written "$g(f)$." Let φ be a CM Hodge structure as above, and M be one of the Mumford-Tate groups intermediate between M_φ and G, so that

$$\mathfrak{m}_\varphi \subseteq \mathfrak{m} \subseteq \mathfrak{g}.$$

We have the Lie algebra representation maps

$$\mathfrak{m} \times T^{m,n}V \to T^{m,n}V$$

whose right kernel defines $\mathrm{Hg}_M^{m,n}$ for each m, n. Since $\mathfrak{m} \supset \mathfrak{m}_\varphi$, these are necessarily subspaces of $(T^{m,n}V)^{(0,0)}$, where the type is with respect to $\mathrm{Ad}\,\varphi$. (In particular, $\mathrm{Hg}_\varphi^{1,1} = E_\varphi$.) An *effective generating set* $\boldsymbol{\tau} = \{\tau_\alpha\}$ for $\mathrm{Hg}_M^{\bullet,\bullet}$ is a set of $\mathbb{Q}$-tensors $\tau_\alpha \in \mathrm{Hg}_M^{m_\alpha,n_\alpha}$ including Q, such that the left-kernel of

$$\boldsymbol{\rho}_{\boldsymbol{\tau}} : \mathfrak{g} \times \mathbb{C}\langle\boldsymbol{\tau}\rangle \to \bigoplus_{m,n} T^{m,n}V$$

is precisely $\mathfrak{m}$. For *any* set of M-Hodge tensors the left-kernel of such a map is a sub-Lie algebra and sub-Hodge structure (with respect to $\mathrm{Ad}\,\varphi$) of $\mathfrak{g}$. So in general, we only need to verify that $\boldsymbol{\rho}_{\boldsymbol{\tau}}$'s left-kernel is not one of the larger $\mathfrak{m}$'s in our classification.

We begin with the most interesting cases IV(a) and IV(f), which are the only ones whose generating sets of Hodge tensors must involve tensors of degree > 2. For IV(a), $K \times K \cong \mathrm{Hg}_\varphi^{1,1}$ together with $Q \in \mathrm{Hg}_\varphi^{0,2}$ cut out only $\langle \varepsilon_1, \varepsilon_2 \rangle$. Now defining

$$\mathrm{Sym}^4 V_{\mathbb{C}} \subseteq T^{4,0}V_{\mathbb{C}}$$

to be the span of tensors of the form

$$\omega_{i_1}\omega_{i_2}\omega_{i_3}\omega_{i_4} := \frac{1}{4!}\sum_{\sigma \in \ell_4} \omega_{i_{\sigma(1)}} \otimes \omega_{i_{\sigma(2)}} \otimes \omega_{i_{\sigma(3)}} \otimes \omega_{i_{\sigma(4)}},$$

we observe that

$$\begin{aligned}
&(\mathrm{Sym}^4 V_{\mathbb{C}}) \cap (T^{4,0}V_{\mathbb{C}})^{(0,0)} \\
&\quad = \mathbb{C}\left\langle \omega_1^2\omega_4^2, \omega_1\omega_3^3, \omega_1\omega_2\omega_3\omega_4, \omega_2^3\omega_4, \omega_2^2\omega_3^2 \right\rangle \\
&\quad = \mathbb{C}\underbrace{\left\langle \omega_1^2\omega_4^2, \frac{\omega_1\omega_3^3 + \omega_2^3\omega_4}{2}, \omega_1\omega_2\omega_3\omega_4, \frac{\omega_2\omega_3^3 - \omega_2^3\omega_4}{2\sqrt{-d}}, \omega_2^2\omega_3^2 \right\rangle}_{\mathbb{Q}\text{-tensors}} \\
&\quad = (\mathrm{Sym}^4 V \cap \mathrm{Hg}_\varphi^{4,0})_{\mathbb{C}}.
\end{aligned}$$

---

[7] In the present notation, $\mathrm{Aut}(V, Q)$ consists of all $\gamma \in \mathrm{GL}(V)$ with $(\boldsymbol{\rho}^{0,2}(\gamma))Q = Q$.

As ε_1 sends $\{\omega_1^2\omega_4^2, \omega_1\omega_2\omega_3\omega_4, \omega_2^2\omega_3^2\} \mapsto 0$ and $\omega_1\omega_3^3 \mapsto \omega_1\omega_3^3$, $\omega_2^3\omega_4 \mapsto -\omega_2^3\omega_4$, we see that any $\mathbb{Q}$-linear combination $\mathfrak{t}$ of the five 4-tensors just listed will suffice to complete our effective generating set provided the coefficients in $\mathfrak{t}$ of $\omega_1\omega_3^3$ and $\omega_2^3\omega_4$ are not 0. We write

$$\boldsymbol{\tau}_{\text{IV(a)}} = \{Q, K \times K, \mathfrak{t}\}.$$

Now let $\mathfrak{m}$ (and hence $M = \overline{\exp(\mathfrak{m})}$) be as in IV(f). Since $\varepsilon_4 + \mu\varepsilon_6 = \widehat{\omega}_{43} + \Xi\widehat{\omega}_{21} + \mu\widehat{\omega}_{32}$ sends (via ad)

$$\left.\begin{array}{lll} \widehat{\omega}_{11} & \mapsto & \Xi\widehat{\omega}_{21} \\ \widehat{\omega}_{22} & \mapsto & \mu\widehat{\omega}_{32} - \Xi\widehat{\omega}_{21} \\ \widehat{\omega}_{33} & \mapsto & \widehat{\omega}_{43} - \mu\widehat{\omega}_{32} \\ \widehat{\omega}_{44} & \mapsto & -\widehat{\omega}_{43} \end{array}\right\} \begin{array}{l}\text{span a} \\ \text{3-dimensional} \\ \text{space,}\end{array}$$

$\text{Hg}_M^{1,1}$ is just the span of id:$= \sum_{j=1}^4 \widehat{\omega}_{jj}$ (which corresponds to $(1,1) \in K \times K$). We claim that there is (up to scale) a unique choice of $\mathfrak{t} \in \text{Hg}_\varphi^{4,0}$ belonging also to $\text{Hg}_M^{4,0}$, and that this $\mathfrak{t}$ completely cuts out M. Writing

$$\mathfrak{t} = A\omega_1\omega_2\omega_3\omega_4 + B\omega_1^2\omega_4^2 + C\omega_2^2\omega_3^2 + (D\mu + E)\omega_1\omega_3^3 + (D\overline{\mu} + E)\omega_2^3\omega_4,$$

we must find A, B, C, D, E so that[8]

$$\begin{aligned} 0 &= (\varepsilon_4 + \mu\varepsilon_6)\mathfrak{t} \\ &= \{A + 2\Xi B\}\omega_1\omega_2\omega_4^2 + \{\Xi A + 2C + \overbrace{3\mu\overline{\mu}}^{4\Xi} D + 3\mu E\}\omega_2^2\omega_3\omega_4 \\ &\quad + \{\mu A + 3\mu D + 3E\}\omega_1\omega_3^2\omega_4 + \{2\mu C + \Xi\mu D + \Xi E\}\omega_2\omega_3^3. \end{aligned}$$

Setting the bracketed quantities to 0 and solving, we get[9]

$$\text{(VII.E.2)} \qquad \mathfrak{t} = -3\omega_1\omega_2\omega_3\omega_4 + \tfrac{3}{2\Xi}\omega_1^2\omega_4^2 - \tfrac{\Xi}{2}\omega_2^2\omega_3^2 + \mu\omega_1\omega_3^3 + \overline{\mu}\omega_2^3\omega_4.$$

(This ties existence of $\mathfrak{t}$ to existence of μ, as it should.) Moreover, every other Hodge-Lie algebra in our list contains a vector $\langle \varepsilon_1, \varepsilon_2 \rangle$ other than $3\varepsilon_1 + \varepsilon_2$. As

$$\varepsilon_1(\mathfrak{t}) = \mu\omega_1\omega_3^3 - \overline{\mu}\omega_2^3\omega_3 \neq 0,$$

$\mathfrak{t}$ can only cut out $\mathfrak{m}$. We conclude that $\text{Hg}_M^{4,0} = \mathbb{Q}\langle \mathfrak{t} \rangle$ and

$$\boldsymbol{\tau}_{\text{IV(f)}} = \langle Q, \mathfrak{t} \rangle.$$

[8] Once this is satisfied, $(\varepsilon_3 + \overline{\mu}\varepsilon_5)\mathfrak{t} = 0$ is automatic.

[9] Up to scale: this particular $\mathfrak{t}$ may not be rational, but some multiple of it is. We will not worry about the difference (but see Section VII.H for a precise example).

VII.F COMPLETION OF THE CLASSIFICATION FOR WEIGHT 3

We first characterize the remaining types by (effective generators of) their Hodge tensors. The identification $\beta : L \xrightarrow{\cong} V_{\mathbb{Q}}$ leads to an embedding

$$\boldsymbol{\beta} : L \hookrightarrow \operatorname{End}(V_{\mathbb{Q},\varphi}) =: E_{\varphi}$$
$$[\boldsymbol{\beta}(\ell)]_{\omega} = \operatorname{diag}\{\theta_i(\ell)\}_{i=1}^{4},$$

which in the irreducible cases (I) and (II) is an isomorphism. As in Section VI.D we can consider, for any (normal[10]) subfield $K \subset L$, the commutant $M_K := G^{\boldsymbol{\beta}(K)}$ — the orbit of φ under which yields a "weak-CM-by-K" Mumford-Tate domain. Its Lie algebra $\mathfrak{m}_K$ is simply the left-kernel of $\rho_{\boldsymbol{\beta}(K)}$, which is to say (since $\boldsymbol{\beta}(K) \subset T^{1,1}V_{\mathbb{Q}}$)

$$\mathfrak{m}_K = \bigcap_{k \in K} \ker(\operatorname{ad}(\boldsymbol{\beta}(k))) \subset \mathfrak{g}.$$

This completely describes the types I(a)–II(d).

Case (I). Other than $\mathbb{Q}$, the only proper subfield of L is the real quadratic one $L_0 (= L^{\rho})$. We have

$$\mathfrak{m}_{L_0} = \ker \operatorname{ad} \begin{pmatrix} \mu & & & \\ & \varsigma & & \\ & & \varsigma & \\ & & & \mu \end{pmatrix}$$

for general $\mu (= \theta_1(k))$, $\varsigma (= \theta_2(k))$, which gives I(b).

Case (II). This time (since $\mathbb{Z}_2 \times \mathbb{Z}_2$ has 3 order-2 subgroups) L has subfields $L_0 = \mathbb{Q}(\xi + \bar{\xi})$, $K_1 = L^{\eta} = \mathbb{Q}(\xi + \tilde{\xi})$, and $K_2 = L^{\eta\rho} = \mathbb{Q}(\xi + \bar{\tilde{\xi}})$. Once again $\mathfrak{m}_{L_0}$ is II(d), while

$$\mathfrak{m}_{K_1} = \ker \operatorname{ad} \begin{pmatrix} \mu & & & \\ & \mu & & \\ & & \varsigma & \\ & & & \varsigma \end{pmatrix} \text{ resp. } \mathfrak{m}_{K_2} = \ker \operatorname{ad} \begin{pmatrix} \mu & & & \\ & \varsigma & & \\ & & \mu & \\ & & & \varsigma \end{pmatrix}$$

yields II(c) resp. II(b). Notice that II(c) lives in $\mathfrak{g}^{(-1,1)} \oplus \mathfrak{g}^{(0,0)} \oplus \mathfrak{g}^{(1,-1)}$, hence it is "classical" while II(b) is not. Continuing on, we consider

Cases (III) and (IV). Observe that for the types III(a) and IV(a), we have $V = V' \oplus V''$ as a (CM) Hodge structure, where $V'_{\mathbb{C}} = V^{(3,0)} \oplus V^{(0,3)}$ resp. $V''_{\mathbb{C}} = V^{(2,1)} \oplus V^{(1,2)}$ are spanned by ω_1, ω_4 resp. ω_2, ω_3. This splitting corresponds to $\mathfrak{m}$ killing $\operatorname{diag}\{1,0,0,1\}$ and $\operatorname{diag}\{0,1,1,0\}$. In fact, from

[10] This is automatic since $\mathcal{G}$ is abelian.

(VII.C.4) we see that $\varepsilon_1, \varepsilon_2, \varepsilon_5, \varepsilon_6, \varepsilon_9, \varepsilon_{10}$ *all* kill (commute with) these two matrices; and so types III(b), (c), (d) and IV(b), (d), (g) are *all* Mumford-Tate Lie algebras of Hodge structure on V with the same decomposition (but with one or both sub-Hodge structures no longer CM). To deal with the remaining types, first note that both $\text{diag}\{\sigma_1(k), \sigma_1(k), \sigma_2(k), \sigma_2(k)\}$ respectively $\text{diag}\{\sigma_1(k), \sigma_2(k), \sigma_1(k), \sigma_2(k)\}$ $(k \in K)$ belong to E_φ. Their commutants in $\mathfrak{g}$ are types IV(e) resp. IV(c), which are therefore both of "weak-CM-by-K" type.

We have essentially identified generating sets of Hodge tensors for all types in Section VII.C, and recast these findings in the following

(VII.F.1) THEOREM: *Let D classify Hodge structures on V with $h^{3,0} = h^{2,1} = h^{1,2} = h^{0,3} = 1$ and polarized by a given nondegenerate alternating form Q. The Mumford-Tate Noether-Lefschetz locus component types with generic polarized Hodge structure* **irreducible** *are:*[11]

(i) D

(ii) [I(b), II(d)] *polarized Hodge structures with weak CM by a real quadratic field*

(iii) [II(c), IV(e)] *polarized Hodge structures with weak CM by an $\underbrace{\textit{imaginary quadratic field}}_{IQF}$ (unconstrained)*

(iv) [II(b), IV(c)] *polarized Hodge structures with weak CM by an IQF (constrained)*

(v) [IV(f)] *polarized Hodge structures with a certain Hodge* 4*-tensor* (VII.E.2)

(vi) [I(a)] *CM Hodge structure with* $\text{Gal}(L/\mathbb{Q}) \cong \mathbb{Z}_4$

(vii) [II(a)] *CM Hodge structure with* $\text{Gal}(L/\mathbb{Q}) \cong \mathbb{Z}_2 \times \mathbb{Z}_2$.

Mumford-Tate Noether-Lefschetz locus component types with generic polarized Hodge structure $V =_{HS} \underset{\substack{(3,0)\\+\\(0,3)}}{V'} \oplus \underset{\substack{(2,1)\\+\\(1,2)}}{V''}$ **reducible** *are:*

(viii) [III(d), IV(g)] V', V'' *both general*

(ix) [III(b), IV(b)] V' *general,* V'' *CM (by IQF)*

[11] *Indicated in brackets is the Lie algebra type of the corresponding Mumford-Tate group.*

(x) [III(c), IV(d)] V' *CM (by IQF),* V'' *general*

(xi) [III(a)] V', V'' *CM by distinct IQFs*

(xii) [IV(a)] V', V'' *CM by same IQF.*

(VII.F.2) **Remark:** The theorem classifies types of Noether-Lefschetz locus *components*; i.e., Mumford-Tate domains D_M. In the *full Noether-Lefschetz loci*, types (iii) and (iv) get mixed, as do types (ix) and (x). For D_M of type (iii) *or* (iv) (resp. (ix) or (x)), $\breve{\mathrm{NL}}_M$ consists of *two components each* of types (iii) and (iv) (resp. (ix) and (x)). In cases:

- (ix) or (x): one component of each type meets D (the set of Q-polarized Hodge structures);
- (iii): the two type (iii) components (only) meet D;
- (iv): one type (iv) component is contained entirely in D (and the others do not meet D).

Associated to the Mumford-Tate domain/group-types in Theorem (VII.F.1) are the following data:

(VII.F.3)

type	unconstrained? /Herm. symm.?	ht(M)	M	$M(\mathbb{R})^0$
(i)	no/no	2	Sp_4	$\mathrm{Sp}_4(\mathbb{R})$
(ii)	no/yes	2	$\mathrm{Res}_{\mathbb{Q}(\sqrt{d})/\mathbb{Q}}\,\mathrm{SL}_{2,\mathbb{Q}(\sqrt{d})}$	$\mathrm{SL}_2(\mathbb{R})\times\mathrm{SL}_2(\mathbb{R})$
(iii)	yes/yes	2	$U_{\mathbb{Q}(\sqrt{-d})}(V,\mathfrak{Q})$	$\left\{ U(1,1)\cong U(1)\times\mathrm{SL}_2(\mathbb{R}) \right.$
(iv)	no/no	2	$U_{\mathbb{Q}(\sqrt{-d})}(V,\mathfrak{Q})$	$U(2)$
(v)	yes/yes	4	SL_2	$\mathrm{SL}_2(\mathbb{R})$
(vi)	yes/yes	2	$\mathrm{Res}_{L_0/\mathbb{Q}}\,U_L$	$U(1)\times U(1)$
(vii)	yes/yes	2	$\mathrm{Res}_{L_0/\mathbb{Q}}\,U_L$	$U(1)\times U(1)$
(viii)	no/yes	2	$\mathrm{SL}_2\times\mathrm{SL}_2$	$\mathrm{SL}_2(\mathbb{R})\times\mathrm{SL}_2(\mathbb{R})$
(ix)	no/yes	2	$U_{\mathbb{Q}(\sqrt{-d})}\times\mathrm{SL}_2$	$U(1)\times\mathrm{SL}_2(\mathbb{R})$
(x)	yes/yes	2	$U_{\mathbb{Q}(\sqrt{-d})}\times\mathrm{SL}_2$	$U(1)\times\mathrm{SL}_2(\mathbb{R})$
(xi)	yes/yes	2	$U_{\mathbb{Q}(\sqrt{-d'})}\times U_{\mathbb{Q}(\sqrt{-d''})}$	$U(1)\times U(1)$
(xii)	yes/yes	4	$U_{\mathbb{Q}(\sqrt{-d})}$	$U(1)$

where "M" means "isomorphism class as a $\mathbb{Q}$-algebraic group" (and $M(\mathbb{R})^0$ is the real Lie group). For a CM field L (with involution ρ and fixed field L_0 as in (VII.B.1), U_L is the L_0-algebraic group with L_0-points $\{l \in L \mid l \cdot \rho(l) = 1\}$. In (iii)–(iv), $\mathfrak{Q} : V \times V \to \mathbb{Q}(\sqrt{-d})$ is a $\mathbb{Q}(\sqrt{-d})$-Hermitian form with $Q = \mathrm{tr}_{\mathbb{Q}(\sqrt{-d})/\mathbb{Q}}(\mathfrak{Q})$, and the resulting unitary group is algebraic over $\mathbb{Q}$. (On the Lie group side, $U(1)$ is just the unit circle in $\mathbb{C}^*$.)

Perhaps more naturally associated to the Noether-Lefschetz-locus types are the remaining invariants:

(VII.F.4)

type	$\dim(\mathrm{NL}_M)$	$\pi_0(\mathrm{NL}_M)$	$\pi_0(\breve{\mathrm{NL}}_M)$	E_φ
(i)	4	1	1	$\mathbb{Q}$
(ii)	2	2	2	$\mathbb{Q}(\sqrt{d})$
(iii)/(iv)	1	2/1	4	$\mathbb{Q}(\sqrt{-d})$
(v)	1	1	1	$\mathbb{Q}$
(vi)	0	2	8	L
(vii)	0	2	8	L
(viii)	2	2	2	$\mathbb{Q} \times \mathbb{Q}$
(ix)/(x)	1	2	4	$\mathbb{Q} \times \mathbb{Q}(\sqrt{-d})$
(xi)	0	2	8	$\mathbb{Q}(\sqrt{-d'}) \times \mathbb{Q}(\sqrt{-d''})$
(xii)	0	1	2	$\mathbb{Q}(\sqrt{-d}) \times \mathbb{Q}(\sqrt{-d})$.

In this, the π_0-columns are computed via Theorem (VI.B.12), and E_φ means the endomorphism algebra for a *generic* $\varphi \in \mathrm{NL}_M$.

(VII.F.5) **Remark:** As in Section VI.D, we can define the field of definition $\mathbb{Q}(\check{D}_M)$ of the component $\check{D}_M = \check{M}(\mathbb{C})\varphi$ of $\breve{\mathrm{NL}}_M$, as the fixed field of all elements $\sigma \in \mathrm{Gal}(\mathbb{C}/\mathbb{Q})$ preserving $\check{D}_M$. Taking an L-CM point $\varphi_0 \in \check{D}_M$, this is easily computed as the fixed field of all $\sigma \in \mathrm{Gal}(L^c/\mathbb{Q})$ preserving the $M(\mathbb{C})$-conjugacy class of the co-character $\chi_0 = \varphi_0 \circ \mu^{-1}$. This yields the results

type	$\mathbb{Q}(\check{D}_M)$
(i)	$\mathbb{Q}$
(ii)	$\mathbb{Q}$
(iii)/(iv)	$\mathbb{Q}(\sqrt{-d})$
(v)	$\mathbb{Q}$
(vi)	L
(vii)	L
(viii)	$\mathbb{Q}$
(ix)/(x)	$\mathbb{Q}(\sqrt{-d})$
(xi)	$\mathbb{Q}(\sqrt{-d'}, \sqrt{-d''})$
(xii)	$\mathbb{Q}(\sqrt{-d})$

VII.G THE WEIGHT 1 CASE

We begin from the following list of non-zero $\mathcal{V} \subset \mathfrak{g}$ satisfying

- $\mathcal{V} = \mathcal{V}^1 \oplus \mathcal{V}^0 \oplus \mathcal{V}^{-1}$, where $\mathcal{V}^i \subset \mathfrak{g}^{(i,-i)}$ and $\overline{\mathcal{V}^1} = \mathcal{V}^{-1}$;
- $[\mathcal{V}, \mathcal{V}] \subseteq \mathcal{V}$; and
- $\mathcal{V} \supseteq \langle \varepsilon_1 + \varepsilon_2 \rangle$:

1. $\langle \varepsilon_1 + \varepsilon_2 \rangle$
2. $\langle \varepsilon_1, \varepsilon_2 \rangle$
3. $\langle \alpha\varepsilon_3 + \overline{\alpha}\varepsilon_4 + ir\varepsilon_1, \varepsilon_1 + \varepsilon_2 \rangle$, $r \in \mathbb{R}$ and $\alpha \in \mathbb{C}^*$
4. $\big\langle \varepsilon_3 + \beta\varepsilon_1, \varepsilon_4 - \overline{\beta}\varepsilon_1, \varepsilon_1 + \varepsilon_2 \big\rangle$, $|\beta|^2 = \Xi$
5. $\langle \varepsilon_9, \varepsilon_{10}, \varepsilon_1, \varepsilon_2 \rangle$
6. $\langle \varepsilon_5, \varepsilon_6, \varepsilon_1, \varepsilon_2 \rangle$
7. $\langle \alpha\varepsilon_9 + \varepsilon_7 - \overline{\alpha}\varepsilon_5, \overline{\alpha}\varepsilon_{10} + \varepsilon_8 - \alpha\varepsilon_6, \varepsilon_1 + \varepsilon_2 \rangle$, $\alpha \in \mathbb{C}$
8. $\langle \varepsilon_7, \varepsilon_8, \varepsilon_1, \varepsilon_2 \rangle$
9. $\langle \varepsilon_9 + \gamma\varepsilon_5, \varepsilon_{10} + \overline{\gamma}\varepsilon_6, \varepsilon_1 + \varepsilon_2 \rangle$, $|\gamma|^2 = 1$
10. $\big\langle \varepsilon_9 + \frac{\overline{\alpha}}{\alpha}\varepsilon_5, \varepsilon_{10} + \frac{\alpha}{\overline{\alpha}}\varepsilon_6, \alpha\varepsilon_3 + \overline{\alpha}\varepsilon_4, \varepsilon_1 + \varepsilon_2 \big\rangle$, $\alpha \in \mathbb{C}^*$
11. $\Big\langle \frac{ir}{\overline{\alpha}}\varepsilon_9 + \varepsilon_7 + \frac{ir}{\alpha}\varepsilon_5, \frac{-ir}{\alpha}\varepsilon_{10} + \varepsilon_8 - \frac{ir}{\overline{\alpha}}\varepsilon_6, \alpha\varepsilon_3 + \overline{\alpha}\varepsilon_4 + \frac{2i|\alpha|^2}{r}\varepsilon_1, \varepsilon_1 + \varepsilon_2 \Big\rangle$, $r \in \mathbb{R}^*$ and $\alpha \in \mathbb{C}^*$
12. $\langle \varepsilon_5, \varepsilon_6, \varepsilon_9, \varepsilon_{10}, \varepsilon_1, \varepsilon_2 \rangle$
13. $\langle \varepsilon_9 + \alpha\varepsilon_7, \varepsilon_5 - \overline{\alpha}\varepsilon_7, \varepsilon_{10} + \alpha\varepsilon_8, \varepsilon_6 - \alpha\varepsilon_8, \varepsilon_1 + \overline{\alpha}\Xi^*\varepsilon_3 - \alpha\Xi^*\varepsilon_4, \varepsilon_1 + \varepsilon_2 \rangle$, $\alpha \in \mathbb{C}$
14. $\langle \varepsilon_9 + \alpha\varepsilon_7, \varepsilon_5 - \overline{\alpha}\varepsilon_7, \varepsilon_{10} + \overline{\alpha}\varepsilon_8, \varepsilon_6 - \alpha\varepsilon_8, \varepsilon_3 - 2\alpha\varepsilon_1, \varepsilon_4 + 2\overline{\alpha}\varepsilon_1, \varepsilon_1 + \varepsilon_2 \rangle$, $4|\alpha|^2 = \Xi$
15. $\mathfrak{g}$.

To narrow these down to the possible Lie algebras about CM points of type (I)–(III) (cf. (VII.B.5), $n = 1$), we impose

- closure under the action of $\mathcal{G}$ (which is the same as for $n = 3$, e.g., (VII.C.10) for case (I)),
- minimality of $\mathcal{V}^0 \supset \langle \varepsilon_1, \varepsilon_2 \rangle$ (or $\langle \varepsilon_1 + \varepsilon_2 \rangle$ in case (II)), and
- that $\mathcal{V}^i$ be defined over L.

For case (I) and (III) the results are *the same* as in Section VII.C: I(a)–(b) and III(a)–(d) exactly as for $n = 3$. In case (II) (which is really the $n = 1$ analog of case (IV)) we find:

$$\begin{array}{|ll|}\hline
\mathrm{II(a)} & \langle \varepsilon_1 + \varepsilon_2 \rangle \\
\mathrm{II(b)} & \langle \varepsilon_9, \varepsilon_{10}, \varepsilon_1, \varepsilon_2 \rangle \\
\mathrm{II(c)} & \langle \alpha\varepsilon_9 + \varepsilon_7 - \overline{\alpha}\varepsilon_5, \overline{\alpha}\varepsilon_{10} + \varepsilon_8 - \alpha\varepsilon_6, \varepsilon_1 + \varepsilon_2 \rangle, \alpha \in K \\
\mathrm{II(d)} & \langle \varepsilon_5, \varepsilon_6, \varepsilon_1, \varepsilon_2 \rangle \\
\mathrm{II(e)} & \langle \varepsilon_9 + \gamma\varepsilon_5, \varepsilon_{10} + \overline{\gamma}\varepsilon_6, \varepsilon_1 + \varepsilon_2 \rangle, \gamma \in K \text{ with } \gamma\overline{\gamma} = 1 \\
\mathrm{II(f)} & \big\langle \varepsilon_9 + \alpha\varepsilon_7, \varepsilon_5 - \overline{\alpha}\varepsilon_7, \varepsilon_{10} + \overline{\alpha}\varepsilon_8, \varepsilon_6 - \alpha\varepsilon_8, \\
 & \quad \varepsilon_1 + \overline{\alpha}\Xi^*\varepsilon_3 - \alpha\Xi^*\varepsilon_4, \varepsilon_1 + \varepsilon_2 \big\rangle, \alpha \in K, \\ \hline
\end{array}$$

where $K = \mathbb{Q}(\sqrt{-\delta})$ is the imaginary-quadratic field of definition of the CM point φ.

Now the isomorphism $M_2(K) \cong E_\varphi$ is given explicitly by the map

$$\begin{pmatrix} A & B \\ C & D \end{pmatrix} \\ \mapsto \underbrace{A\widehat{\omega}_{11} + B\widehat{\omega}_{12} + C\widehat{\omega}_{21} + D\widehat{\omega}_{22} + \overline{A}\widehat{\omega}_{44} + \overline{B}\widehat{\omega}_{43} + \overline{C}\widehat{\omega}_{34} + \overline{D}\widehat{\omega}_{33}}_{=:\mu}$$

and the endomorphism algebra in each subcase ($\mathfrak{m} =$) II(a)–(f) can be computed by finding which μ's bracket trivially with $\mathfrak{m}$. As subrings of $M_2(K)$ the results are:

II(a) $\quad M_2(K)$;

II(b) $\quad \left(\begin{smallmatrix} a & 0 \\ 0 & D \end{smallmatrix}\right)$, $a \in \mathbb{Q}$ and $D \in K$;

II(c) $\quad \left(\begin{smallmatrix} A & \Xi^*\overline{C} + 2\alpha i \operatorname{Im}(A) \\ C & \overline{A} - 2\operatorname{Re}(\alpha C) \end{smallmatrix}\right)$, $A, C \in K$;

II(d) $\quad \left(\begin{smallmatrix} A & 0 \\ 0 & d \end{smallmatrix}\right)$, $A \in K$ and $\alpha \in \mathbb{Q}$;

II(e) $\left(\begin{smallmatrix} a & b\bar{\eta} \\ c\eta & d \end{smallmatrix}\right)$, $\eta \in K$ with[12] $\xi/\bar{\eta} = \gamma$, $a, b, c, d \in \mathbb{Q}$; and

II(f) $\left(\begin{smallmatrix} a & \bar{\alpha}\Xi^*(a-d) \\ \alpha(a-d) & d \end{smallmatrix}\right)$, $a, d \in \mathbb{Q}$.

So for example, the II(e) endomorphism algebra is $\cong M_2(\mathbb{Q})$; while for II(f) we have $\mathbb{Q} \times \mathbb{Q}$ if $\Delta := 1 + 4|\alpha|^2\Xi^*$ is a square in $\mathbb{Q}$ ("$\Delta = \square$") and the *real quadratic field* $\mathbb{Q}(\sqrt{\Delta})$ otherwise. But II(c) requires a bit more thought.

Recall that the *rational quaternion algebras* $(x, y)_{\mathbb{Q}} =: \mathcal{Q}$ (here x and y are rational numbers) have generators $1, i, j, k$ subject to relations

$$i^2 = x, \quad j^2 = y, \quad ij = k = -ji$$

(which imply $k^2 = -xy$). Their isomorphism classes are the same as those of the quadratic forms over $\mathbb{Q}$ given by the norm

$$N_{\mathcal{Q}}(a + bi + cj + dk) = a^2 - xb^2 - yc^2 + xyd^2. \tag{VII.G.1}$$

In particular, $(x, y)_{\mathbb{Q}} \cong M_2(\mathbb{Q})$ (is split) if, and only if, (VII.G.1) represents zero over $\mathbb{Q}$. Indeed, this observation is valid over any field, and if $(x, y)_{\mathbb{R}} := (x, y)_{\mathbb{Q}} \otimes_{\mathbb{Q}} \mathbb{R}$ is split while $(x, y)_{\mathbb{Q}}$ is not, then the latter is said to be *indefinite*. It turns out that one can write down an explicit isomorphism between the II(c) endomorphism algebra above, and

$$\mathcal{Q} := (-\delta, |\alpha|^2 + \Xi^*)_{\mathbb{Q}},$$

which always splits over $\mathbb{R}$. It is split over $\mathbb{Q}$ exactly when $|\alpha|^2 + \Xi^*$ is the norm of an element of K, which can be decided as in (VII.D.4).

Everything is summarized neatly in the following result:

(VII.G.2) PROPOSITION: *The Mumford-Tate-Noether-Lefschetz locus-types for weight-1 rank-4 Hodge structure with polarizing form Q are (irreducible cases first):*

(i) D;

(ii) [I(b); II(f) *if* $\Delta \neq \square$] *polarized Hodge structures with weak CM by a real quadratic field* K';

(iii) [II(c) *if* $|\alpha|^2 + \Xi^* \notin \mathrm{Im}(N_{K/Q})$] *polarized Hodge structures with endomorphisms by an indefinite quaternion algebra* $\mathcal{Q}$;

[12] Such an η clearly always exists, either when $K = \mathbb{Q}(i)$ or $\mathbb{Q}(\zeta_3)$ (where there are nontrivial possibilities for γ) or otherwise (when $\gamma = \pm 1$).

(iv) [I(a)] *CM-Hodge structure with* $\mathrm{Gal}(L/\mathbb{Q}) \cong \mathbb{Z}_4$;

and (reducible) polarized Hodge structures of the form $V_1 \oplus V_2$ *with*

(v) [III(d); II(f) *if* $\Delta = \square$] *both general*;

(vi) [II(e); II(c)*if* $|\alpha|^2 + \Xi^* \in \mathrm{Im}(N_{K/\mathbb{Q}})$] $V_1 \cong V_2$ *(otherwise general)*;

(vii) [II(b), II(d), III(b), III(c)] V_1 *or* V_2 *CM by IQF* K *(the other general)*;

(viii) [III(a)] V_1 *and* V_2 *CM by distinct IQF's* K_1, K_2;

(ix) [II(a)] $V_1 \cong V_2$, *with CM by same IQF* K.

All are unconstrained and Hermitian symmetric, with $\pi_0(\mathrm{NL}_M) = 1$ and $ht(M) = 2$ in each case — i.e., E_φ and Q always give a generating set for the Hodge tensors (where φ is now generic in NL_M). We have the additional data:

type	dim	π_0	M	$M(\mathbb{R})$	E_φ	$\mathbb{Q}(\check{D}_M)$
(i)	3	1	Sp_4	$\mathrm{Sp}_4(\mathbb{R})$	$\mathbb{Q}$	$\mathbb{Q}$
(ii)	2	1	$\mathrm{Res}_{K'/\mathbb{Q}}\,\mathrm{SL}_{2,K'}$	$\mathrm{SL}_2(\mathbb{R}) \times \mathrm{SL}_2(\mathbb{R})$	K'	$\mathbb{Q}$
(iii)	1	1	$U_{\mathcal{Q}^{\mathrm{opp}}}$	$\mathrm{SL}_2(\mathbb{R})$	$\mathcal{Q}$	$\mathbb{Q}$
(iv)	0	4	$\mathrm{Res}_{L_0/\mathbb{Q}}\,U_L$	$U(1) \times U(1)$	L	L
(v)	2	1	$\mathrm{SL}_2 \times \mathrm{SL}_2$	$\mathrm{SL}_2(\mathbb{R}) \times \mathrm{SL}_2(\mathbb{R})$	$\mathbb{Q} \times \mathbb{Q}$	$\mathbb{Q}$
(vi)	1	1	SL_2	$\mathrm{SL}_2(\mathbb{R})$	$M_2(\mathbb{Q})$	$\mathbb{Q}$
(vii)	1	2	$\mathrm{SL}_2 \times U_K$	$\mathrm{SL}_2(\mathbb{R}) \times U(1)$	$\mathbb{Q} \times K$	K
(viii)	0	4	$U_{K_1} \times U_{K_2}$	$U(1) \times U(1)$	$K_1 \times K_2$	$K_1 K_2$
(ix)	0	2	U_K	$U(1)$	$M_2(K)$	K

where "dim"$=\dim(NL_M)$, "π_0"$=\pi_0(\widetilde{\mathrm{NL}}_M)$, and $U_{\mathcal{Q}^{\mathrm{opp}}}$ is a ($\mathbb{Q}$-) algebraic group with[13] $U_{\mathcal{Q}^{\mathrm{opp}}}(\mathbb{Q}) \cong \ker(N_{\mathcal{Q}})^{\mathrm{opp}}$. □

In fact, types (i)–(iv) could just have been read off of Example (2.7) in [Mo1],[14] but with some loss of detail. In case (ii), NL_M should be viewed as a pre-Hilbert modular surface.

(VII.G.3) **Remark:** It is well-known [Mi2] that for $n = 1$, the field of definition $\mathbb{Q}(\check{D}_M)$ coincides with that of the Shimura variety $\mathrm{Sh}_{K_f}(D, D_M)$, which contains an arithmetic quotient of D_M as an irreducible component.

[13]Here by $\ker(N_{\mathcal{Q}})$ we mean the elements of norm 1.

[14]Cases (ii), (iii), (iv) correspond to type (I), (II), and (IV) in the Albert classification.

The obvious similarities between the $n = 1$ and $n = 3$ classifications beg some discussion. Writing X for the $n = 1$ period domain from this section, and D for the $n = 3$ domain studied previously, there is a natural (but non-holomorphic) fibration

$$\begin{array}{ccc} D & \subset & \check{D} \\ \downarrow{\scriptstyle \pi} & & \downarrow{\scriptstyle \check{\pi}} \\ X & \subset & \check{X}. \end{array}$$

This is given by sending any co-character $\chi : \mathbb{G}_m \to \mathrm{Aut}(V, Q)$ with eigenvalues z^3, z, z^{-1}, z^{-3} to the one with eigenvalues z, z^{-1}, z, z^{-1} on respective 1-dimensional subspaces. Indicating weight by a subscript, one can show that $\check{\pi}$ restricts to 2:1-maps[15] of $\check{\mathrm{NL}}$-loci from type

$$\begin{aligned} (\mathrm{ii})_3 &\mapsto (\mathrm{ii})_1 \\ (\mathrm{vi})_3 &\mapsto (\mathrm{iv})_1 \\ (\mathrm{viii})_3 &\mapsto (\mathrm{v})_1 \\ (\mathrm{ix/x})_3 &\mapsto (\mathrm{vii})_1 \\ (\mathrm{xi})_3 &\mapsto (\mathrm{viii})_1 \end{aligned}$$

and an *isomorphism* from type

$$(\mathrm{xii})_3 \mapsto (\mathrm{ix})_1.$$

(This is discovered by letting the permutation of the $\{\varepsilon_i\}$ induced[16] by $\omega_2 \leftrightarrows \omega_3$ act on the Lie algebras I(a)–IV(g), replacing $\mathcal{V}^0$ by $\langle \varepsilon_1 + \varepsilon_2 \rangle$ and taking closure under $[\ ,\ ]$.) Outside those cases, the situation is less straightforward: e.g., for type (v)$_3$, sending a generic $\varphi \in D_M$ to $\pi(\varphi)$ destroys "Hodgeness" of the 4-tensor $\mathfrak{t}$. Consequently, $\pi(D_M)$ is not a Mumford-Tate domain at all, and $M_{\pi(\varphi)} = \mathrm{Sp}_4$. For type (vii)$_3$, $\check{\mathrm{NL}}_M$ maps (2-to-1) onto *two distinct* Noether-Lefschetz loci of type (ix)$_1$, corresponding to enlarging $E_\varphi \cong L$ to $M_2(L^\eta)$ or $M_2(L^{\eta\rho})$. Most bizarre of all, given a type (iii)$_3$ $\check{\mathrm{NL}}_M$, the components with trivial IPR map 2:1 onto a component of type

$$\begin{cases} (\mathrm{vi})_1 & \text{if } -\Xi^* \in \mathrm{im}(N_{\mathbb{Q}(\sqrt{-d})/\mathbb{Q}}) \\ (\mathrm{iii})_1 & \text{otherwise;} \end{cases}$$

while the components with nontrivial IPR are contracted to a (0-dimensional) locus of type (ix)$_1$. What is true in *all cases* where $\pi(\check{D}_M)$ is a Mumford-Tate domain, i.e., excluding (v)$_3$, is that $\mathbb{Q}(\check{D}_M)$ is a field extension of $\mathbb{Q}(\check{\pi}(\check{D}_M))$.

[15]Two connected components to one; isomorphisms on individual components.

[16]This also exchanges $\Xi \leftrightarrow -\Xi$.

VII.H ALGEBRO-GEOMETRIC EXAMPLES FOR THE NOETHER-LEFSCHETZ-LOCUS TYPES

The weight-one polarized Hodge structure-types delineated in (VII.G.2), are the H^1 of obvious irreducible abelian surfaces (types (i)–(iv)) and products of pairs of elliptic curves (types (v)–(ix)). In particular, cases (ii), (iii), and (iv) correspond respectively to abelian varieties of type (I), (II), and (IV) in the Albert classification [Mo1]. The weight-three types are not "self-generating" in this way, so we will give some examples of CY 3-folds whose third cohomologies realize most of them.

Type (i): We first turn to the existence of (families of) motivic $(1,1,1,1)$-Hodge structure with the largest possible Mumford-Tate group. As discussed in [GGK], polarized $\mathbb{Z}$-VHS of weight 3 and rank 4 (with $h^{3,0}=h^{2,1}=1$) over a smooth curve $\mathcal{S}$ have three possible classes of *nontrivial unipotent* monodromy transformations T_p about points $p \in \overline{\mathcal{S}}\backslash\mathcal{S}$:

I) $(T_p - I)^3 \neq 0 = (T_p - I)^4$;

IIa) $(T_p - I)^2 = 0$, rank $(T_p - I) = 1$;

IIb) $(T_p - I)^2 = 0$, rank $(T_p - I) = 2$.

C. Doran and J. Morgan [DM] showed that for $\mathcal{S} = \mathbb{P}^1\backslash\{0,1,\infty\}$ there are 111 or 112 such VHS subject to the condition (roughly speaking) that T_0 be of type **I** and T_1 of type **IIa**; furthermore, at least 21 (but conjecturally no more than 23) of them are produced by H^3 of families of toric complete-intersection CY 3-folds.

We claim that, considered as curves in $\mathrm{Sp}_4(\mathbb{Z})\backslash D$, none of these VHS can lie in a Mumford-Tate subdomain quotient, with the consequence that $M = \mathrm{Sp}_4$ for almost all fibres of each. Here are the possible monodromy/LMHS types for VHS into a D_M of each of the types in Theorem (VII.F.1):

type	possible T's
(i)	**I, IIa, IIb**
(ii)	**IIa**
(iii)	**IIb**
(iv)	none
(v)	**I**
(vi), (vii)	none
(viii)	**IIa**
(ix)	none
(x)	**IIa**
(xi), (xii)	none.

Type (i) is the only one with both **I** and **IIa** possible, proving our assertion.

Types (ii) and (iv): Unfortunately, the only motivic polarized Hodge structures we know of in Mumford-Tate domains of these types, are CM-Hodge structures hence contained in strictly smaller (point) subdomains. One way to motivically produce a polarized Hodge structure of type (iv) could be to first find a surface U satisfying:

- $H^2(U)$ has a sub-Hodge structure V_0 with Hodge numbers $(2, 0, 2)$;
- V_0 has weak CM by $K = \mathbb{Q}(\sqrt{-d})$ with eigenspaces V_0^+, V_0^-;
- $V_{0,\mathbb{C}}^+ \cap V_0^{(2,0)}$ has rank 1.

Then by the half-twist construction, if E is an elliptic curve with CM by K, then $V_0 \otimes H^1(E) \subset H^3(U \times E)$ has a $(1, 1, 1, 1)$-sub-Hodge structure V with weak CM by K such that $V_{\mathbb{C}}^+ = V^{(3,0)} \oplus V^{(1,2)}$.

Type (v): Let $\mathbb{Z}_2 \times \mathbb{Z}_2 \cong \{1, \iota_1, \iota_2, \iota_3\}$ act on the product of 3 elliptic curves $E_1 \times E_2 \times E_3$, with ι_j fixing the jth factor and involuting the other two. The resulting quotient has CY resolutions with $H^3 \cong \bigotimes_{i=1}^{3} H^1(E_i)$ [Bo], and if $E_1 = E_2 = E_3 = E \cong \mathbb{C}/\mathbb{Z}\langle 1, \tau\rangle$ then this contains $V := \text{Sym}^3(H^1(E))$ as a sub-Hodge structure with Hodge numbers $(1, 1, 1, 1)$. Representing elements by

$$\eta_1\eta_2\eta_3 := \frac{1}{6} \sum_{\sigma \in \mathfrak{S}_3} \eta_{\sigma(1)} \otimes \eta_{\sigma(2)} \otimes \eta_{\sigma(3)} \qquad (\eta_i \in H^1(E)),$$

we have Hodge basis

$$\begin{aligned}
\omega_1 &= \omega^3 = \beta^3 - 3\tau\beta^2\alpha + 3\tau^2\beta\alpha^2 - \tau^3\alpha^3 \\
\omega_2 &= \omega^2\bar{\omega} = \beta^3 - (\bar{\tau} + 2\tau)\beta^2\alpha + (2|\tau|^2 + \tau^2)\beta\alpha^2 - \tau|\tau|^2\alpha^3 \\
\omega_3 &= \omega\bar{\omega}^2 = \beta^3 - (\tau + 2\bar{\tau})\beta^2\alpha + (2|\tau|^2 + \bar{\tau}^2)\beta\alpha^2 - \bar{\tau}|\tau|^2\alpha^3 \\
\omega_4 &= \bar{\omega}^3 = \beta^3 - 3\bar{\tau}\beta^2\alpha + 3\bar{\tau}^2\beta\alpha^2 - \bar{\tau}^3\alpha^3
\end{aligned}$$

if $H^1(E, \mathbb{Q}) = \mathbb{Q}\langle\alpha, \beta\rangle$ and $\omega = \beta - \tau\alpha$ generates $H^{1,0}(E)$. By direct computation (e.g., using PARI), one checks that for any $\tau \in \mathfrak{h}$

$$\frac{-3\omega_1\omega_2\omega_3\omega_4 + \frac{1}{2}\omega_1^2\omega_4^2 - \frac{3}{2}\omega_2^2\omega_3^2 + 2\omega_1\omega_3^3 + 2\omega_2^3\omega_4}{(\tau - \bar{\tau})^6}$$

is rational of type $(0, 0)$ in $\text{Sym}^4 V \subset T^{4,0}V$. This is, of course, a multiple of (VII.E.2) with $\Xi = 3$ and $\mu = 2$. (In fact, that $\Xi = 3$ for $\text{Sym}^3(H^1(E))$ is easy to check directly.) Hence, for τ general, V is of type (v). As Mumford-Tate domains of type (v) only contain CM points of type (xii) (by Section VII.C), "general" simply means "not a quadratic irrationality."[17]

Type (iii): Repeat the Borcea CY construction with $E_1 = E_2$ having CM by $K = Q(\sqrt{-d})$ and E_3 *general.* Then $H^1(E_1) \otimes H^1(E_2)$ has a sub-Hodge structure V_0 of type $(1, 0, 1)$ with CM by K, and $V := V_0 \otimes H^1(E_3) \subset \bigotimes_{i=1}^{3} H^1(E_i) = H^3(\mathrm{CY})$ has K-eigenspaces $V^+ = V^{(3,0)} \oplus V^{(2,1)}$ and $V^- = \overline{V^+}$ as required.

Type (vi): A small resolution of the standard quotient of the Fermat quintic $\left\{\sum_{i=0}^4 z_i^5 = 0\right\} \subset \mathbb{P}^4$ by $(\mathbb{Z}/5\mathbb{Z})^3$, yields a CY X with $H^3(X)$ of Hodge type $(1, 1, 1, 1)$ and having CM by $L = \mathbb{Q}(\zeta_5)$. This is easier to see from the model of X as a smooth toric compactification of

$$\left\{x_1 + x_2 + x_3 + x_4 + \frac{1}{x_1 x_2 x_3 x_4} = 0\right\} \subset (\mathbb{C}^*)^4,$$

which has an obvious automorphism induced by $x_i \mapsto \zeta_5 x_i$ $(\forall i)$.

Type (vii): Similar to type (iii), except taking E_3 to have CM by $K' = \mathbb{Q}\sqrt{-d'}) \not\cong K$. It is easy to see (by considering the Galois action on a Hodge basis) that V does not split, hence has CM by KK'.

Type (xii): Borcea construction one more time, with $E_1 = E_2 = E_3 = E$ having CM by $K = \mathbb{Q}(\sqrt{-d})$, so that $V = \mathrm{Sym}^3(H^1(E)) \subset H^3(X)$ splits. A CY with H^3 exactly (rather than just having a sub-Hodge structure) of type (xii), is given in Section 6.1 of [Yui] as (the resolution of) a $(2, 2, 2, 2)$-complete intersection in $\mathbb{P}^7$.

Types (viii)–(xi): For the remaining split types we will just discuss the $(3, 0)+(0, 3)$-sub-Hodge structure V'. Arithmetic geometers interested in modularity[18] of *rigid* CY's (those with $h^{2,1} = 0$) over $\mathbb{Q}$ have built up a considerable database of examples, e.g., [Yui], [Me], some with H^3 of CM type and some evidently not. In the latter category lies the CY in example (VIII.A.9) (see the discussion there and following (VIII.B.4)). For a CM example, set $E = \mathbb{C}/\mathbb{Z}\langle 1, \zeta_3\rangle$ and consider the construction of Beauville [Be]: quotienting $E^{\times 3}$ by the diagonal multiplication action by ζ_3, then minimally resolving the 27 singular points, yields a CY 3-fold with Hodge numbers $(1, 0, 0, 1)$ and CM by $\mathbb{Q}(\zeta_3)$. The same Hodge structure also arises as H^3 of a small resolution

[17] Naively, it looks like taking $\tau = \zeta_5$ might give a CM-Hodge structure of type (vi), but this is not the case.

[18] Now proven in [GY].

of the self-fibre product of the rational elliptic surface given by the universal elliptic curve over $X(3)$.

Remark: In independent work (and from a different point of view) [Ro2], J. C. Rohde classified the possible generic Mumford-Tate groups of 1-parameter families of CY 3-folds with $h^{2,1} = 1$; $M(\mathbb{R})$ can only be $\mathrm{Sp}_4(\mathbb{R})$, $U(1,1)$, or $\mathrm{SL}_2(\mathbb{R})$. That $\mathrm{SL}_2(\mathbb{R}) \times \mathrm{SL}_2(\mathbb{R})$ (type (ii)) cannot occur is essentially a consequence of the Bryant-Griffiths result that for a maximal family of CY 3-folds, F^2 must be generated by F^3 and ∇F^3. Rohde also showed that the Doran-Morgan examples all have $M \cong \mathrm{Sp}_4$.

Additionally, in a forthcoming work of M. Kerr with A. Clingher and C. Doran, a class of examples of maximal CY families is constructed (using isotrivial families of Picard rank 18 K3's) with H^3 having weak CM by an imaginary quadratic field.

Chapter VIII

Arithmetic of Period Maps of Geometric Origin

Recall that a period domain D has a differential ideal $\mathcal{I} \subset \Omega^\bullet(D)$ that must pull back to zero under the local lifting of any VHS. We refer to $\mathcal{I}$ as the infinitesimal period relation (IPR). A period domain on which $\mathcal{I}$ is zero, is called *unconstrained.* More generally, a Mumford-Tate domain on which the pullback of $\mathcal{I}$ is zero is also called *unconstrained.* A key point is that the quotients, by arithmetic groups Γ, of unconstrained period and Mumford-Tate domains, including unconstrained Mumford-Tate subdomains of constrained period domains, will yield quasi-projective (Shimura[1]) varieties, with the projective embedding given by automorphic functions.[2]

Put differently, the whole idea of automorphic functions and forms in the unconstrained case is to provide the highly transcendental passage from periods/Hodge structures corresponding to the points of D, to coefficients of the defining equations of algebraic varieties with these Hodge structures — e.g., $g_4(\tau), g_6(\tau)$ for Weierstraß elliptic curves, or the inverse period map for $H \oplus E_8 \oplus E_7$-polarized K3 surfaces in [CD]. What gives conceptual backing to the nonexistence of holomorphic sections of $K_{\Gamma\backslash D}^{\otimes \bullet > 0}$ in the constrained case, then, is that only a "dense foliation" of D corresponds to algebraic geometry at all.

Related to this last observation are questions that are trivial for classical D and probably deep otherwise: *Does every unconstrained Mumford-Tate domain parametrize a variation of Hodge structure of geometric origin?*[3] *What are the Mumford-Tate groups of maximal VHS's (into D, resp. some D_M) of algebro-geometric origin?* A constrained D (resp. D_M) is, of course, the uncountable union of maximal IPR-integral submanifolds, but those arising as lifted images of "motivated" period maps are rigid, by transversality. Hence, the family $\mathcal{X} \to S$ motivating the period map may be "spread down to $\overline{\mathbb{Q}}$" without consequence, which tells us both that the integral submanifolds com-

[1]An unfortunate coincidence with our terminology of "Shimura domains" for orbits of a Mumford-Tate group cut out by 1- and 2-tensors.

[2]This will be true, more generally, for Hermitian symmetric Mumford-Tate domains, but the automorphic functions will not have the right algebro-geometric meaning for us.

[3]See the beginning of VIII.B for an interesting example.

ing from algebraic geometry are countable in number and arise from algebraic geometry/$\overline{\mathbb{Q}}$. The same sort of argument is what implies that the regulators of $K_3^{\text{ind}}(\mathbb{C})$ are the same as those of $K_3^{\text{ind}}(\overline{\mathbb{Q}})$. Our question is whether there is any meaningful sense in which Mumford-Tate is a *strict* subgroup of G (resp. M) *more often* for the manifolds in this countable subset than for all maximal integral submanifolds. A related issue of differing flavor is: What groups can occur as Mumford-Tate groups of variation of Hodge structures coming from algebraic geometry?

Another possibility is that motivic variations of Hodge structure might distinguish themselves by their behavior with respect to subdomains of D (resp. D_M). Indeed, by the main theorem of [CDK], the pullbacks of quotients of Noether-Lefschetz loci and of Mumford-Tate domains under a VHS $\Phi : S \to \Gamma\backslash D$ are algebraic subvarieties of S. In the motivated setting, one may inquire as to their field of definition, keeping in mind that already in the classical weight 1 case for CM points this requires the full machinery of the class field theory and the Shimura-Taniyama theory of complex multiplication. Also, while CM points are dense for maximal unconstrained variations of Hodge structure, in the constrained maximal setting it is an interesting question whether there are any! One might add that for families of Calabi-Yau 3-folds this is quite relevant to string theory, where mirror pairs of CY's with CM by the same field are conjectured to yield a rational conformal field theory [GV].

All of these questions will be refined here, where we consider four conjectures on motivic variations of Hodge structure and their consequences. The first three of these are variants or generalizations of Hodge $\Rightarrow$ absolute Hodge, Grothendieck, and André-Oort.

VIII.A BEHAVIOR OF FIELDS OF DEFINITION UNDER THE PERIOD MAP — IMAGE AND PREIMAGE

Let Y be a smooth projective variety defined over a subfield $k \subset \mathbb{C}$, and $\sigma \in \mathrm{Aut}(\mathbb{C})$. By identifying

$$F^m H^{2m}_{dR}(Y^{(an)}, \mathbb{C}) \cong \mathbb{H}^{2m}_{\text{zar}}(Y_{\bar{k}}, \Omega^{\bullet \geq m}_{Y_{\bar{k}}}) \otimes_{\bar{k}} \mathbb{C},$$

we have induced maps

$$F^m H^{2m}_{dR}(Y, \mathbb{C}) \underset{\sigma_{dR}}{\longrightarrow} F^m H^{2m}_{dR}({}^\sigma Y, \mathbb{C}), \quad H^{2m}_{\text{ét}}(Y_{\bar{k}}, \mathbb{Q}_\ell) \underset{\sigma_{\text{ét}}}{\longrightarrow} H^{2m}_{\text{ét}}({}^\sigma Y_{\bar{k}}, \mathbb{Q}_\ell).$$

(VIII.A.1) **Definition:** We say a Hodge class $\zeta \in \mathrm{Hg}^m(Y)$ is *absolute Hodge* (AH) if

$$\sigma_{dR}\zeta \in \mathrm{Im}\{\mathrm{Hg}^m({}^\sigma Y) \hookrightarrow F^m H^{2m}_{dR}({}^\sigma Y, \mathbb{C})\} \qquad \text{for all } \sigma,$$

and *strongly absolute Hodge* (SAH) if

$$(\sigma_{dR}\zeta, \sigma_{\text{ét}}\zeta) \in \operatorname{Im}\{\operatorname{Hg}^m({}^\sigma Y) \hookrightarrow F^m H^{2m}_{dR}({}^\sigma Y, \mathbb{C}) \times H^{2m}_{\text{ét}}({}^\sigma Y_{\bar{k}}, \mathbb{Q}_\ell)\}.$$

One has the famous "Hodge ⇒ absolute Hodge" criterion of Deligne ([DMOS], also see [Vo]):

(VIII.A.2) CONJECTURE: (i) *for all* (Y, ζ), ζ *is strongly absolute Hodge*; (ii) *for all* (Y, ζ), ζ *is absolute Hodge.*

By a theorem of Deligne (again [DMOS]), both (i) and (ii) are known for abelian varieties. A new exposition of the proof is contained in [CS].

Now let X/k be smooth and projective, $V = H^n(X^{an}_{\mathbb{C}}, \mathbb{Q})$ with accompanying Hodge structure φ, and $M = M_\varphi$. Noting that $V_{\mathbb{Q}_\ell} \cong H^n_{\text{ét}}(X_{\bar{k}}, \mathbb{Q}_\ell)$, consider the absolute Galois action

$$\iota : \operatorname{Gal}(\bar{k}/k) \to \operatorname{GL}(V_{\mathbb{Q}_\ell}).$$

Assuming (VIII.A.2)(i), where Y is taken to be various powers of X, the Galois action takes Hodge tensors to Hodge tensors, and so

$$\iota(\operatorname{Gal}(\bar{k}/k)) \subset N_{\operatorname{GL}(V)}(M, \mathbb{Q}_\ell). \qquad \text{(VIII.A.3)}$$

(VIII.A.4) **Remark:** In fact, there is an ℓ-adic analog Π_ℓ of the geometric monodromy group, obtained by taking the identity connected component of the Zariski closure of the left-hand side of (VIII.A.3). If one assumes the Hodge conjecture, then the finite set of Hodge classes determining M are represented by cycles defined over a finite extension of k, which would show that $\Pi_\ell \subset M_{\mathbb{Q}_\ell}$. Under this assumption, one could also ask whether $\Pi_\ell = M_{\mathbb{Q}_\ell}$ holds, which is a generalization of the so-called Mumford-Tate conjecture [M2, §4].

Next we look at a consequence of (VIII.A.2)(ii), in the following setting:

$$\text{(VIII.A.5)} \qquad \begin{cases} \mathcal{X} \underset{\pi}{\longrightarrow} S & \text{smooth and proper algebraic family defined over } k, \\ \Phi : S \to \Gamma \backslash D & \text{the period map associated to a subquotient of some } R^n\pi_*\mathbb{Z}. \end{cases}$$

Let $\mathcal{D} \subset S$ be an irreducible component of the preimage of the quotient of a Mumford-Tate domain D_M. Consider the $\bar{k}$-spread $\mathcal{P}$ of an arbitrary point $p \in \mathcal{D}(\mathbb{C})$, which is the Zariski-closure of the set of points $q \in \mathcal{D}(\mathbb{C})$ such that $X_q = {}^\sigma X_p$ for some $\sigma \in \operatorname{Gal}(\mathbb{C}/\bar{k})$. These $\{\sigma\}$ produce a continuous family of isomorphisms $H^n_{dR}(X_p) \xrightarrow{\cong} H^n_{dR}(X_q)$ inducing, by (VIII.A.2)(ii), isomorphisms defined over $\mathbb{Q}$ of spaces of Hodge tensors. Hence, the Hodge-tensor spaces must be constant with respect to the $\mathbb{Q}$-Betti structure, and we conclude that $\mathcal{P}$ lies in the preimage of NL_M, hence (being irreducible) in $\mathcal{D}$. So assuming (VIII.A.2)(ii), we get that

$$\text{(VIII.A.6)} \qquad \mathcal{D} \text{ is defined over } \bar{k}.$$

A special case is that, if *varieties* with Hodge structure of CM type (on their cohomology groups in some degree) are isolated in the family, then *they* are defined over $\bar{k}$. Moreover, an exact analog of (VIII.A.6) in the context of VMHS, is the conjecture [KP1, §4] that the zero locus of a k-motivated normal function is defined over $\bar{k}$.

Now suppose that $k = \overline{\mathbb{Q}}$ but S is defined over $\mathbb{Q}$; can we say anything more about the actual fields of definition of Mumford-Tate preimage components, starting with the CM points in S and perhaps proceeding to the special "weak-CM-loci" studied in Section VI.D? To give some idea of how this might go, we recall a bit of the theory of complex multiplication for abelian varieties.

Let L be a CM field of degree $2g$, Ψ a CM-type for L, and $\mathfrak{a} \in \mathfrak{J}(L)$ a fractional ideal; we have (cf. Section V.B) $A_{\mathfrak{a}}^{(L,\Psi)} = \mathbb{C}^g/\Psi(\mathfrak{a}) \in \mathfrak{Ab}(L,\Psi)$, the set of abelian varieties with CM by $(\mathcal{O}_L, \Psi)$. Denote the composition

$$\mathrm{Gal}(\mathbb{C}/L') \xrightarrow{|_{H_{L'}}} \mathrm{Gal}(H_{L'}/L') \xrightarrow[\cong]{\phi^{-1}} C\ell(L') \xrightarrow{\mathcal{N}'(\cdot)} C\ell(L)$$

by $[\mathfrak{A}(\cdot)]$, where $H_{L'}$ is the Hilbert class field of the reflex field L', ϕ is the Artin map, and $\mathcal{N}'(\mathfrak{b}) := N_{\Psi'}(\mathfrak{b})\mathcal{O}_L$ is the reflex norm. Then the easiest part of the main theorem of CM is

$${}^{\sigma}A_{\mathfrak{a}}^{(L,\Psi)} \cong A_{[\mathfrak{A}(\sigma)]^{-1}\mathfrak{a}}^{(L,\Psi)},$$

for σ fixing L'. If σ does not fix L', then ${}^{\sigma}A_{\mathfrak{a}}^{(L,\Psi)} = A_{\mathfrak{b}}^{(L,\sigma\Psi)}$ for some $\mathfrak{b} \in \mathfrak{J}(L)$.

So if, in (VIII.A.5), π is a family of abelian varieties, with S defined over $\mathbb{Q}$ and $\mathcal{X}, \pi$ defined over $\overline{\mathbb{Q}}$ and $X_p = A_{\mathfrak{a}}^{(L,\Psi)}$, then

- $\sigma \in \mathrm{Gal}(\mathbb{C}/\mathbb{Q}) \implies X_{\sigma(p)}$ has CM by $\mathcal{O}_L$;
- $\sigma \in \mathrm{Gal}(\mathbb{C}/L') \implies X_{\sigma(p)}$ has CM-type $(\mathcal{O}_L, \Psi)$;
- $\sigma \in \mathrm{Gal}(\mathbb{C}/H_{L'}) \implies X_{\sigma(p)} \cong X_p$;

and if σ fixes various ray class fields containing $H_{L'}$, then the isomorphism in the last bullet respects marked torsion, of various orders, as well. The field of definition of p would then be contained in $\mathbb{Q}, L', H_{L'}$, respectively a ray class field according to whether X_p is the only member of the family with CM by $\mathcal{O}_L$, of CM type $(\mathcal{O}_L, \Psi)$, of its isomorphism class as an abelian variety, respectively as a marked abelian variety.

Provided one works with integral Hodge structure, assumes the Hodge conjecture, and replaces "$X_{\sigma(p)}$ has..." by "$H^n(X_{\sigma(p)})$ has..." etc., the same story works as far as L', as defined in Section V.A using partitions, for a more general (e.g., a CY-) fibration π. The point is that once the CM-endomorphisms are induced by correspondences, the n-orientation can be read off from the eigenvalues of the action on $H^q_{\text{Zar}}(\Omega^p_{X_{\overline{\mathbb{Q}}}}) \otimes_{\overline{\mathbb{Q}}} \mathbb{C}$, and is acted on as required by σ. What we do not know is whether σ fixing $H_{L'}$ would produce an isomorphism of Hodge structure $H^n(X_p^{(an)}) \cong H^n(X_{\sigma(p)}^{(an)})$. This could perhaps be treated by looking at the various Jacobians produced by the types in Remark (V.A.2).

Note that if $\text{Gal}(\mathbb{C}/\mathbb{F})$ fixes p for some $\mathbb{F}$, then it also stabilizes the component of the pullback of any "weak-CM-locus" (cf. Section VI.D) through it, assuming (VIII.A.2)(ii). Even in the abelian variety case, we do not know if class field theory says anything about how the field of definition should relate to K in Section VI.D. However, if X is a *universal*[4] family of abelian varieties over $\mathbb{Q}$, then the field of definition of $\mathcal{D}$ will be an abelian extension of that of $\check{D}_M$ (by the theory of Shimura varieties).

If the reader will indulge us a bit of speculation, another way of looking at the question of fields of definition of Mumford-Tate Noether-Lefschetz loci (assuming AH) is as follows. Consider the setting (VIII.A.5) with $k = \mathbb{Q}$, Φ maximal, $\mathcal{D}$ a component of $\Phi^{-1}(\Gamma'\backslash D_M^0)$ ($\Gamma' = M(\mathbb{R})^0 \cap \Gamma$), and $L' :=$ the field of definition of $\check{D}_M$. By (VIII.A.6) $\mathcal{D}$ is defined over a number field $L^\#$, and we define $\mathcal{D}^\#$ to be its $\text{Gal}(L^\#/L')$-orbit. Then into what sort of orbit of $\Gamma'\backslash D_M^0$ does Φ map $\mathcal{D}^\#$?

One intriguing possibility (which would only partially answer this question) is suggested by the theory of Shimura varieties, and by the fact that general Mumford-Tate domains can also be written as adelic quotients [Ke, remark 4.17]. Expressing Γ' as $K_f \cap M(\mathbb{Q})^0$, for some compact open subgroup $K_f \leq M(\mathbb{A}_f)$, let $\mathcal{C}$ denote a system of representatives for the (finite) double-coset[5] $M(\mathbb{R})_+\backslash M(\mathbb{A}_f)/K_f$. The Mumford-Tate analog of the union that gives

[4]i.e., Φ is an isomorphism.

[5]$M(\mathbb{Q})_+$ denotes the preimage of $M^{\text{ad}}(\mathbb{R})^0$ in $M(\mathbb{R})$, and $M(\mathbb{Q})^0 := M(\mathbb{R})^0 \cap M(\mathbb{Q})$.

a Shimura variety[6] is

$$\text{(VIII.A.7)} \qquad \coprod_{g \in \mathcal{C}} \Gamma_g \backslash D_M^0 ,$$

where $\Gamma_g := gK_f g^{-1} \cap M(\mathbb{Q})^0$. One can then at least ask the question: *is* $\Phi(\mathcal{D}^{\#})$ *contained in a union complex-analytically isomorphic to* (VIII.A.7)*?*

We turn to a version of the "Grothendieck conjecture"; see [A2] for a more thorough treatment. The general concept is "transcendentality of special values of a transcendental map." Write $\rho : D \twoheadrightarrow \Gamma \setminus D$.

(VIII.A.8) CONJECTURE: *In the setting* (VIII.A.5) *with* $k = \overline{\mathbb{Q}}$, *let* $p \in S(\overline{\mathbb{Q}})$ *and suppose that* $\varphi \in D$ *satisfies* $\rho(\varphi) = \Phi(p)$. *Then* φ *is very general in* $D_{M_\varphi} = M_\varphi(\mathbb{R})\varphi$, *i.e., it is a point of maximal transcendence degree in the complex projective variety* $\check{D}_{M_\varphi}$.

Note that the countability of maximal motivic variations of Hodge structure and of $\overline{\mathbb{Q}}$-points means that the transcendental numbers occurring should have arithmetic meaning.

The classical example is elliptic curves: the period ratio τ for an elliptic curve defined over $\overline{\mathbb{Q}}$ satisfies $[\mathbb{Q}(\tau) : \mathbb{Q}] = 2$ or ∞, where 2 is the CM case in which the Mumford-Tate group is not SL_2.[7] Here is probably another:

(VIII.A.9) **Example:** Consider the rigid CY 3-fold ($h^{3,0} = 1, h^{2,1} = 0$) W, defined by taking a minimal resolution of the 125 double-points on the quintic

$$\breve{W} = \{z_0^5 + z_1^5 + z_2^5 + z_3^5 + z_4^5 - 5z_0z_1z_2z_3z_4 = 0\} \subset \mathbb{P}^4.$$

Its period ratio τ is the quotient of two linear combinations, with coefficients involving ζ_{10} and $\{\Gamma(k/5)\}_{k=1,2,3,4}$, of ${}_4F_3$-special-values; it is essentially the τ on pp. 128–129 of [GGK]. It seems likely that this is not only non-quadratic but transcendental. □

We remark that this conjecture also has a nice VMHS analog: the higher normal function associated to a family of $\mathrm{CH}^n(X_s, n)$ Milnor cycles has well-defined "periods" against topological cycles, in $\mathbb{C}/\mathbb{Z}$. One conjectures that, for $\mathcal{X} \xrightarrow[\pi]{} S/\overline{\mathbb{Q}}$ and $p \in S(\overline{\mathbb{Q}})$, the special values of these periods at p that are not torsion, are transcendental, and have arithmetic meaning in the context of the Beilinson conjectures.

[6] In the Shimura variety context, the theory of canonical models shows how to define (VIII.A.7) as an algebraic variety over L'.

[7] Important work of Tretkoff, Shiga and Wolfart [SW] extends this result to Shimura varieties of Hodge type, showing that the period point of a non-CM abelian variety over $\overline{\mathbb{Q}}$ is transcendental.

VIII.B EXISTENCE AND DENSITY OF CM POINTS IN MOTIVIC VHS

In his thesis, J. Rohde [Ro1] constructed many interesting families of CY varieties with dense subsets of CM points. By showing that they map *onto* quotients of Hermitian symmetric subdomains of a non-Hermitian symmetric D, he is able to pull density of CM points in D_M back to S. (For example, M might be $U(n,1)$ with D_M the complex ball parametrizing weight-3 CY-Hodge structures having weak CM by $\mathbb{Q}(\sqrt{-d})$ with $V^{3,0} \oplus V^{2,1}$, $V^{1,2} \oplus V^{0,3}$ as eigenspaces.) When the infinitesimal period relation (IPR) gets in the way, the situation is completely different; we have the following generalization of the André-Oort conjecture. A similar, less general, conjecture was made by Zhang [Zh]. We shall again assume the situation (VIII.A.5), $k = \overline{\mathbb{Q}}$. Recall that by [So], $\Phi(S)$ has the structure of a quasi-projective variety over $\mathbb{C}$.

(VIII.B.1) CONJECTURE: *Consider a set of CM points $\mathcal{E} \subset D$, with $\rho(\mathcal{E}) \subset \Phi(S)$. Then the $\mathbb{C}$-Zariski-closure of $\rho(\mathcal{E})$ in $\Phi(S)$ is a union of ρ-images of (unconstrained) Hermitian symmetric Mumford-Tate domains.*

This would mean that if $\Phi(S)$ meets no image of an Hermitian symmetric domain in dimension ≥ 1, then it can only meet finitely many Γ-equivalence-classes of CM points. A special case of this is the conjecture of de Jong and Oort, that there are only finitely many CM points in the VHS arising from H^3 of the mirror quintic family of CY 3-folds.

In an early outline of this paper, in view of Abdulali's result that CM-Hodge structures are motivic, and because of the density of CM points in *all* (even non-Hermitian symmetric) domains, we had conjectured

$$\text{(VIII.B.2)} \quad \left\{ \begin{array}{l} \text{If } \Phi : S \to \Gamma \setminus D_M \text{ lifts to a maximal IPR-integral} \\ \text{submanifold of } D_M\text{, then } \Phi \text{ is of algebro-geometric origin} \\ \iff \Phi \text{ passes through a CM point.} \end{array} \right.$$

This had the following implications:

$$\text{(VIII.B.3)} \quad \left\{ \begin{array}{l} \text{A nonabelian subgroup } M \leq G \text{ with } M(\mathbb{R}) \\ \text{compact, cannot be the Mumford-Tate group} \\ \text{of a (V)HS coming from algebraic geometry.} \end{array} \right.$$

$$\text{(VIII.B.4)} \quad \left\{ \begin{array}{l} \text{Suppose that } D \text{ classifies rigid Hodge structures;} \\ \text{i.e., no two non-zero } h^{p,q}\text{'s are adjacent. Then if} \\ \varphi \in D \text{ is motivic, } \varphi \text{ is CM.} \end{array} \right.$$

Indeed, if $M = M_\varphi$ for $\varphi \in D$ and $M(\mathbb{R})$ is compact, then by [Mo1, (1.18)], the Weil operator C_φ for φ lies in $Z(M)$. But then $\mathrm{Ad}(C_\varphi)$ fixes every $\varphi' \in$

D_M, so that $V_\varphi^+ = V_{\varphi'}^+$ and $V_\varphi^- = V_{\varphi'}^-$. Any direction in $T_\varphi D_M$ thus violates transversality, so φ itself is maximal, and by (VIII.B.2) would be CM so that M is abelian, a contradiction. This gives (VIII.B.3); the argument for (VIII.B.4) is just the last sentence again, without the contradiction.

The problem is that (VIII.B.4), i.e., "rigid implies CM", is wrong, from Example 5.13 of [Sch], which shows for $X = W$ from (VIII.A.9), that $\Pi_\ell \supseteq \mathrm{SL}(V_{\mathbb{Q}_\ell})$ so that $\iota(\mathrm{Gal}(\overline{\mathbb{Q}}/\mathbb{Q}))$ cannot normalize a torus. Therefore, *assuming* (VIII.A.2)(ii), and referring to (VIII.A.3), M cannot be a torus; it must be SL_2. More concretely, it is probably not hard to show that, for the specific τ mentioned in (VIII.A.9), $[\mathbb{Q}(\tau) : \mathbb{Q}] \neq 2$.

Furthermore, once one has this counterexample in hand, one can disprove (VIII.B.2) for maximal variations of positive dimension. For instance, consider the weight four 1-parameter variation of Hodge structure defined by $H^1_{\mathcal{E}/S} \otimes H^3(W)$, where $\mathcal{E} \to S$ is a non-isotrivial family of elliptic curves. Since this is of type $h^{4,0} = 1 = h^{3,1}, h^{2,2} = 0$, having 1 parameter it is maximal; and as the Mumford-Tate group of any $H^1(E_s) \otimes H^3(W)$ surjects onto both $M_{H^1(E_s)}$ and $M_{H^3(W)} \cong \mathrm{SL}_2$, it cannot be abelian for any ι.

What remains is the half of (VIII.B.2) that says maximal integral submanifolds through CM points come from algebraic geometry, which seems plausible, and (VIII.B.3), for which we do not have a counterexample.

Bibliography

[Ab] S. Abdulali, Hodge Structures of CM type, *J. Ramanujan Math. Soc.* **20** (2005), no. 2, 155–162.

[Ad] J. F. Adams, *Lecture Notes on Lie Groups*, W. A. Benjamin, New York, 1969. Reprinted: University of Chicago Press, 1982.

[ACT] D. Allcock, J. A. Carlson, and D. Toledo, Orthogonal complex hyperbolic arrangements, *Sympos. in Honor of C. H. Clemens* (Salt Lake City, UT, 2000), 1–8, *Contemp. Math.* **312**, Amer. Math. Soc., Providence, RI, 2002.

[Ag] I. Agricola, Old and new on the exceptional group G_2, *Notices Amer. Math. Soc.* **55** (2008), 922–929.

[A1] Y. André, Mumford-Tate groups of mixed Hodge structures and the theorem of the fixed part, *Compositio Math.* **82** (1992), 1–24.

[A2] ———, Galois theory, motives, and transcendental numbers, available at http:// arxiv.org/abs/0805.2569.

[Be] A. Beauville, Some remarks on Kähler manifolds with $c_1 = 0$, in *Classification of Algebraic and Analytic Manifolds, Prog. Math.* **39** (1983), 1–26.

[BHR] D. Blasius, M. Harris, and D. Ramakrishnan, Coherent cohomology, limits of discrete series, and Galois conjugation, *Duke Math. J.* **73** (1994), 647–685.

[Bo] C. Borcea, Calabi-Yau 3-folds and complex multiplication, in *Mirror Symmetry*. I (S.-T. Yau, ed.), 431–444, *AMS/IP Stud. Adv. Math.* **9**, Amer. Math. Soc., Providence, RI, 1998.

[Bor] A. Borel, *Linear Algebraic Groups, Grad. Texts in Math.* **126**, Springer-Verlag, New York, 1991.

[BEKPW] J. Brown, K. Eisentrager, K. Klosin, J. Pineiro, O. Watson, Arizona Winter School project (under supervision of J. de Jong), 2002, available at http://math.arizona.edu/ ~swc/aws/2002/notes.html#deJong

[Br] R. Bryant, Élie Cartan and geometric duality, in *Journées Élie Cartan 1998 et 1999*, **16** (2000), 5–20, Institut Élie Cartan (Nancy, France).

[BH] R. Bryant and L. Hsu, Rigidity of integral curves of rank 2 distributions, *Invent. Math.* **114** (1993), 435–461.

[C1] H. Carayol, Limites dégénérées de séries discrètes, formes automorphes et variétés de Griffiths-Schmid: le case du groupe $U(2,1)$, *Compos. Math.* **111** (1998), 51–88.

[C2] ______, Quelques relations entre les cohomologies des variétés de Shimura et celles de Griffiths-Schmid (cas du group $\mathrm{SU}(2,1)$), *Compos. Math.* **121** (2000), 305–335.

[C3] ______, Cohomologie automorphe et compactifications partielles de certaines variétés de Griffiths-Schmid, *Compos. Math.* **141** (2005), 1081–1102.

[CN] H. Carayol and A. W. Knapp, Limits of discrete series with infinitesimal character zero, *Trans. Amer. Math. Soc.* **359** (2007), 5611–5651.

[Ca] E. Cartan, Les systèmes de Pfaff à cinq variables et les équations aux dérivées partielles du second ordre, *Ann. Éc. Norm.* **27** (1910), 109–192.

[C-MS-P] J. Carlson, S. Mueller-Stach, and C. Peters, *Period Mappings and Period Domains*, Cambridge Univ. Press, 2003, 559pp.

[CGG] J. Carlson, M. Green, and P. Griffiths, Variations of Hodge structures considered as an exterior differential system: Old and new results, *SIGMA* **5**, 087, 40 pp., available at http://www.emis.de/journals/SIGMA/2009/087/

[CDK] E. Cattani, P. Deligne, and A. Kaplan, On the locus of Hodge classes, *J. Amer. Math. Soc.* **8** (1995), 483–506.

[CKS] E. Cattani, A. Kaplan, and W. Schmid, Degeneration of Hodge structures, *Ann. of Math.* **123** (1986), 457–535.

[CS] F. Charles and C. Schnell, Notes on absolute Hodge classes, preprint, 2011, available at http://arxiv.org/abs/1101.3647.

[CD] A. Clingher and C. Doran, Lattice polarized $K3$ surfaces and Siegel modular forms, preprint, available at http://arXiv.org/abs/1004.3503

[De1] P. Deligne, La conjecture de Weil pour les surfaces $K3$, *Invent. Math.* **15** (1972), 206–226.

[De2] ______, Travaux de Shimura, in *Séminaire Bourbaki, Exposé 389, Février 1971, Lect. Notes Math.* **244**, Springer-Verlag, New York, 1971, 123–165.

[De3] ______, Variétes de Shimura: interprétation modulaire, et techniques de construction de modeles canoniques, in *Automorphic Forms, Representations, and L-Functions* (A. Borel and W. Casselman, eds.), *Proc. of Symp. in Pure Math.* **33**, Amer. Math. Soc., 1979, 247–290.

[De4] ______, Théorie de Hodge II, *Inst. Hautes Etudes Sci. Publ. Math.* **40** (1971), 5–57.

[De5] ______, Un theoreme de finitude pour la monodromie, in *Discrete Groups in Geometry and Analysis, Progr. Math.* **67**, 1–19, Birkhauser, 1987.

[De-M1] P. Deligne and G. D. Mostow, Monodromy of hypergeometric functions and non-lattice integral monodromy, *Publ. Math. I.H.E.S.* **63** (1986), 5–89.

[DMOS] ______, Hodge cycles on abelian varieties (Notes by J. S. Milne), in *Hodge Cycles, Motives, and Shimura Varieties, Lect. Notes in Math.* **900**, Springer-Verlag, New York, 1982, 9–100.

[DR] M. Dettweiler and S. Reiter, Rigid local systems and motives of type G_2, *Compos. Math.* **146** (2010), 929–963.

[Do] B. Dodson, On the Mumford-Tate group of an Abelian variety with complex multiplication, *J. Algebra* **111** (1987), 49–73.

[DM] C. Doran and J. Morgan, Mirror symmetry and integral variations of Hodge structure underlying one-parameter families of Calabi-Yau threefolds, in *Mirror Symmetry* V, pp. 517–537, AMS/IP *Stud. Adv. Math.* **38**, AMS, Providence, RI, 2006.

[FHW] G. Fels, A. Huckelberry, and J. A. Wolf, *Cycle Spaces of Flag Domains, A Complex Geometric Viewpoint*, *Progr. Math.* **245**, (H. Bass, J. Oesterlé, and A. Weinstein, eds.), Birkhäuser, Boston, 2006.

[FL] R. Friedman and R. Laza, Semi-algebraic horizontal subvarieties of Calabi-Yau type, preprint, 2011, available at http://arxiv.org/abs/1109.5632.

[FH] W. Fulton and J. Harris, *Representation Theory — A First Course*, *Grad. Texts in Math.* **120**, Springer-Verlag, Dordrecht-Heidelberg-London-New York, 2011.

[EGW] M. Eastwood, S. Gindikin, and H. Wong, A holomorphic realization of $\bar{\partial}$-cohomology and constructions of representations, *J. Geom. Phys.* **17**(3) (1995), 231–244.

[Gi] S. Gindikin, Holomorphic language for $\bar{\partial}$-cohomology and representations of real semisimple Lie groups, in *The Penrose Transform and Analytic Cohomology in Representation Theory* (South Hadley, MA, 1992), *Contemp. Math.* **154**, Amer. Math. Soc., Providence, RI, 1993, pp. 103–115.

[G1] P. A. Griffiths, Periods of integrals on algebraic manifolds I, *Amer. J. Math.* **90** (1968), 805–865.

[G2] ______, Periods of integrals on algebraic manifolds: Summary of main results and discussion of open problems, *Bull. Amer. Math. Soc.* **76** (1970), 228–296.

[GS] B. H. Gross and G. Savis, Motives with Galois group G_2: an exceptional theta correspondence, *Compositio Math.* **114** (1998), 153–217.

[GW] R. Goodman and N. Wallach, *Symmetry, Representations and Invariants*, Springer-Verlag, Dordrecht-Heidelberg-London-New York, 2009.

[GY] F. Gouvêa and N. Yui, Rigid Calabi-Yau threefolds over $\mathbb{Q}$ are modular: a footnote to Serre, preprint, 2009, available at http://arxiv.org/abs/0902.1466.

[GGK] M. Green, P. Griffiths, and M. Kerr, Néron models and boundary components for degeneration of Hodge structure of mirror quintic type, in *Curves and Abelian Varieties*, *Contemp. Math.* **465**, Amer. Math. Soc., Providence, RI, 2008.

[GS] P. Griffiths and W. Schmid, Locally homogeneous complex manifolds, *Acta Math.* **123** (1969), 253–302.

[GV] S. Gukov and C. Vafa, Rational conformal field theories and complex multiplication, *Comm. Math. Phys.* **246** (2004), no. 1, 181–210.

[HC1] Harish-Chandra, Discrete series for semisimple Lie groups, I, *Acta Math.* **113** (1965), 241–318.

[HC2] ———, Harish-Chandra, Discrete series for semisimple Lie groups, II, *Acta Math.* **116** (1966), 1–111.

[H] M. Harris, Automorphic forms and the cohomology of vector bundles on Shimura varieties, in *Automorphic Forms, Shimura Varieties, and L-Functions* (L. Clozel and J. Milne, eds.), 41–91, Academic Press, Inc., 1990.

[HNY] J. Heinloth, B.-C. Ngo, and Z. Yun, Klooserman sheaves for reductive groups, 2010, available at http://arxiv.org/abs/1005.2765.

[IvG] E. Izadi and B. van Geemen, Half-twists and the cohomology of hypersurfaces, *Math. Z.* **242** (2002), no. 2, 279–301.

[Ka] N. Katz, *Exponential Sums and Differential Equations*, *Ann. of Math. Studies* **124**, Princeton University Press, Princeton, NJ, 1990. xii+430pp.

[Ke] M. Kerr, Shimura Varities: a Hodge-theoretic perspective, preprint, 2011, available at http://www.math.wustl.edu/~matkerr/SV.pdf .

[KP1] M. Kerr and G. Pearlstein, An exponential history of functions with logarithmic growth, to appear in *Proc. MSRI Conf. on Topology of Stratified Spaces*.

[KP2] ———, Normal functions and the GHC, to appear in *RIMS Kokyuroku*.

[Kl1] B. Klingler, The Andre-Oort conjecture, preprint, 2006, available at http://people. math.jussieu.fr/~klingler/papiers/KY2010Last.pdf.

[Kl2] ______, Some remarks on non-Abelian Hodge theory, preprint, available at http://people.math.jussieu.fr/~klingler/papiers/hodge.pdf.

[Kl3] ______, On the cohomology of Kaehler groups, preprint, 2010, available at http://people.math.jussieu.fr/~klingler/papiers/coh1.pdf

[Kl4] ______, Local rigidity for complex hyperbolic lattices and Hodge theory, *Invent. Math.* **184** (2011), 405–498.

[K] A. Knapp, *Lie Groups: Beyond an Introduction, Progr. in Math.* **140** (2005), Birkhäuser, Boston.

[KMRT] M.-A. Knus, A. Merkurjev, M. Rost, and J.-P. Tignol, The Book of Involutions (English summary) with a preface in French by J. Tits, *Amer. Math. Soc. Colloq. Publ.* **44**, Amer. Math. Soc., Providence, RI, 1998. xxii+593pp.

[Ku] M. Kuga, Fibre variety over symmetric space whose fibres are abelian varieties, *Proc. U.S.-Japan Seminar on Differential Geometry*, Kyoto, Japan, 1965, Nippon Hyoronsha, 1966.

[KU] K. Kato and S. Usui, *Classifying Spaces of Degenerating Polarized Hodge Structure, Ann. of Math. Studies* **169**, Princeton Univ. Press, Princeton, NJ, 2009.

[La] S. Lang, *Complex Multiplication*, Springer-Verlag, New York, 1983.

[Me] C. Meyer, Modular Calabi-Yau threefolds, *Fields Inst. Monograph* **22** (2005).

[Mi1] J. S. Milne, The action of an automorphism of $\mathbb{C}$ on a Shimura variety and its special points, in *Arithmetic and Gemoetry, Pap. dedic. I. R. Shafarevich* (M. Artin and J. Tate, eds.), *Prog. Math.* **35**, 239–265, Birkhauser, Boston.

[Mi2] ______, Introduction to Shimura varieties, available at http://www.jmilne.org/math.

[M1] D. Mumford, Families of abelian varieties, in *Algebraic Groups and Discontinuous Subgroups, Proc. Sympos. Pure Math.* **9**, Amer. Math. Soc., Providence, RI, 1966, 347–351.

[M2] ———, A note of Shimura's paper "Discontinuous groups and abelian varieties", *Math. Ann.* **181** (1969), 345–351.

[Mo1] B. Moonen, Notes on Mumford-Tate groups, preprint, 1999, available at http://staff. science.uva.nl/~bmoonen/index.html#NotesMT.

[Mo2] ———, An introduction to Mumford-Tate groups, 2004, notes, available at http:// staff.science.uva.nl/~bmoonen/index.html#NotesMT.

[Mor] S. Morel, On the cohomology of certain non-compact Shimura varieties, *Annals Math. Studies* (2009).

[O1] A. Otwinowska, Composantes de petite codimension du lieu de Noether-Lefschetz: un argument asymptotique en faveur de la conjecture de Hodge pour les hypersurfaces, *J. Algebraic Geom.* **12** (2003), 307–320.

[O2] ———, Composantes de dimension maximale d'un analog du lieu de Noether-Lefschetz, *Compositio Math.* **131** (2002), 31–50.

[PS] C. Peters and J. Steenbrink, *Mixed Hodge Structures*, *Ergeb. Math. Grenzgebiete* **52**, Springer-Verlag, New York, 2008, 470 pp.

[Ro1] J. Rohde, *Cyclic Coverings, Calabi-Yau Manifolds, and Complex Multiplication*, *Lect. Notes Math.* **1975**, Springer-Verlag, New York, 2009.

[Ro2] ———, Calabi-Yau manifolds and generic Hodge groups, preprint, 2010, available at http://arxiv.org/abs/1001.4239.

[R] V. Rotger, Shimura varieties and their canonical models, notes on the course, Centre de Recerca Matemàtica, Bellaterra, Spain, Sept. 20–23, 2005.

[Sa] I. Satake, Holomorphic imbedding of symmetric domains into a Siegel space, *Am. J. Math.* **87** (1965), 425–461.

[Sch] C. Schoen, Varieties dominated by product varieties, *Internat. J. Math.* **7** (1996), 541–571.

[Schm1] W. Schmid, Variation of Hodge structure: The singularities of the period mapping, *Invent. Math.* **22** (1973), 211–319.

[Schm2] ______, On a conjecture of Langlands, *Ann. of Math.* **93** (1971), 1–42.

[Schm3] ______, L^2-cohomology and the discrete series, *Ann. of Math.* **103** (1976), 375–394.

[Schm4] ______, Discrete series, *Proc. Symp. Pure Math.* **61** (1997), 83–113.

[Schm5] ______, Homogeneous complex manifolds and representations of semisimple Lie groups, in *Representation Theory and Harmonic Analysis on Semisimple Lie Groups, Math. Surveys and Monographs* **31**, 223–286, (P. Sally Jr. and D. A. Vogan, Jr., eds.), Amer. Math. Soc., Providence, RI, 1989.

[Schw] J. Schwermer, The cohomological approach to cuspidal automorphic representations, *Contemp. Math.* **488** (2009), 257–285.

[Se] Jean-Pierre Serre, Propriétés conjecturales des groupes de Galois motiviques et des représentations ℓ-adiques, *Proc. Symposia Pure Math.* **58**, Amer. Math. Soc., Providence, RI, 1994.

[Sh1] G. Shimura, On analytic families of polarized abelian varieties and automorphic functions, *Ann. of Math.* **78** (1963), 149–192.

[Sh2] ______, Moduli of abelian varieties and number theory, *Proc. Sympos. Pure Math.* **9**, Amer. Math. Soc., Providence, RI, 1966, pp. 312–332.

[Simp1] C. Simpson, Higgs bundles and local systems, *Publ. I.H.E.S.* **75** (1992), 5–95.

[So] A. Sommese, Criteria for quasi-projectivity, *Math. Ann.* **217** (1975), 247–256.

[SW] H. Shiga and J. Wolfart, Criteria for complex multiplication and transcendence properties of autormorphic forms, *J. Reine Angew. Math.* **464** (1995), 1–25.

[vG] B. van Geemen, Half-twists of Hodge structures of CM type, *J. Math. Soc. Japan* **53** (2001), 813–833.

[Vo] C. Voisin, Hodge loci and absolute Hodge classes, *Compos. Math.* **143** (2007), 945–958.

[W] H. C. Wang, Closed manifolds with homogeneous structure, *Amer. J. Math.* **76** (1954), 1–32.

[WW] R. O. Wells, Jr. and J. A. Wolf, Poincaré series and automorphic cohomology on flag domains, *Ann. Math.* **105** (1977), 397–448.

[W1] J. A. Wolf, The action of a real semisimple group on a complex flag manifold, I: Orbit structure and holomorphic arc components, *Bull. A.M.S.* **75** (1969), 1121–1237.

[W2] ———, The Stein condition for cycle spaces of open orbits on complex flag manifolds, *Ann. of Math.* **136** (1992), 541–555.

[Ya] H. Yanai, On the rank of CM type, *Nagoya Math. J.* **97** (1985), 169–172.

[Yui] N. Yui, Update on the modularity of Calabi-Yau varieties, in *Calabi-Yau Varieties and Mirror Symmetry, Fields Inst. Comm.* **38** (2003), 307–362.

[Yun] Z. Yun, Motives with exceptional Galois groups and the inverse Galois problem, preprint, available at http://arxiv.org/abs/1112.2434.

[Z1] Y. G. Zarhin, Hodge groups of K3 surfaces, *J. Reine Angew. Math.* **341** (1983), 193–220.

[Z2] ———, Weights of simple Lie algebras in the cohomology of algebraic varieties, *Math. USSR-Izv.* **24** (1985), 245–281.

[Zh] Y. Zhang, Some results on families of Calabi-Yau varieties, in *Mirror Symmetry V*, 361–378, *AMS/IP Stud. Adv. Math.* **38**, Amer. Math. Soc., Providence, RI, 2006.

Index

GPSR Authorized Representative: Easy Access System Europe - Mustamäe tee 50, 10621 Tallinn, Estonia, gpsr.requests@easproject.com

www.ingramcontent.com/pod-product-compliance
Ingram Content Group UK Ltd.
Pitfield, Milton Keynes, MK11 3LW, UK
UKHW011920260325
456773UK00010B/176